From 5G to 6G

From 5G to 6G

Technologies, Architecture, AI, and Security

Abdulrahman Yarali
Murray State University
Murray, KY, USA

WILEY

Library of Congress Cataloging-in-Publication Data

Names: Yarali, Abdulrahman, author.
Title: From 5G to 6G : technologies, architecture, AI, and security /
  Abdulrahman Yarali.
Description: Hoboken, New Jersey : Wiley, [2023] | Includes index.
Identifiers: LCCN 2023013868 (print) | LCCN 2023013869 (ebook) | ISBN
  9781119883081 (hardback) | ISBN 9781119883098 (adobe pdf) | ISBN
  9781119883104 (epub)
Subjects: LCSH: Mobile communication systems–Technological innovations. |
  Mobile communication systems–Security measures. | Artificial
  intelligence.
Classification: LCC TK5103.2 .Y3657 2023 (print) | LCC TK5103.2 (ebook) |
  DDC 621.3845/6–dc23/eng/20230405
LC record available at https://lccn.loc.gov/2023013868
LC ebook record available at https://lccn.loc.gov/2023013869

Cover Design: Wiley
Cover Image: © Rob Nazh/Shutterstock

Set in 9.5/12.5pt STIXTwoText by Straive, Pondicherry, India
Printed and bound by CPI Group (UK) Ltd, Croydon, CR0 4YY
C9781119883081_140723

Dedication

To my children

Fatemeh Zahra and Sadrodin Ali

Contents

About the Author

Abdulrahman Yarali, PhD, has led research teams working on wireless mobile communications systems design, implementation, and optimization for organizations such as AT&T, Nortel, and Sprint PCS. He is a faculty member at the School of Engineering and Cybersecurity and Network Management program at Murray State University in Murray, Kentucky, USA.

Preface

The advancement of technology never stops because the demands for improved internet and communication connectivity are increasing. With the advancements in digital circuitry and microprocessors making sensors small, lighter, and more powerful, a digital spillway has been opened. With wireless technology seeing exponential growth in the last decades, it is no wonder why it is becoming one of the largest and fastest-growing industries in this technical revolution promoting economic advancement. Competition drives this high adoption through access, affordability, and flexibility. These changes have impacted virtually all parts of life, society, and industries, fulfilling the communication needs of humans and intelligent machines. They have led to many new challenges and increased new horizons that keep the industry on its toes.

With every generation, new advancements are being made. These improvements have dramatically changed the international communication sector, improving how information is transmitted.

Wireless communications began about five decades ago and have expanded very rapidly. The transition from 1G to 5G has occurred rapidly for mobile communications. Today 5G is a strong technology intensively developed; however, it has limitations with respect to future needs. Just as 5G networks, with their key enablers technologies of M-MIMIO, small cells, and millimeter wave spectrum, are rolling out, allowing for many more devices to be connected to help pave the way for IoT devices like autonomous cars, smart devices, and many other things., the world has begun to talk about sixth-generation networks (6G). Wireless generations of 4G, 5G, and 6G are iterative steps rather than huge leaps. There are intermediate steps between each large release. For example, LTE-Advanced Pro was before 5G implementation started. There will be 5G-Advanced before it is time for 6G. Although 5G is starting to pave the road for vehicle automation and smart cities, 6G should create a platform for all device connectivity driven by the applications such as multisensory XR, robotic and autonomous systems, wireless brain-computer interactions, blockchain, and database ledger technologies. 6G is expected to outperform 5G in multiple specifications, such as latency of less than 1ms and .1ms for C-plane and U-Plane, respectively, with greater data rate than 1Tbps for individual and peak data rates greater, mobility up to 1000 Km/h and also spectral efficiency of 100bps/Hz. Although 6G will not replace 5G, 6G opens a new frontier of connectivity with the network built on the infrastructure put down for 5G extending the performance of existing 5G applications. Sixth-generation wireless technology will be integrated with a set of previously disparate technologies with an extreme focus on short-range communication, very high network heterogeneity, transformed radio topology, and AI, inherently expanding the scope of capabilities of new and innovative applications across the realm of wireless cognition, sensing, and imaging. With higher data rates, higher reliability, higher density, and higher intelligence, 6G is all about networks, human and machine with ubiquitous hyper-connectivity, efficient, and intelligent systems, and will be characterized in many ways as substantively advanced as compared to 5G in prior versions of wireless evolution. The key features of 6G are high security, secrecy, and privacy by applying blockchain and quantum key; operational, environmental, and service intelligence; energy efficient network via harvesting energy from sunlight, wireless power, micro-vibration, and ambient radio frequency signals; and multiband ultrafast transmission speed.

Newer efficient technologies such as—Enhanced mobile broadband (eMBB-Plus), massive machine type communications (mMTC), 3D integrated communications (3D-Integrated), and secure ultra-reliable low latency communications (SURLLC)—support a host of applications across a diverse set of industries. 6G is expected to expand on vectors like holographic-type communications using unconventional data communications technology (UCDC) for fully immersive 3D experiences and tactile Internet for real-time remote operation with audio, visual, and haptic feedback, health, autonomous vehicles, and beyond. The perpetual demand for more data bandwidth pushes researchers to explore underutilized spectrums in the sub-THz frequency bands. On the security side, asymmetric cryptography will become more prevalent in

6G technology. Radio access network (RAN) enhancements will lead to a large number of base stations that can be in different forms, such as all bands, BT, NFC, satellite, cloud-based WIFI, etc.

In the era of 6G, several key technologies, such as AI to develop an intelligent network, big data analytics, and computing, will converge, and we will likely see extreme experiences with alternative forms of secure credential storage and trustworthiness in sustainable and global service coverage. As public mobile networks are expected to exchange information with a broadening spectrum of parties, given the variety of use cases, services, devices, and users, security controls will have to be adaptable and dynamic in nature, supported by machine learning (ML) and artificial intelligence (AI). Advanced AI systems could be used as a form of penetration testing for cybersecurity issues. Using AI for this purpose will allow systems to be tested far more thoroughly to detect vulnerabilities in systems. To address the security demands of 6G networks, SDN and NFV should be complemented with AI techniques for proactive threat discovery, intelligent mitigation techniques, and self-sustaining networks. Enhancements in services will also lead to re-consideration of complete key management, mobility security, and considerations of confidentiality, integrity, and availability of communication networks. An issue for the Telecom industry is the fewer regulations that come with more tech and more use cases across the world. The priority list for many is the need for constant security advancements for device connectivity and the infrastructure involved. With the applications, these technologies are being used in medicine, manufacturing, banking, agriculture, and other important aspects of the economy. Security absolutely must be taken seriously to avoid a collapse in the economy on a global scale. The sixth generation has a significantly higher cost associated with its production, and the current network infrastructure cannot support its implementation. There are some challenges, such as increased hardware complexity, intelligent wireless energy harvesting, the coexistence of multiple RAT, and dynamic radio resource allocation, that scientists, researchers, and the industry need to overcome. Since its predecessor, the fifth generation, is still developing, it may take a while before the sixth generation is integrated into modern daily life. Due to its higher cost and global coverage, the sixth-generation networking and connectivity will see itself deployed into integrated space, aerial, ground, and undersea communications systems for ubiquitous 3D coverage.

We are very pleased that the technology, academic, and industry communities discuss this important and fast-growing future 5G-to-6G-transition networking. We are certain that this book's content will shed some light on these fast-growing technologies with their impacts on our daily life and economy. The chapters presented in this book discuss the enabling technologies, design paths, implementation requirements, and solutions with AI and ML roles to pave the road for a higher generation of wireless technologies. The challenges and issues faced in providing applications and services to ubiquitously and securely user experiences are presented.

1

Technologies and Development for the Next Information Age

1.1 Introduction

The advancement of technology never stops because the demands for improved internet and communication connectivity keep increasing. Just as 5G networks are rolling out, the world has begun to talk about sixth-generation networks (6G). The semantics of 6G is more or less the same as 5G networks because they strive to boost speeds, machine-to-machine (M2M) communication, and latency reduction. However, some of the distinctive focuses of 6G include optimization of networks of machines through super speeds and innovative features. This chapter discusses many aspects of technologies, architectures, challenges, and opportunities of 6G wireless communication systems. We will discuss super-smart societies, extended reality, wireless brain-computer interactions, haptic communication, smart healthcare, five-sense information transfer, the encompassing world of the Internet of Everything (IoE), and cybersecurity.

As we enter a new age of high-speed data transfer and the implementation of new technologies that have been in our minds for decades, we begin to see the light that illuminates the beginning of the tunnel. This is quite different from the light at the end of the tunnel because of the vast advances that will take place when we start the journey to high-speed data transfer. That will enable us to enter the proverbial tunnel, a virtual rabbit hole continuing to advancements at a speed that has never before been possible. With the advancements in digital circuitry and microprocessors making sensors small, lighter, and more powerful, a digital spillway has been opened. This paves the way for artificial intelligence (AI) and the Internet of Things (IoT) that have already changed our behaviors in life [1] – to the point that even the recent term IoT has a new umbrella term being applied that will be discussed later in this book. In this chapter, I will cover several topics that will be leading the way in advancement in the next decade. The advent of the 5G network began this process but will be short-lived without 6G rearing its head to turn imagination into realization. In this chapter, we will discuss super-smart societies, extended reality, wireless brain-computer interactions, haptic communication, smart healthcare, five-sense information transfer, the encompassing world of the IoE, and cybersecurity.

1.2 Roadmap to 6G

The telecommunications sector is comprised of companies that deal with communication globally. These organizations range from cell phones, internet services, and airwave cables to other wired and wireless dealers. Telecom organizations have introduced data information of words, video, audio, and voice to the world through their improved basic infrastructure. The telecommunications industry consists of various subsectors such as wireless communication, communication equipment, processing systems, and products. In addition, subsectors such as long-distance carriers, domestic telecom services, foreign telecom services, and diversified communication services also constitute the telecommunication industry.

Telecommunication achievements have led to dramatic changes in humans' lives and interactions with the inevitable advancement in technology associated with it. These changes have led to many challenges new to the industry and increased new horizons that keep the industry on its toes. The telecommunications sector consists of companies that enable communication globally; there can be multiple sources to achieve this, ranging from cell phones, internet services, airwaves, or cables, to other wired and wireless devices.

For example, Wi-Fi facilitates access by low-income individuals and countries and uses the network to empower its people.

The journey of wireless communication only began to develop about 51 years ago. Wireless technologies give an excellent opportunity for economic development, ensuring that services are delivered to developing nations. Such technological connectivity can play critical roles, considering how fast they facilitate access to better knowledge and acquisition of information.

Due to these advancements, the industry is experiencing several trends that will change the face of the industry in general:

- The change in the network from 4G to 5G has been implemented in some regions of the world. The 5G advent will ensure increased network speed, increased efficiency, and less latency. Through this, the innovation of IoT has been the outcome.
- Technology enabling autonomous vehicles is improving, and this could change the face of the world.
- There is the potential for fewer regulations globally, and telecom companies would be exposed to global markets with more freedom and competition.
- Security and privacy will be critical areas of concern for many internet users with the advancement in technology.
- There would be an increase in cross-industry partnerships associated with the evolution of new technological experiences and demands from clients worldwide.

The wireless mobile industry has seen large, exponential growth in the past decade. Wireless communications have brought dramatic changes in the international communication sector, improving informational infrastructure that uses radio waves in place of wires to transmit information. Wireless technologies give an excellent opportunity for economic development. Ensuring that services are delivered in developing nations can play critical roles, considering how fast they facilitate access to better knowledge and acquisition of information.

Over the span of five decades, wireless networks underwent various developments: wireless personal area networks (WPAN), wireless local area networks (WLAN), ad hoc networks (Adhoc), or wireless mesh networks, metropolitan area networks (WMAN), wireless wide area network (WWAN), cellular networks that spans from first-generation 1G to 5G, and space networks. Broadband communication has gone through continuous and massive developments in recent decades, which has seen it become a more efficient and reliable industry for the function of people globally. The evolution of each generation of wireless technology has been a decade trend to cope with the end-users demands.

Technologies such as Global Systems for Mobile (GSM) and CDMA (Coe Division Multiple Access) also offered data roaming services. Third generation (3G) was a huge milestone in transmitting the bulk of data and speed of transmission compared to 2G. 3G was introduced in the late 1990s with wider data bandwidth and faster internet connectivity. It is well applicable to a wireless voice telephone, internet access, wireless internet service, video calls, and mobile television technologies. This platform provided faster and better uplink/download packet access. It includes evolved HSPA (high-speed packet access) and long-term evolution-advanced (LTE-A). Advanced third generation, 3.5G, is a cellular telephone grouping and data technology designed to provide more efficient performance than the 3G network, as an interim advancement toward realizing full fourth-generation (4G) capabilities.

The 4G cellular network is supposed to provide an environment for dynamic data access, HD video streaming, and global roaming regarding service delivery. Such services were not prevalent in previous mobile network technologies. 4G technology uses a combination of standards of IEEE802.11 and IEEE802.16 radio technologies ratified by IEEE.

5G network has demonstrated improved and advanced services with connectivity is considered to have a bandwidth of higher seed than the other four network generations. In addition, it has improved signal efficiency, significantly lower latency than 4G, and enhanced spectral reliability and efficiency.

Advancements in wireless communications as well as foreign and domestic telecom services have dramatically changed people's lives across the world, but they also pose new challenges and open new horizons to the industry players. While 4G was 3G but faster, 5G and 6G represent different iterations of wireless connectivity. Many predictions expect 6G will be reserved for business, military, and industrial purposes, with some consumer use such as immersive entertainment. It will not be practical initially to have every device streaming with 6G – but other advances may change that. Figure 1.1 shows the evolutionary roadmap of wireless generations.

Several new technologies are being tested to support the future generation of wireless communications (beyond 5G) B5G or 6G, which are expected to deliver high data rates for enhanced mobile broadband (eMBB), support ultra-reliable, and low-latency services, and accommodate a massive number of connections.

The 5G standard has gone through two releases, Rel-15 and Rel-16, that will complete the development of 5G wireless communication. The standard will also include nonorthogonal multiple access (NOMA), ultra-reliable low-latency communication, vehicle-to-X communication, unlicensed band operation, integrated access, backhaul, terminal power saving, and positioning. Figure 1.2 shows various 3GPP 5G toward 6G releases. The standard will also include NOMA,

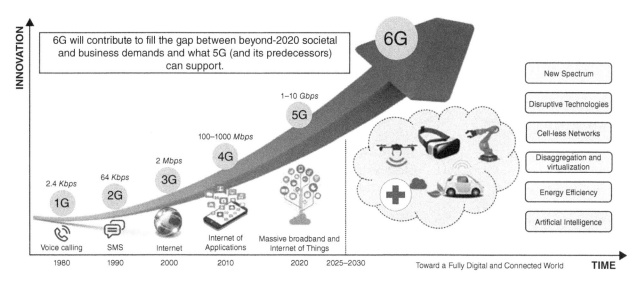

Figure 1.1 Evolution of wireless technologies from 1G to 6G [2].

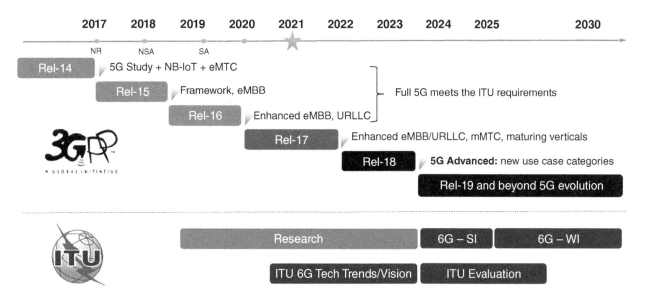

Figure 1.2 Features for various 3GPP releases [3].

ultra-reliable low-latency communication, vehicle-to-X communication, unlicensed band operation, integrated access, and backhaul, terminal power saving, and positioning.

Every 10 years, wireless (or mobile) communication undergoes a generation change. The next generation, 6G, has already started. The International Telecommunication Union (ITU) Focus Group Technologies for Network 2030 has listed several aspects of 6G, including new holographic media, services, network architecture, and internet protocol (IP). Many researchers and industries have started discussions on use cases, deployment scenarios, and performance requirements of 6G. These were compared with previous generations of mobile communication, and new network architectures were discussed. Technology maturity levels vary in 6G, with many technologies in their scientific exploration stage.

Technologies being considered include holographic radio, terahertz communication, large intelligent surface (LIS), orbital angular momentum, advanced channel coding/modulation, visible light communication, and advanced duplex. The goal of 6G is to provide the information society with services by around 2030. The 6G network will support many challenges, such as intelligence of network elements and network architecture, the intelligence of connected objects (terminal devices), and information support of intelligent services. The 6G network should be able to deliver 10 Tb per second and should have lower latency, higher reliability, and higher energy efficiency.

Several key features of 6G are terahertz communications, visible light communications, very large-scale antenna, advanced channel coding, and space-air-ground-sea integrated communications. Some technologies in the next generation of mobile communications are difficult to categorize and may eventually become part of 5G. Revolutionary technologies are still in the exploratory stage and might not reach maturity before 2030. In order to improve the performance of the physical layer, we need to develop new technologies such as photonics-defined radios and holographic radios. Optical holography can be used to record the electromagnetic field in space based on the interference principle of electromagnetic waves, and radio frequencies are electromagnetic waves, so a continuous aperture antenna array can be used to record the electromagnetic field in space. The terahertz band is the last span of the radio spectrum and can be used for 6G, but it is not yet commercially available. The characteristics of the terahertz spectrum are unique, including wide bandwidth and low attenuation, but the spectrum is also prone to shadows and blocking due to the weak diffraction effect at short wavelengths and the short time for channel stabilization. There are many challenges in the engineering implementation of THz, including ultra-high processing power and very large-scale antenna, which could lead to a superfast-speed broadband processing chip with high power consumption. Commercial products for THz communications are now available, and it is expected that these products will allow high data rates in the 6G era. Micro-scale networks can be classified into two categories: indoor and outdoor scenarios.

1.2.1 Society 5.0

The first topic is a term that has been thrown around without many noticing what was being stated. The process started with the term *smart society* and progressed to a *super-smart society*. Many have even mentioned Society 5.0, with most people thinking it was referring to an upgrade so that Apple's iOS systems use numbers; this was a slight overlook in my option. The term is used to prepare us for a society enriched with technology that will help push us into a more efficient life. Over the past 10 years, we have used search engines to increase human abilities and as a way to retrieve the information that will make users more productive. This requires us to initiate the sequence to get the process started by first looking for the information and then acting on it. In realizing the new Society 5.0 [4], we will not have to find the information to solve our daily problems, but instead, the information will find us the solution. This has been considered the fourth industrial revolution and will be implemented into our daily lives. This will contribute to more sustainable and comfortable living environments in people's lives. We have already progressed as a society through several of these without considering that it was propelling us forward. We started this process with the following stages, beginning with the Hunter-Gatherer society (Society 1.0), Agrarian society (Society 2.0), Industrial society (Society 3.0), and Information society (Society 4.0) that we are currently in and starting to enter Society 5.0 without a coined name. With the collection of Big Data by IoT devices, the information will be converted into new types of intelligence by AI machines that will mine the data-making informed decisions with only minimal human intervention if any at all. As AI devices are learning patterns, they will make people's lives more comfortable and sustainable. The participants will be provided with only the needed products and services, in the amounts needed and at the time they are needed. One of the areas that will be greatly affected by societal change is healthcare. This will provide rapid solutions for the connection and sharing of medical records. This will also initiate the implementation of remote medical care services. Using AI and robots at nursing facilities will lessen a load of healthcare workers to focus on less mundane tasks. This jump in society will also increase mobility for everyone, including the elderly and disabled. This will promote autonomous driving taxis and buses for public transportation, increasing accessibility in the country's rural areas.

Also, increasing the distribution and logistics in our supply chain by using unmanned-following-vehicle systems and drones to deliver products. We will be able to increase efficiency in our infrastructure using sensors, AI, and robots to inspect and maintain roads, bridges, tunnels, and dams. This will ensure that places that need repair can be detected at an early stage. Doing so will cut down on unexpected accidents and the time spent on construction while increasing safety and productivity. One of the last large categories will be the financial technology (FinTech) category. This will include using blockchain technology for monetary transfers, introducing open application programming interfaces (API) to firms and banks, and promoting cashless payment systems. The implementation of blockchain technology alone will reduce time and cost while increasing security in global business transactions.

1.2.2 Extended Reality

The second topic to discuss is the exciting world of extended reality. This covers three main areas in itself. We will start with the first area of extended reality (XR), an umbrella term for immersive technologies [5]. The ones that are currently in use today are augmented reality (AR), virtual reality (VR), and mixed reality (MR), plus new ones that are still to be created.

AR is virtual information and objects that are overlaid on the real-world background. This is accessed through items such as AR glasses or smart devices. This experience enhances the real world with digital details such as images, text, and animation. The next area is VR, a virtual reality experience that which users are fully immersed within a simulated digital environment. This requires participants to put on a VR headset or head-mounted display to get the full experience of a 360° view of an artificial world that tricks their brains into believing they are in the place or world the developers have created. The last area is MR, mixing reality, digital, and real-world objects that coexist while interacting in real time. This is the latest in immersive technology and requires an MR headset and a lot more processing power than the less-advanced VR or AR devices.

1.2.3 Wireless Brain-Computer

The third topic is a wireless brain-computer interaction demonstrated by BrainGate researchers with the first human use of a wireless transmitter capable of delivering high-bandwidth neural signals [6]. With this, the traditional cables are no longer needed. The wireless signal could be recorded and transmitted with similar fidelity, which means the same decoding algorithms are used as with wired systems. The main difference is that people do not have to be tethered to the equipment. This opens up a world of new possibilities in how the systems can be used. This development will be a step forward to a fully implantable system that helps restore the independence of people who have lost the ability to move. This is a huge achievement considering that the brain interacts with technology without new algorithms, sophisticated programs, or attached wires.

1.2.4 Haptic Communication

Haptic communication is also referred to as kinesthetic communication and 3D touch [7]. This is any technology that can create an experience of touch by applying forces, vibrations, or motions to the user. This technology can also be combined with VR to create virtual objects in a computer simulation to control objects and enhance the remote control of machines and devices. Haptic technology helps investigate how the human sense of touch works by creating controlled haptic virtual objects. The different ways of implementation are listed here with a brief description [8, 9]:

- *Vibration*. This is created using an eccentric rotating mass (ERM) actuator, which is an unbalanced weight attached to a motor shaft. When the shaft spins, the unbalanced weight causes the attached device to shake.
- *Force feedback*. Devices use motors to manipulate the movement of an item that the user holds. This is best known in racing games that use a steering wheel that, when turned, gives the feel of resistance or pull in the opposite direction. This allows the haptic simulation of objects, textures, recoil, momentum, and objects' perceived physical presence in games.
- *Air vortex rings*. A donut-shaped air pocket is made up of concentrated gusts of air. These gusts can blow out a candle or move papers from a few yards away and can be used to deliver noncontact haptic feedback.
- *Ultrasound*. A focused beam can be used to create a sense of pressure on a finger without touching an object. The beams can also give sensations of vibration and give users the ability to feel virtual 3D objects.

The advancement of ultrasound technology has been integrated into many areas such as personal computers (PCs), mobile devices, virtual reality (VR), teleoperators and simulators, robotics, medicine, dentistry, art, aviation, space, automobiles, and teledildonics. The technology has been used for a while now but has gone under the radar for most customers [9].

1.2.5 Smart Healthcare

The fifth topic is the most expansive with the new age data transfer and is the one that will grow the fastest as 6G emerges. It is the smart healthcare topic of discussion. The increase in population and the recent COVID-19 pandemic have taken the forefront in the technology age. With the extension of telemedicine and internal systems of the hospitals, the workload has become overwhelming. Smart healthcare covers services, medical devices, connectivity technologies, system management, applications, and end users.

When it comes to architecture and the requirements needed, there is a big demand placed on the systems. One of the broader demands comes from hardware and software requirements. With this, all areas have to be balanced and in the best performance abilities. There is a need for quality of service, low power usage, high efficiency, high system reliability, and form factor, interoperability with other systems, excellent connectivity, high speed, ambient intelligence, and sufficient memory. There is also a whole category of deployment with some that need to be further enhanced. There is a need for reliable Wi-Fi/WLAN, MEMS, BLE, RFID, WPAN/6LoWPAN, GPS, and WSN. Even with this technology implemented, there is still space for improvement [10].

The IoT aspect of smart healthcare has to be mentioned. With some of the enormous attributes such as identification, location, sensing, and connectivity that come with the IoT, it is a huge component of the smart healthcare system. IoT can be broadly implemented, starting from calibrating medical equipment to the personalized monitoring system. IoT plays one of the most significant parts of the smart healthcare system, in my opinion, with much more to come. Big Data has taken a strong stance in the smart healthcare industry with the increase in smart sensors, social networks, and web services. Mobile devices are estimated to be generating more than 2.5 quintillion bytes per day [11]. This data collected by the smart systems have to be consistent in every way. The ability to offer medical services to users despite their geographic location results from cloud assistance. This gives the professionals the ability to offer services from any distance. With every good technological advancement, there are also challenges and vulnerabilities. With a large amount of personal data with smart healthcare, a higher bar has been set. Some of the issues start with patient privacy challenges. This includes but is not limited to data confidentiality, data privacy, data eavesdropping, identity threats, data freshness, service availability, data confidentiality, integrity, location privacy, authentication, self-healing, access control, and unique identification, and resiliency. The final point in this category is one of the most interesting to me personally. It is the implementation of nano-smart Healthcare, which consists of nanotechnologies to diagnose from the inside out. With the use of microimage sensors, processors, encapsulated battery, LED light, antenna, RF transmitter, and electrogram, a pill camera can be made for the patient to take that allows the physician to do an endoscopy or colonoscopy without the traditional methods that take longer to diagnose and treat. It makes the procedure as simple as swallowing a pill to retrieve more information than could be captured with a traditional diagnosis.

1.2.6 Five-Sense Information

The sixth topic outlined is five-sense information transfers. This is a way to experiment with the world around using 6G communication. Since humans use their five senses, the systems will remotely transfer the data obtained from the senses. This is used to detect the sensations from the human body and environment and uses the body effectively without the environment and local circumstances.

1.2.7 The Internet of Everything

The seventh topic in this chapter is the IoE, which is not the same as IoT in many ways. IoE is the intelligent connection of people, processes, data, and things. The IoE is the world where billions of devices have sensors to detect and assess the states of the environment. These are all connected over a public or private network using protocols both standard and private. The main characteristics of IoE are people, data, processes, and things. The IoT is the network of physical objects accessed through the internet. The main difference is that IoE has four distinguishable characteristics, while IoT has one. You will hear the term IoE as much or more than IoT shortly due to the vast range and expansion of the high-speed data transfer and accompanying technology.

1.2.8 5G to 6G

Wireless communication has been developing since the 1980s with significant changes and advancements with each generation. We have recently entered the fifth generation of wireless communication (5G). The adoption and deployment of this newest generation are underway but will likely take three to five more years to see fully realized deployment. 5G has brought about the move to the cloud for software-based networking, bringing about on-demand, automated learning management of networking functions. But even with these advancements, researchers have already begun to look at the next generation (6G) and how the networking landscape will change with its adoption.

Through a variety of technologies, 6G is speculated to bring about a shift to completely intelligent network orchestration and management. The current evolution of 5G networking is helping to visualize the architectural framework of 6G. Heterogeneous cloud infrastructure should be an expected part of 6G as the existing cloud infrastructure and that of future generations will not simply disappear with the introduction of a new infrastructural component. Where it will begin to diverge from the present is the increased variety and flexibility of specialized networks for personal subnetworks and things such as flexible workload offloading.

As with every next step, 6G communication networks will have their own security and privacy concerns. Mobile security continues to be a very significant issue with each generation of wireless communications and seems to only get more difficult with each technological advancement. Past generations have dealt with everything from authentication issues to cloning. With the availability of computing resources and the sophistication of attackers now, it is expected that 6G networks see some of the same attacks such as zero-day attacks and physical layer attacks, but also an increase in advanced attacks such as quantum-based attacks and AI/ML-based intelligent attacks. The addition of AI into the networks of 6G will work for and against its security, it seems. Figure 1.3 shows a summary of the potential and new threats, vulnerability, and corresponding security mechanisms simply by examining the risk exposure of proposed 6G technologies [13].

We must build on fifth-generation security research to gain insight into what will be coming with the later generations of communication. Currently, there is not much insight and research that looks at the holistic picture of 6G security.

Current mobile data network technologies such as 4G and 5G are reaching their physical limit as more and more devices in our lives depend on the internet and mobile network connections. The idea of a 6G network is becoming the new ideal network upon which our society can grow into. My only concern is that the idea of the "6G communication system" will become a monolith that business and education communities will flock around, to a point where the idea of the 6G network becomes a near sci-fi concept that is unreachable or that should be actualized immediately. Simply put, 6G is not some approaching bleeding edge technology but is an aggregate of multiple technologies that is a natural extension of our current mobile networks and wireless communication systems. The idea and fantasies behind 6G are attainable; people just have to keep their heads straight.

In the past few years, the amount of traffic over mobile data networks has drastically increased and shows no signs of stopping. This is due to an increasing number of applications that require connection to mobile data networks. Examples include the IoT, smart automobiles, and smart cities among many other things. Because of these new uses for the existing wireless mobile networks, our current 3G/4G networks struggle to keep up with the demand for these technologies that grow more numerous and more data-hungry as time goes on. A potential solution to the limited capacity of current networks is to broaden available bandwidth to accommodate higher data rates. Even with newer 5G frequencies, more bandwidth is still a highly coveted goal that remains out of reach with the narrow bandwidth below 6 GHz. Technologies like multiple input, multiple output (MIMO) antennas do make the sub 6 GHz communications more efficient, however.

Very mobile devices, like smartphones and wearable devices, also tend to operate on multiple bands, which put more strain on existing networks, especially when these devices become more numerous and more data-hungry with time. Also,

Technology	Risk Level	Primary Cause	Time Frame
AI	Medium	Adversarial manipulation and malicious AI development	<3 years
IT-OT Convergence	High	Lack of cybersecurity designed and deployed in IoT devices	Immediate
Self-Adaptive Networks	Medium	Lack of automation and real-time intelligence processing	>5 years
Quantum Computers	High	Break complex encryption asymmetric algorithms	>10 years

Figure 1.3 Potential threats and novel events, and corresponding security measures [12].

mobile devices, by their nature of being incredibly mobile, are hard to communicate with consistently as they move and change orientation, making full, satisfying coverage harder to achieve. This difficulty of communication results in multiple different types and configurations of antennae.

5G and associated networks also struggle with complete coverage. Being ground-based systems, 5G networks have trouble extending to inhospitable terrain. This problem is trying to be solved by integrating 5G communication systems into space-based 6G systems. These space systems, built mainly on the back of satellite technology, will give rural and remote areas across the world entirely new or greatly improved mobile data network access.

There are some concerns regarding human health when it comes to 5G technologies and transmissions. An important fact is that many existing applications used similar bands and these technologies have had no evidence crop up that could suggest they are harmful to human health. Despite this, the effect of 5G signals, due to not having as many standards as a newer technology, will remain a concern until evidence convincingly proves its safety.

The current goal among relevant circles in business and education is an idea of the 6G system, featuring every aspect of a modern network, perfected. This includes high speed, high capacity, high security, and extensive coverage. The 6G network can support the massive, growing number of devices that require a connection. It will be expected to be very fast for communicating with transportation systems like airlines. Deep learning and AI will provide security insights to the 6G network. The 6G network will be fast enough to support live broadcasts and streaming. It is expected to be ubiquitous, being much larger than any standing 5G network. Finally, the 6G network will have significantly higher bandwidth and thus transmission rates. Many technologies that would allow the goal of 6G to be reached can be derived from existing 5G technologies or are new frontiers currently being explored.

In a society where high-speed data transfer is not just a luxury but also a way of life, we have realized the need and demand for our wireless infrastructure have come under immense pressure to expand. I will cover several characteristics and trends in the mobile industry that show that the need to move from a 5G system to a 6G system is on the horizon. This will be a very condensed summary due to a large amount of information involved. I will briefly cover the prospects, applications, specifications, and requirements, technologies that enable growth, industry standardization, and future challenges. All the topics listed can be vastly expanded on and can occupy pages that cover changes in the network systems.

We will start with the global traffic volume of mobile devices in 2010 consumed 7.5 EB/month of data. This was not the beginning but gives a quick reference to the large jump made in the next decade. The predicted traffic in 2030 was estimated at a whopping 672% increase making data traffic consistent at 5016 EB/month with expectations to increase even greater in the future. This will strain the 5G network that is not fully functional in all areas of North America and will cripple it if the network structure does not change quickly [9]. The 5G network is rapid, including new techniques of managing data, including but not limited to new frequency bands. These changes are being implemented rapidly but not fast enough to keep up with the trends. We are already coming to the limit of 5G, which was stated to max out in 2030. With that being said, the predecessor of the 5G network is the new 6G network that will include all the bells and whistles of the 5G network, such as network densification, higher throughput rate, increased reliability, lowered energy consumption, and massive connectivity. This will pave the way for new services and devices to automate much of our lives. The most important aspect of the 6G network is handling massive amounts of data with very high data rate connectivity per device [14].

With the increased data rates and communication speeds, the new 6G systems will foster a new revolution in the digital age of industrial manufacturing. Some of the key prospects and applications, such as the super-smart society, will accelerate the quality of life for all humanity with smart AI-based communication. The ushering in of extended reality (XR) includes AR, MR, and virtual reality (VR), which all used 3D objects within the real-life environment. This will also lead to advancements in robotics and autonomous systems, wireless brain-computer interaction, haptic communication, smart healthcare, automation in the manufacturing industry, full sense information transfer, and the ever-growing IoT technologies. IoT has been placed under an umbrella term of IoE, which will interconnect the world.

Some of the specifications and requirements that will be changing on the new horizon are the service requirements of the 6G network. This will include eMBB, ultra-reliable low latency communications (URLLC), massive machine-type communication (mMTC), AI communication, higher throughput, increased network capacity, and higher energy efficiency, low backhaul congestion, and – a hot topic the last few years – increased data security. This change will also include newer integrated networks with connected intelligence, seamless integration of wireless information, and energy transfer of super 3D connectivity. This will require fewer general requirements in the network characteristics, including small cell networks, Ultra-dense heterogeneous networks, high-capacity backhaul, and mobile

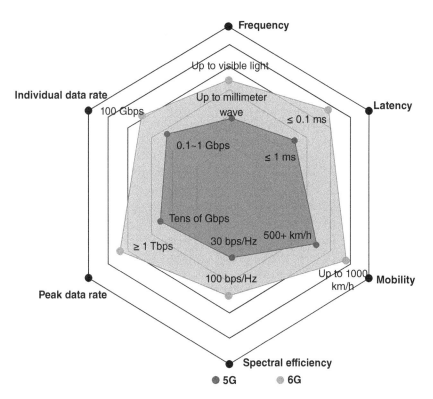

Figure 1.4 5G and 6G wireless characteristics [15].

technologies integrated with radar technology, softwarization with visualization. 6G is expected to outperform 5G in multiple specifications. We highlight six of them in Figure 1.4:

- Frequency
- Individual data rate
- Peak data rate
- Spectral efficiency
- Mobility
- Latency

Some of the key enabling technologies that will help push this high-speed age are:

- AI, with its ability to learn without the need for human interaction
- Terahertz communication which is needed for large data transfers
- Free-space optical (FSO) backhaul network for constant communication with devices and sensors
- MIMO technology for increasing data connection while reducing latency
- Ever-growing blockchain ledger system with its decentralized data system
- Fast retrieval speeds and quantum communication

Big Data analytics must be added to this list to market the expanding technologies to process the growing data retrieval.

There will have to be standardization in protocol and research activities. This has begun with Samsung Electronics leading the way and Finnish 6G Flagship programs that will help maintain consistency in the rapidly changing environment. The ITU has also kicked up while discussing the new 6G wireless network at the summit by the same name.

Some of the future challenges and the headed directions include but are not limited to areas in high propagation and the atmospheric absorption of THz to show the loss as frequency travels through free space. The added complexity in the resource management required for the 3D environments while minimizing hardware constraints. The required need to model sub-mmWave (THz) frequencies to be managed with higher wavelength capabilities. This will include spectrum management to prevent overlap or interference and beam management in the THz communication range.

What this all boils down to is that there is a rapidly changing technological environment coming very soon. It leaves the consumers oblivious to the amount of change needed behind the scenes to bring this to fruition. Looking on the bright side, this will make our lives easier and even better in the future. And for students taking this class, it will bring increased opportunities in the industry.

1.3 AI and Cybersecurity: Paving the Way for the Future

In the twenty-first century, cybersecurity and AI have presented numerous opportunities for the world while facing challenges behind the scenes. There is a battle to counter cyberthreats, turning to AI and machine learning. The US government employs methods to combat a new type of terrorism in a completely different environment. While it might seem that there are many challenges for cybersecurity and AI, there are also many opportunities for advancements in all types of technology.

Through recent advances in AI, we have seen what it is truly possible. Image recognition can analyze an image with a precision that would take hours for a human to do. New AI applications will bring challenges to the community that they have not previously faced. There are various security concerns with these applications because they are still vulnerable. Security will be the most important factor in the future for AI systems and applications.

There are many complexities involved with these systems, and they need to operate in all environments. There are four components in integrated AI systems: perception, learning, decisions, and actions. Each of these components must react independently from its counterparts, and each has unique vulnerabilities.

Since these AI systems can have high risks, trustworthiness must be ensured. Areas that need to be addressed for trustworthy decision-making include defining performance metrics, developing techniques, making AI systems explainable and accountable, improving domain-specific training and reasoning, and managing training data.

Even though AI does well on many tasks, vulnerabilities are still produced from corrupt inputs that produce inaccurate responses. These are current challenges the community is facing. "Modern AI systems are vulnerable to surveillance where adversaries query the systems and learn the internal decision logic, knowledge bases, or the training data. This is often a precursor to an attack to extract security-relevant training data and sources or to acquire the intellectual property embedded in the AI" [1].

You can see how cybersecurity could be affected by AI. Cybersecurity can use AI to increase its awareness, react in real time, and improve its effectiveness. This means that it uses self-adaptation and adjustment to face ongoing attacks and the ability to alter its course of action. It can help highlight weaknesses in an adversary's strategies by analyzing and reacting to them in real time.

1.4 Fusion of IoT, AI, and Cybersecurity

In the world of IoT, people are sometimes reluctant to adapt to the trend of connected products because of security concerns. These devices are more vulnerable to security threats because they are connected over a network with many open paths that can be attacked. Practices are in play involving strong authentication and tracking IoT devices. However, AI and 5G may address some of the concerns people have about IoT devices [14].

IoT has utilized home security, the medical field, utilities, appliances, entertainment, and so much more. With nearly every aspect of our daily lives utilizing technology, there must be strong cybersecurity practices in place. To understand the importance of cybersecurity, you must first understand what cybersecurity is and how you can practice it. Cybersecurity is the practice of protecting networks, data, and devices from unauthorized users. Such unauthorized users may be trying to access your data for financial gain, blackmail, or identity theft. According to statistics from TechJury, 30,000 websites are hacked daily. As of March 2021, 20 million records were breached [22]. The statistics are increasing significantly each year (Figure 1.5).

It is necessary to not only have good security practices but also have knowledge about attacks so you can better defend yourself:

- *Malware attacks* use malicious software to download onto a user's computer system. This attack is typically done through unsecure, fraudulent links.
- *Password attacks* are done when hackers use password generator software to crack a user's password.

Figure 1.5 Real-time display of cyberattacks being performed/requested [21].

- *Phishing frauds* are another cyberattack you must watch out for. These are emails that seem trustworthy and entice the user to click on a link that will unknowingly install malicious malware on their device.
- Structured query language (SQL) injections are when an attacker places a deceptive programming code into SQL statements to gain access to the network database.

All these things are serious threats and can be prevented with knowledge of common attacks and with good security practices. If you have access to the IoT, always make sure to follow cybersecurity practices. Make sure to check for software updates and patches to ensure that all your software is running up to date. To make things easier, there is typically an option for automatic updates that can be applied. A good practice would also be to use a strong password. Strong passwords usually consist of letters, numbers, and characters. You should never use personal information that is easy to guess as to your password. An example of a poor passcode could be the name of your pet. To better secure your password, you can enable two-factor authentication. By enabling this, hackers will more than likely move on from trying to access your account. Hackers generally like to target easy victims.

Another great example of good cybersecurity practices is to run up-to-date antivirus software. In case you are unable to detect the threat and/or virus and its origin for yourself, antivirus software will do it for you and will even remove the threat and/or virus. Also, you want to always be suspicious of unexpected emails. Generally, emails are good at filtering spam, but if the spam email was not filtered out, be extra cautious. A common hacker tactic is phishing emails, and the goal of this tactic is to gain information about you, steal money, or install malware onto your device [16, 22].

One positive effect that 5G is having on the security of IoT devices is the low latency it offers and faster speeds. There is one thing worrying developers about 5G, though: the protocol flaws discovered. Hackers have demonstrated that they can find certain connected devices in a certain area on 5G networks. One can see that 5G has many advantages but also has been shown to have its faults. The future for 5G and IoT devices is still largely unknown, but engineers hope it will pave the way for innovative capabilities.

1.4.1 Where Did AI Begin?

Before 1949, computers lacked a key prerequisite for intelligence. Computers, at the time, could not store commands, but could only execute them. In addition, they were expensive. The average cost of leasing a computer was equivalent to almost $200,000 a month. From 1957 to 1974, AI flourished. Computers were now cheaper and more accessible, and data could finally be stored. However, it was eventually met with a mountain of obstacles. Even though computers could store information, they could not store enough information or process it fast enough. Computers were simply too "weak" at the time. In the 1980s, an expansion of the algorithm toolkit and an increase in funds helped reignite the development of AI. John Hopfield and David Rumelhart popularized "deep learning" techniques that allowed computers to learn user experience.

1.4.2 Role of AI

AI is the use of machines to understand human intelligence. In this concept, intelligence is the computational part of the ability to achieve goals. AI uses algorithms designed to make decisions from real-time data. In order to operate on real-time data, AI must use vast amounts of data. Such data's quality and quantity drive AI effectiveness. The goal of AI is to use little to no human interaction. However, some human interaction is needed to tell machines how to extract features. A common example of AI would be autonomous driving (Tesla is a well-known autonomous vehicle). In autonomous driving, the vehicle receives data visually via cameras and through different sensing technologies. AI uses the data input from these features and produces an output, such as switching lanes or coming to a stop. It completely allows a passenger to ride in the vehicle without having to touch the steering wheel. Another common example of AI is face detection technology. Face detection technology allows the user to unlock a device simply by looking into the camera. The AI technology identifies key features and grants or denies access based on whether those features are met. This is a common feature with iPhone.

1.4.3 Disadvantages of AI

As much as AI has to offer, it does have its disadvantages as well. For example, there is high-cost implementation. Setting up AI-based machines is very costly because of the resources and complexity of engineering that goes into a building. Not only is it expensive to build but it can also be expensive to maintain or repair. Another disadvantage would be that AI does not improve with experience. Unlike humans, machines are not able to develop with age. Machines cannot alter their responses to changing environments. Machines are also unable to determine what is morally right or wrong because they are unaware of what is ethical and legal. If a machine is not programmed for a certain situation, the user cannot be certain about how the machine will respond. Even though the goal of AI is for little to no human interaction, humans simply cannot be replaced altogether. However, it can increase unemployment. AI machines are able to perform repetitive tasks that many humans do. According to a study conducted by McKinsey Global Institute, intelligent agents and robots could replace 30% of the world's current human labor by the year 2030. The study then goes on to state that "automation will displace between 400 and 800 million jobs by 2030, requiring as many as 375 million people to switch job categories entirely" [17].

1.4.4 Advantages of AI

Although reducing labor percentages has its disadvantages, it can also be classified as an advantage in AI. With machines being able to carry out most basic functions, it eliminates the need for companies to pay human employees to do such jobs. As a result, saving companies a significant amount of money in the long haul and/or free up employees' time to work on more complex tasks. According to the Oxford Economics Report in June 2019, more than 2.25 million robots are deployed worldwide. Amazon deploys more than 100,000 of these AI-based machines [18]. AI also helps increase safety. According to the World Health Organization Report, over a million people die in road accidents each year. AI has played a huge role in analyzing traffic patterns, lane discipline, and other driving patterns to develop self-driving cars. With most accidents being because of human fault, self-driving cars have and can improve road safety. For example, a 2018 survey conducted by IHS showed that drivers are 57% more likely to use a cellphone than drivers who answered the same question in 2014. Using a cellphone while driving is extremely dangerous, and, at the time of the survey, took over 385 lives. Another example would be driving while under the influence. Being under the influence can impair your decision-making skills and could eventually lead you to drive, especially if you are not in a comfortable situation. Statistics, according to National Highway Traffic Safety Administration (NHTSA), show that about one-third of car crash fatalities in the US involve drunk drivers. In both examples, if you take away the human driver and let AI take over, the probability of getting into a crash would significantly decrease. It would be like taking a taxi, in which you can be intoxicated or text as much as you want.

The first self-driving car was introduced on the road in 1995. Notice how accidents eventually decline (Figure 1.6). One thing about AI and IoT that has amazed everyone is how advanced it has become in the last decade. Thirty years ago, people could not imagine the advancements we can make today. However, one thing is halting advancement in a certain AI field that has yet to be solved, and it is in the autonomous car industry. Too much data must be processed and stored for these cars to run. If we want to fill the roads with autonomous cars, implementations need to tackle these problems.

The fusion of IoT, 5G, and AI has a long way to go, but continual developments help pave its future in technology. We will soon be living in a world run by connected devices, and AI will pave this road.

Four key aspects of 6G networks are real-time intelligent edge computing, distributed AI, intelligent radio, and 3D intercoms. 6G opens new frontiers of connectivity but 5G does not. 5G has struggled to arrive because of its infrastructure requirements. In contrast, 6G will build on the infrastructure we put down for 5G and enhance connectivity – on land, under the sea, or even in space.

6G applications will have much higher requirements and need for functionality for their networks. Networks may be required to reach Tbps connection speeds while also facilitating connectivity for extremely high counts of devices due to IoT and the plethora of mobile communication devices that utilize these networks. With applications that are extremely sensitive to latency, end-to-end latency of a network will require latency reduction down to microseconds. With these extreme jumps in speed, density, and latency requirements, 6G will still require the network to be 10–100 times more energy efficient than current generations. This may be accomplished through extremely low-power communication in some cases. These all breed causes for concern. Latency-sensitive applications will have to consider how implementation affects system latency, while the extremely large bandwidth of networks will offer up many challenges for ingesting and processing the network's data as it flows. Some applications will likely require processing locally to avoid these issues. As these networks evolve and become less and less homogenous, they will also introduce another security issue with the diversity of devices on a given network and their ability to change their edge networks frequently and quickly. The hypothesized

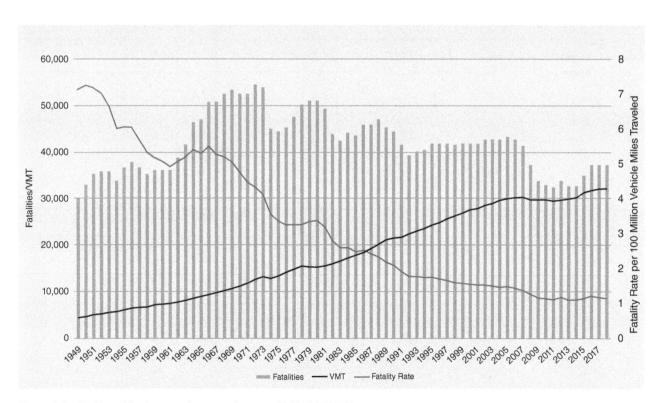

Figure 1.6 Decline of fatal car crashes over the years 1949–2017 [19].

key performance indicators (KPIs) to help address these requirements include protection level, time to resolution (TTR), coverage, autonomy level, AI robustness, security AI model convergence time, security function chain round-trip time (RTT), and cost to deploy security functions.

Additionally, the benefits of intelligent radio and RAN-core convergence offer up further considerations. As these technologies enable more rapid and automated management and deployment of radio networks via AI/ML, they will add an attack vector through AI/ML training manipulation. Detection of these types of exploits will be key to securing things. RAN and core functions being virtualized and able to be implemented close to the edge as well as at the core for higher-layer RAN functions will improve low-latency communications, but also add complexity to securing these new networks.

Edge computing married with AI will be essential, but will also offer up its own challenges. Edge intelligence relies on the hierarchical training of systems. This is to allow analysis and prediction to happen closer to the edge, making way for real-time analysis and prediction, but its algorithms and training are highly data-dependent, meaning that it is very important that the data be aggregated from reliable sources that authenticate and can guarantee data integrity. This may be a great space for the implementation of blockchain to keep consistency and authenticity. Another interesting topic mentioned here is homomorphic encryption.

The specialization of networks is also a potential area of concern. 6G network capabilities will allow for use of specialized subnetworks in many parts of society. The implementation of these smaller-scale specialized networks will require the use of lightweight but well-tested authentication and encryption devices. The creation and utilization of trusted execution environments offer another potentially more feasible solution for some closed networks. As with 5G, the continued use and expansion of the use of open APIs in 6G should also be recognized. These have been found to be vulnerable to multiple types of attack. The flexibility and computational capabilities of 6G networks could also allow for entirely closed-loop networks that are zero-touch and autonomous but will likely still have to consider things like distributed denial of service (DDOS) attacks and man-in-the-middle attacks. Figure 1.7 shows an AI-enabled 6G wireless network and possible applications.

1.4.5 Threats from Hackers

One of the biggest problems in the cybersecurity world is hackers. Many people do not realize how broad the term *hacker* can be. Most people think hackers only perform malicious acts and are dangerous criminals. There are three different types of hackers. Black-hat hackers are people who break laws, spread viruses, and steal information. They are often portrayed on TV and in the media, breaking into people's networks. Next, you have gray-hat hackers, who sometimes venture over into the dark side, but they are kind of neutral [23].

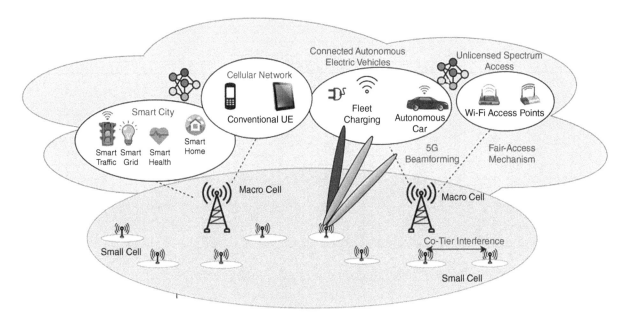

Figure 1.7 AI-enabled 6G wireless network and related applications [20].

They sometimes perform malicious acts, but other times they follow the rules and do not break any laws. Finally, you have the white-hat hackers, and this is where ethical hacking comes into play. They use their skillset for good and try to help people test their network for vulnerabilities and then patch them. If the media portrayed some hackers as good, some people might realize they can make it into an interesting career.

Ethical hackers can become certified and get a job in the community as penetration testers or security defense experts [24]. There are many opportunities in this field, and the type of work environment is limitless. You can travel abroad, and get a job with a private agency, essentially anywhere, allowing you to enjoy all aspects of your career. Ethical hacking is essential for businesses because it can save them money in the end. Sometimes when people build a company from the ground up, they miss a few things. Some people might design their network and never hire an outside firm to test their security.

Anyone can see that it is vital for a business or organization to hire a consultant to test its network. This has created an explosion of ethical hackers trying to be certified to be hired by an agency that can legally hack systems. A new field has arisen solely based on network security, which continues to advance.

Millions of dollars each year are saved by firms patching security flaws that could've been exploited. In the past, companies like Sony and multiple crypto exchanges have been targeted and lost millions because hackers found vulnerabilities in their systems. No telling how hard it was for the hackers to gain access. It could've been something simple or a complex attack. This is why it's important to test every aspect of your security and hire professionals to do it [25].

Another huge thing that companies and organizations need to pay attention to is social engineering. This is one of the easiest tools a hacker has in their arsenal because they can access your system with little or no work. People have been doing it for centuries, and other people are always the easiest targets because they are not as hard to crack as computers; people make mistakes. This can be accomplished with a defined skillset such as shoulder surfing, identity theft, and gaining unauthorized access to closed-off building areas such as the server room.

With the pace of technology never slowing down, the industry needs to keep up with the latest information on systems. Whenever a system is patched, or something new is added, it potentially opens doors to new vulnerabilities. That is why it is important to stay up to date with the security industry to learn about new vulnerabilities and exploits so they can prevent intrusions from happening to their network. With AI, automated systems can detect most of these problems before they happen [26].

1.5 How AI Can Help Solve These Problems

With the introduction of AI in cybersecurity, there are various ways to tackle the problems mentioned above. The world is changing, and so is how devices are connecting. IoT is revolutionizing the world of connected devices and bringing many innovations to the technology around us. One problem that has arisen from connected devices is the security behind the technology. Many companies are seeing the number of threats increasing because more devices must be managed. This is not hurting adoption, but it is a concern within the community.

One fix has been suggested, and that involves the 5G network and AI. One example is an AI company that focuses on user authentication. This provides an extra layer of security that is harder to crack than the application or device's traditional security. The 5G network brings faster speeds and low latency, which provides better management for devices that go a step further than user authentication. Both implementations address people's concerns about IoT-connected devices and will help adoption in the future [27].

Another approach is to analyze traffic patterns and look for an activity that could present as an attack. A type of AI that defends against cyberattacks is called the *decision tree*. A decision tree creates a set of rules based on its training data samples. An example of a decision tree would be detecting DoS attacks by analyzing the flow rate, size, and duration of traffic. If something seems unusual within those areas, the decision tree will categorize it as an attack on the system. Another AI example would be the k-nearest neighbor k-NN$_0$ technique. The k-nearest neighbor technique "learns from data samples to create classes by analyzing the Euclidean[1] distance between a new piece of data and already classified pieces of data to decide what class the new piece should be put in" [17, 22]. This technique has been used to detect attacks such as false data injection attacks.

---

1 The Euclidean distance is defined as the distance between two points ($d = \sqrt{[(X_2 - X_1)^2 + (Y_2 - Y_1)_2]}$).

Nevertheless, as always with benefits, there will also be cons. The 5G network has protocol flaws that could cause security flaws. Security experts have agreed that the 5G network has better security than the previous versions and has some of the same flaws ported over to the new network. 5G is also vulnerable to new attacks that we have not been concerned about with the previous versions. With the increase in data, attackers can go unnoticed because there is more traffic [28].

There is a huge factor at play with all these implementations growing with each other; they are gaining traction together. 5G, IoT, and AI are all helping each other out by implementing their features all under one roof. AI used to be something straight out of a comic book, and now we see it used more every day. IoT is making this possible by allowing data to be shared through devices. AI also enables data to be processed with much greater efficiency than ever before and will only grow as we progress. This is creating smart devices that are essentially learning.

These technologies are creating new platforms for companies. They bring new functions in transport, entertainment, medical, and public services. From a technological standpoint, the applications are limitless, and so are the number of devices that can communicate with each other.

The only problem with this is where exactly the data will be stored and how much we will store. So far, this has not been much of a problem, but it will need to be addressed in the future. However, the benefits outweigh the cons and will have a lasting positive impact on people and the world. Innovations will change the simple things we do.

We already see this effect in the world of virtual reality. We can fly drones and view the world at our leisure and watch live concerts from the comfort of our homes. This technology is creating a great future for everyone and will only get better.

IoT and its implementation in AI will continue to expand how we use technology and how we are connected to the world. There are some security concerns, but the technology itself is helping collect data on how we can fix these problems. The applications are limitless with what IoT, 5G, and AI have in store, and we will see life-changing technology appear faster than ever before.

1.6 Connected Devices and Cybersecurity

IoT is transforming the world, as we know it. As the technology advances, so does the limitless applications that are built on top of it. Data is being collected from thousands of sensors that help improve products' quality, management, and security. IoT brings innovation to the healthcare industry, cities and communities, energy and construction industries, and insurance. These developments come solely from IoT, and anyone can see how much of an impact it has made [29].

IoT has been great for businesses, but its practices are being deployed slower than expected; the reasons for this are security, interoperability, and costs. Many IoT products have not been pushed to production and are still in the concept phase. It is also hard for them to hire experienced staff who can work on these complicated projects. For now, companies are relying on simple IoT projects because they generate value quickly, thus helping the company retain a consumer base.

Even though IoT is slowly growing, great innovations have been made to the cost, security, and technology. There have been many great products developed through IoT, one being by the pharmaceutical industry. IoT solutions can track records of medicines from production to the patient. Introducing these new products offers competitive rates to the consumers. This will essentially develop into more use cases.

These are just a few examples of limitless applications and products from IoT. I believe the future is through IoT and the optimization it brings to the table with AI. We will continue to see this market grow, expand to other industries, and create more competitive pricing models [30].

IoT-connected products are implemented in homes, wearables, the industrial internet, connected cities, and cars. One product that is playing a huge role in its expansion is the use of sensors. They can be used for security, optimization, and autonomous technologies. Data from sensors is even being used to enhance the safety of first responders. Anyone can see how limitless the applications can be and how important they will be for the future.

The use of sensors is creating the next generation of public safety services. Sensors can track data in real-time and send alerts to responders. A high-level architecture is used, allowing various sensors to be interconnected, allowing more than one application is the same architecture. This is huge for optimization in any industry because it creates less hands-on work, and the data is even being processed and stored more efficiently.

The current state of technology is readily developed for a future state. It will be connected over broadband internet, have multi-media/supplemental data, and provide situational awareness in the future state. There are many considerations for this future technology, such as an open ecosystem for applications and developers, data security and privacy, and a national movement toward open government data.

This technology provides way more benefits than cons. At first, developing some of these products could be pricey, but as technology advances, the pricing will get more competitive, reducing the cost. IoT-connected devices are here to stay, and we will see many applications in our life-giving us a better quality of life [31].

We recognize connected devices as mobile phones and PCs, virtual assistants, and GPS, but now we have automated cars and radio frequency identification (RFIDs). This presents many security challenges. With the emergence of cloud-based operations, there are increasing challenges, making analyzing the data a lot harder.

With cloud computing, it's much harder for forensics professionals to analyze what's on the hard drive because it is sometimes potentially unrecoverable. More complex procedures for investigating must be used because it is much harder to analyze this type of device data. Sometimes it's not necessarily harder to come across the data, but it becomes cumbersome. Often there is just so much data that needs to be analyzed, and it takes a lot of extra work.

The devices that are connected come with weak security because they are so new to the market. It is hard to test for all vulnerabilities if the item has not been made public to be tested in real-time. The attack surface for hackers is becoming a lot wider with all the new products coming to market. All the sensors, automated systems, and infrastructures can be easily targeted because of their unknown vulnerabilities. Attention must be paid to all of these new devices because of the high workloads they are creating and the data they are processing.

This can be combated by properly seizing data, storing data, extracting, and analyzing it. As of now, there is no procedure that covers the world of IoT. This must change as we continue to push out connected devices and build the world of IoE [32].

1.7 Solutions for Data Management in Cybersecurity

For IoT to go well, there must be a solution for data management. Data being generated with IoT devices is hard to process, and the speed of the data may be slow. 5G must face challenges in network latency, bandwidth usage, security, and governance.

Cloudera Dataflow has proven that it has come up with a solution to these problems. It tackles major problems that most IoT devices were experiencing. Its management, processing, and analysis all help make this possible. It is also open-source, which is a game-changer for products in the IoT field.

With all the devices being connected, something must be done with their data. This is where data management comes into play, which has a key focus on automation. By using data management, businesses will become more productive, accelerate IoT initiatives, and help deliver a comprehensive view of IoT-related data.

Other vendors are also tapping into this new market and offering data management solutions. Google, Amazon, and Microsoft all offer data management solutions using a vast computing and storage infrastructure. Data management vendors are constantly releasing open source projects so other companies can help build on the technology. Several hardware vendors have designed systems used for IoT.

1.8 Conclusion

There has been a large expansion of ways to collect, manipulate, organize, and mine data. With this comes greater responsibility for the networks and connected devices to take into consideration security issues. This not only includes software, password, and hardware security but now encompasses site security as well. This will require large amounts of data to be stored and more data space for redundancy to make sure security measures are at the maximum they can be.

We have discussed the role of AI in cybersecurity and its impact on our everyday lives. From security to smart devices, there are multiple uses for AI. Most of the problems in cybersecurity can be addressed by the use of AI. We have discussed how intrusion detection systems can provide more security for your network and how AI can prevent and deter hackers. Data management is a problem that will continue to be discussed until we find a permanent solution to the problem. Another problem it faces is the constant connectivity to smart devices and sensors. With everything being connected to the internet now, it is more vital than ever to have AI deployed in more places than one.

References

1 Yang, K. (2017, April 3). AI is a mechanism for creating diversity. In: *Speeding the Arrival of a Super Smart Society*. Hitachi: Inspire the Next https://www.hitachi.com/rd/sc/ai/001.html.

2 Vaigandla, K.K., Bolla, S.R., and Karne, R.K. (2021). A Survey on Future Generation Wireless Communications-6G: Requirements, Technologies, Challenges and Applications. *International Journal of Advanced Trends in Computer Science and Engineering* 10 (5): 3067–3076. doi: 10.30534/ijatcse/2021/211052021.

3 Soldani, D. (2021). 5G evolution, 6G vision, security controls, and assurance. Webinar at AISA. https://youtube/S9215UdnJs4.

4 Powered by NewsPicks Brand design. Realizing Society 5.0 ROBOT (n.d.). https://www.japan.go.jp/abenomics/_userdata/abenomics/pdf/society_5.0.pdf.

5 Marr, B. (2019). What is extended reality technology? A simple explanation for anyone. Forbes. https://www.forbes.com/sites/bernardmarr/2019/08/12/what-is-extended-reality-technology-a-simple-explanation-for-anyone/?sh=2dceca047249.

6 Stacey, K. (2021, March 31). Researchers demonstrate the first human use of a high-bandwidth wireless brain-computer interface. *News from Brown* https://www.brown.edu/news/2021-03-31/braingate-wireless.

7 Haptic Communication (2018, November 15). Communication Theory. https://www.communicationtheory.org/haptic-communication.

8 Haptic communication (2019). Psychology Wiki. https://psychology.wikia.org/wiki/Haptic_communication.

9 Wikipedia contributors (2019, March 13). Haptic technology. https://en.wikipedia.org/wiki/Haptic_technology.

10 Sundaravadivel, P., Kougianos, E., Mohanty, S., and Ganapathiraju, M. (2018). Everything you wanted to know about smart healthcare. http://www.smohanty.org/Publications_Journals/2018/Mohanty_IEEE-CEM_2018-Jan_Smart-Healthcare.pdf.

11 Zaman Chowdhury, M., Shahjalal, M., Ahmed, S., and Jang, Y. (n.d.). 6G wireless communication systems: applications, requirements, technologies, challenges, and research directions. https://arxiv.org/ftp/arxiv/papers/1909/1909.11315.pdf.

12 Menting, M. (2021). Conceptualizing security in a 6G world. ABI research. *6G World* https://www.6gworld.com/conceptualizing-security-in-a-6g-world-3.

13 Soldani, D. (2021). 6G fundamentals: vision and enabling technologies: from 5G to 6G trustworthy and resilient systems. *Journal of Telecommunications and the Digital Economy* 9 (3): https://jtde.telsoc.org/index.php/jtde/article/view/418.

14 Banafa, A. (2016, August). The Internet of Everything (IoE). https://www.bbvaopenmind.com/en/technology/digital-world/the-internet-of-everything-ioe.

15 Yang Zhao, Wenchao Zhai, Jun Zhao et al. (2021). A comprehensive survey of 6G wireless communications. arXiv:2101.03889v2 [eess.SP].

16 Cybersecurity & Infrastructure Security Agency (2019, February 1). What is cybersecurity? https://www.cisa.gov/uscert/ncas/tips/ST04-001.

17 Das, A. (2019, November 25). 5 disadvantages of AI. https://www.proschoolonline.com/blog/what-are-the-disadvantages-of-ai.

18 Pedamkar, P. (2019, October 1). Benefits of artificial intelligence. Benefits of artificial intelligence. https://www.educba.com/benefits-of-artificial-intelligence.

19 Collisionweek Editor (2019). U.S. roadway fatalities decline for second straight year. https://collisionweek.com/2019/10/23/u-s-roadway-fatalities-decline-second-straight-year.

20 Yang, Z., Zhao, J., Zhai, W. et al. (2021). *A Survey of 6G Wireless Communications: Emerging Technologies*. SpringerLink.

21 Acura, M. (2017, May 13). Global real-time cyber-attack infects thousands of computers worldwide. PakistanToday. https://archive.pakistantoday.com.pk/2017/05/13/global-realtime-cyber-attack-infects-thousands-of-computers-worldwide.

22 Kuzlu, M. (2020). Role of artificial intelligence in the internet of things (IoT) cybersecurity. *Discover the Internet of Things* 1: 7.

23 Gupta, S. (2019). *Exploitation Attack. Ethical Hacking – Learning the Basics*. Apress. http://dx.doi.org/10.1007/978-1-4842-4348-0_11.

24 Ethical Hacking (n.d.). https://repo.zenk-security.com/Magazine%20E-book/EN-Ethical%20Hacking.pdf.

25 Ashvin, G. (n.d.). Ethical hacking tutorial. https://www.academia.edu/32432762/Ethical_hacking_tutorial.

26 Ethical Hacking. (n.d.). https://www.tutorialspoint.com/ethical_hacking/ethical_hacking_tutorial.pdf.

27 Porter, M.E. (2014, November). How smart, connected products are transforming competition. Harvard Business Review.

28 Sterlite Technology Limited (n.d.). Don't just change, Transform with Intelligent Connectivity. Sterlite Tech. https://www.stl.tech/mwc19/pdf/01_Intelligent_Connectivity_Whitepaper_16_01_19_web.pdf.

29 Yarali, A., Ramage, M.L., and May Manu Srinath, N. (2019). Uncovering the true potentials of the Internet of Things (IoT). In: *2019 Wireless Telecommunications Symposium (WTS)*. New York: IEEE https://doi.org/10.1109/wts.2019.8715545.

30 MacDermott, A., Baker, T., and Shi, Q. (2018). IoT Forensics: Challenges for the IoA Era. *2018 9th IFIP International Conference on New Technologies*, Mobility and Security (NTMS). https://doi.org/10.1109/ntms.2018.8328748.

31 Cloudera (2019). Cloudera dataFlow: IoT data management from edge to cloud. https://www.cloudera.com/content/dam/www/marketing/resources/analyst-reports/forrester-now-tech-iot-data-management.pdf?daqp=true.

32 Simon Segars, Advisor to the CEO, Arm, 2020. AI, 5G, and a secure IoT: The fifth wave of computing.

2

Networks of the Future

2.1 Introduction

Telecommunication sectors consist of companies that allow communication to happen globally. These companies have created the standard infrastructure that we know today. They consist of these subsectors: wireless communications, communications equipment, processing systems and products, long-distant carriers, domestic telecom services, foreign telecom services, and diversified communication services.

Advancements within the telecommunication sector have caused a dramatic change in how people live across the globe. This change has also opened new doors, posing challenges and new opportunities. The first was the launching of 5G in 2018, offering more speed, efficiency, and less latency for the future. Another opportunity is the creation of autonomous vehicles. The possibility of fewer global regulations also opens up the market for more freedom and competition. Cross-industry partnerships are another area that may see new growth, as they are reliant on the telecommunications industry. Finally, security and privacy will be a higher priority as demand continues to grow.

With wireless technology seeing exponential growth in the last decade, it is no wonder that it is becoming one of the largest and fastest-growing industries in this technical revolution. With every generation, new advancements are being made. These improvements have had a dramatic change on the international communication sector, allowing for greater improvements in the way information is transmitted. The television and radio have been around for as long as the new generations can remember, and even these have been improved with the rapid expansion of wireless technology. These advancements have also ensured that the cost of connectivity is cheaper than ever before, providing access to many different areas. This journey for wireless communications has only happened for around 50 years. In that time frame, the standard wireless network has undergone significant development: wireless personal area networks (WPAN), wireless local area network (WLAN), ad hoc network (adhoc), or mesh networks, metropolitan area networks (WMAN), wireless wide area network (WWAN), cellular networks, and space networks.

Broadband telecommunications have also undergone massive development over the past decade, evolving to become more efficient with their broadband usage than ever before. In the early 1970s, in its first generation, also known as 1G, the telecommunications standards via analog only allowed voice services to its subscribers. In the 1980s, 2G came out with the first iteration of digital signals. 2G also saw the creation of text messages for those mobile phones and even began roaming data services. 3G would emerge in the late 1990s, allowing for large data bandwidth and faster internet connectivity. The unfortunate part is that the actual network infrastructure was not rolled out the way it was marketed, but that's where 3.5G came in, making the 3G network even more efficient than it originally was, allowing for fast and better uplink/downlink packet access.

Then came the fourth generation, 4G. Its purpose was to allow internet access with speeds of up to 100 Mbps. Due to the specifications of the 4G network, many different service providers and users could access it. Based on the standards created by the International Telecommunications Union-Radio communications (ITU-R), the data rate would also vary, depending on the facilities and equipment used. Those who connected with those experiencing high mobility, such as people in cars and trains, would have access to data rates varying from 100 Mbps or higher, while those in low mobility, such as pedestrians and stationary devices, would experience around 1 Gbps. These policies were created by the International Mobile Communication Advanced specifications (IMT-Advanced). 4G networks could also be used in network infrastructures like WLANs and WANs.

From 5G to 6G: Technologies, Architecture, AI, and Security, First Edition. Abdulrahman Yarali.
© 2023 The Institute of Electrical and Electronics Engineers, Inc. Published 2023 by John Wiley & Sons, Inc.

With 4G, a plethora of standards ratified by IEEE also came into effect after meeting with ITU-R to ensure they meet the peak speed of at least 100 Mbps. The goals of this were as follows: To be a completely IP-based integrated system, to provide communication indoor and outdoor with speeds of between 100 Mbps and 1 Gbps, to have 4G combine the usage of Wi-Fi and WiMAX (Worldwide Interoperable Microwave Access) technologies, and to establish fourth generation long-term evolution (4G-LTE). 4G LTE is a standardized mobile technology that meets the minimum requirements of 4G that consists of a wide range of advancements and features that were not available for previous versions of both LTE and 3G. To allow 4G LTE to work in all locations, it combined network types with 3G to meet specific requirements.

Finally, we get to 5G, which is as unique as the technologies used to create it. It is expected to increase both the range and capacity of telecommunications. 5G will no longer be limited to broadband networks, as it will be built for various systems that need higher speeds. However, with 5G not yet fully implemented, policies and other implications must be analyzed and considered well before its launch. Safety and security are also being discussed and debated by many experts with the current global focus against terrorism.

5G is set to surpass 4G networks in both its higher bandwidth speeds and internet connectivity. The 5G spectrum also supposedly can hold between 3 GHz up to 300 GHz. With 5G's cellular networks, while being similar to 4Gs, will have the advantage of being based on the Orthogonal Frequency Division Multiplexing scheme. This allows for smaller technologies to be able to enhance the quality and capacity of data rates. 5G will also have access to major features such as significantly lower latency, improvement of IoT, M2M, and device-to-device communications, and reduced energy consumption, increasing battery life in general for the device. 5G also boasts several characteristics that are important for high performance. Some of these include improved reliability and security, network traffic prioritization, and seamless communication via the Internet of Things (IoT).

The present generation has seen many changes in technology that have transformed our lives, daily work, and how we communicate with each other. Information and communication technology (ICT), in all its variants, has been incorporated into our social lives, and we are dependent on this technology in our education, work, and business sectors. In this chapter, we will discuss Wi-Fi, LTE-A, 4G, 5G, HetNet (heterogeneous networks), and 6G, along with the discussions of third-generation partnership project (3GPP). The reader will learn the importance of 3GPP and its contribution to wireless cellular networks and communications. Also discussed in this chapter is how the energy consumption of utilizing variant ICT usages, applications, and mobile communication networks has become an issue and concern for the research and service providers of communication systems. Research has focused on reducing network energy consumption as devices migrate to higher generations and hence higher data consumption and possible solutions that can reduce energy intake by the networks, devices, and data centers are being explored. Moving toward green networks for the future can help to reduce CO_2 emissions caused by wireless mobile communication systems, for example. We begin our chapter by discussing some of the areas where energy consumption in ICT can be reduced.

2.2 The Motive for Energy-Efficient ICTs

As of late, the National Academy of Engineering has discovered many disputes in the twenty-first century. Out of these issues, three of them concentrate on energy-related problems. All the challenges exhibit the significance of energy issues, and they have a common objective –minimizing the cost and the emission of carbon dioxide. In the last two decades, the growth in ICTs, which comprise business, commercial, and educational sectors, has exploded. In the past, the performance and price of the ICT were the main criteria that researchers and developers focused on; power utilized and what kind of impact it will have on the environment were given less priority. However, the current tendency is the focus on the rising price of electricity or power resources, as well as how carbon-dioxide-mitigating technologies might affect our energy needs. Here we concentrate on energy utilized by present product network technologies such as wired and wireless.

As of now, we have seen a serious shift in the importance of information technology. This change and prominence to bolster the ICT used by utilizing only efficient energy networks and reducing the carbon dioxide emitted in different parts like business, educational, and industrial are being moved by different factors, which also include the following:

- Environmental problems like global warming
- Rise in demand for more electricity to bolster present and upcoming ICT devices
- Rise in the price of the energy
- Rise in the consciousness of national security

A study shows that the growth in the utilization of energy and the carbon dioxide discharge, which is caused by ICT devices, is almost doubling every five years. The total effect of the cost of the energy utilized by ICT devices has been exhibited, and it is shown that educational institutions are consuming 30–50% of the total energy. Present ICT is the biggest consumer of energy in household and business sectors [1]. Many researchers have published their work on green networking, paving the road toward efficiency of the bandwidth, coverage range of communication, dynamic channel allocation to users, convert-and-deliver, and file sharing scenarios.

ICT produces carbon dioxide indirectly by utilizing the power from fossil fuels, which leads to the discharge of CO_2. The energy-saving ICT plans distributed until now have been pointed at increasing energy utilization and managing ICT devices. There is a hope that these plans will minimize the effect on the environment of ICT usage but also may help the energy providers, suburban, and business users to increase their profitability in the long term. Currently, telecommunications infrastructure and server farms utilize almost 3% of the total world's power, and this is growing abundantly, up to 16–20% each year [2]. Cellular networks utilize almost 0.5% of the world's power, with the network and the terminals of end users using a high percentage of cellular energy usage. Researchers foresee that in the next seven years, the energy cost to run the cellular network will almost triple, with 80% of the energy used by the sites in the base stations [3]. The rise in the usage of the internet is also the main factor for worsening energy usage. Researchers have said that the CO_2 emissions related to global computing were responsible for a greater share of ICT discharges than it was thought [4]. As the user wants the device to be more portable and the companies to meet the expectations of users for services from almost anywhere irrespective of time, cellular network service providers are competing to support this type of communication demand. The invention of 4G (fourth generation) and 5G networks is also a factor in the rise in energy usage. Massive connectivity of IoT devices is estimated to add about 30% of more energy consumption. However, the efficiency of the networks will be enhanced by about 60% with the implantation of artificial intelligence (AI) and machine learning (ML) in the future 5G and 6G networks when fully deployed.

2.2.1 Approaches

Developers have been working on energy-efficient networking for many years. Increase in the usage of the internet and wired and wireless networks, studies have drilled down the energy efficiency of protocols, networks, and the applications, which run on them. Different types of energy-related problems have been checked vigorously, covering most topics like media access control (MAC), allocation of spectrum, and routing. This section depicts a few examples of recent progress, especially in increasing the energy efficiency of product-based networks like ethernet and wireless LAN.

Ethernet is one of the most prominent and famous wired technologies used for local area networks (LANs), which are widely used worldwide by 3 billion people. Lumen, AT&T, Spectrum Enterprise, Verizon, Comcast Business, and Cox Business are the top six companies to qualify for a rank of having 4% or more of the US retail ethernet services market [5]. Ethernet is mostly used in commercial, business, and residential areas. Almost every personal computer, laptop, and server has at least one network adapter of ethernet. Nowadays, ethernet interfaces are also used in home appliances.

Many relevant technical advancements such as GbE and wireless 802.11ac contribute to the evolution of ethernet usage in the industry for IIoT (Industrial Internet of Things). Data transmission is rapidly increasing; in the past, it was 10 Mbps, and now it is increased to 100 Gbps (100 GbE), replacing fast ethernet. Present network adapters that support high data transfer rates utilize the most power; for instance, a 10 GbE Base-T transceiver uses nearly 10 times more than 1000 Base-T.

Researchers and developers have shown great interest in improving energy efficiency in the last five years. Their hard work resulted in the growth of standard Energy Efficient Ethernet (EEE), which was accepted in 2010. The change made by this EEE introduced the concept of the low power idle (LPI). The sender should send the idle signals, which are used to maintain the position from the sender to the receiver when no data is transmitted. Using a traditional idle signal, we can save at least 50% of the energy during this time. Approaches like packet coalescing near the sender before sending the data to the receiver they send in a break are investigated and exhibited to reduce the energy usage of EEE. At the same time, they keep the data packets in a predefined jump for applications that are based on the internet. It is believed that the use of EEE has resulted in saving energy, which is crossing almost $1 billion globally.

One more famous effort to save energy for ethernet technology resulted in present ethernet coding plans and a novel energy sensible encoding that is easy to apply. This suggested technique returns almost 18% and 60% betterment in encoding and the transmission of the circuit energy.

Recent studies have found very positive feedback about ICTs that they will minimize the amount of power used and reduce the emission of CO_2, which helps the environment. Some of the positive impacts that can be gained with ICTs are

- Making electricity procedures more efficient
- Enabling technologies that improve business operations
- Like fiber optics, promoting the development of low carbon outcomes

The ICT department is developing abundantly worldwide, and the improving speed will grow in the upcoming years as more ICT devices are used in various sectors of everyday life.

High-energy performance and minimizing energy usage is the basic requirement of 5G. It reduces the ownership cost and extends the network connectivity to almost everywhere, and the network access is a very bearable and resource-efficient way. The main technology to finish the ultra-lean model and separation of user's data on the radio interference. 5G is very costly when compared to other data plans. Its functioning model is different from 3G and 4G.

The structure of 5G plays an important role in energy saving while the data is transmitted. The device does not transmit the data until a user data transfer occurs. This concept saves a lot of energy. 5G networks will handle more antennas and frequency bands with a faster data transfer rate. The main two design principles of this technology are

1) Being active only when the transmission is required.
2) Being active only where transmission is required.

It is an ultra-lean model whose mechanism "only transmit when needed" endeavors to reduce transmissions unrelated to the data delivery. This kind of model is for all types of positions, including macro.

6G needs to be a very energy-efficient communication structure. Energy and battery efficiency are both topics of interest for future phones to pair with 6G, as the future 6G applications need to have almost zero latency and an extremely high data rate. One of the technologies to accomplish this efficiency is that THz waves are very high-powered and can utilize directional power beams. Improving mobile nodes will also help efficiency by lessening the power needed to operate them.

We can conclude from this section of the chapter that reducing processing energy usage only will not be enough to maintain steadiness. A comprehensive method is needed that covers the whole product life cycle, not energy consumption alone. Information and communication technologies are very powerful catalysts that can help to speed up efficiency and productivity in many departments. If we distribute the usage of power and the network load, we can primarily improve the network quality, and it makes work easier. We will apply the feasible aggregation plans in the upcoming years, where we can talk about real and unreal aggregation plans to make the suggested algorithm appropriate for most applications.

2.3 Wireless Networks

Wireless mobility is a large part of society in today's world. Cellphones, tablets, computers, and watches are just a few names that interconnect our lives to the internet. We need the internet for work, school, entertainment, and sometimes to monitor critical aspects of our lives. Many different technologies allow us to interconnect to the world of the internet.

As new advancements are made daily, cellular advancement has also become constant because of the demands of the customers and users. Figure 2.1 [5] shows the rapid growth of global mobile data traffic forecast by ITU (International Telecommunication Union) during the past few years. Overall, mobile data traffic is estimated to grow at an annual rate of around 55% in 2020–2030 to reach 607 Exabytes (EB) in 2025 and 5016 EB in 2030 [6].

The journey starts with 1G (first generation), and now 5G (fifth generation) is being introduced and implemented in most areas. The second generation was introduced in three phases – 2G, 2.5G, 2.75G – and then 3G (third generation), 4G (fourth generation), were introduced. We have started working on the latest generation, namely 6G. Although there have been no ratified standards, research and scientist have discussed and published many applications for this Terahertz technology.

In 1979, the Japanese company introduced the 1G (first generation) with a speed of 10 MHz and frequencies of 800 and 900 MHz. Still, due to analog signaling and technical limitations, it was not very successful. GSM, also known as the global mobile communication system, was introduced in the second generation and was a key feature for further advancement. Digital wireless systems, email and SMS services, voice transmission (coded), and relatively high security were the other main features of the second generation. Then in 2001, 3G was presented in Japan and in the United States. In 2002, Verizon extended the redundancy range into the 2.1 GHz band, permitting phones to work at higher rates. MIMO (multiple inputs, multiple outputs) was the primary leap forward during this time since it supported the network information transmission

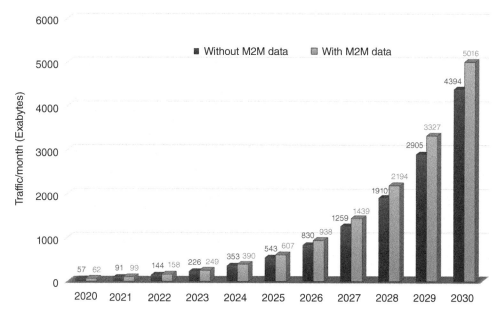

Figure 2.1 The rapid growth of global mobile data traffic forecast.

and gadget limit [7]. 3G could hypothetically arrive at paces of 40 Mbps, and its quicker information move rates empowered now-standard components like versatile web perusing, photograph sharing, and GPS tracker.

4G was presented in 2009, with frequencies in the 600 MHz, 700 MHz, 1.7 GHz, 2.1 GHz, and 2.5 GHz ranges. 4G network velocities may surpass 400 Mbps, considering superior quality video gaming, continuous video web-based, and video conferencing. Most firms currently utilize 4G LTE, which has diminished torpidity and expanded capability. (LTE, long-term evolution, demonstrates the way to 4G.) Millimeter wave-based 5G is anticipated to give surprisingly massive levels of apex downlink throughput execution, going from 4 Gbps on the low finish to 20 Gbps on the better quality. These rates address an enormous danger in each business and vertical. Expect more regular checking, low-dormancy correspondence in remote spots, and the advancement of "urban regions" in the not very far future. Table 2.1 shows the evolution phases of mobile network technology [8].

This mobile technology is based on LTE, and it is in its early stages of development, intending to revolutionize mobile communication by coming up with systems that adequately eliminate current problems experienced with the existing mobile systems. The following are other features that 5G is aiming to implement better than its predecessor:

- Reduced battery consumption
- Increased range of coverage and higher data rate throughout the cell
- High number of concurrent data transmission paths and handover

Table 2.1 Summary of comparison between the five mobile communication generations.

Technology	1G	2G	3G	4G	5G
Deployment	1970–1980	1990–2004	2004–2010	now	2020
Throughput	2 kbps	64 kbps	2 Mbps	1 Gbps	Higher than 1 Gbps
Technology	Analog Cellular Technology	Digital Cellular Technology	CDMA 2000 (1xRTT, EVDO) UMTS, EDGE	Wi-Max LTE Wi-Fi	WWWW (coming soon)
Service	Mobile telephony (voice)	Digital voice, SMS, higher capacity packetized data	Integrated high-quality audio, video, and data	Dynamic information access, wearable devices	Dynamic information access, wearable devices with AI capabilities
Multiplexing	FDMA	TDMA, CDMA	CDMA	CDMA	CDMA
Switching	Circuit	Circuit, packet	Packet	All packets	All packets
Core network	PSTN	PSTN	Packets N/W	Internet	Internet

- Interactive multimedia services such as voice calls, video calls, the internet, and other broadband services that are more effective for bidirectional traffic statistics
- Mobility data rate averaged at over 1 Gbps and a higher broadcast capacity of over 65,000 connections at a time
- Improved security features such as better cognitive software development radio (SDR)
- Worldwide wireless web (WWWW)-based applications that have full multimedia capability
- AI-aided applications
- High resolution for cellphone users and bidirectional large bandwidth shaping

2.3.1 Wi-Fi

For many people, Wi-Fi plays a very important role in our day-to-day operations. As soon as we arrive at a location of any sort, we pull out our mobile devices and look for public Wi-Fi. Most technical devices are equipped with Wi-Fi-capable devices. Even things such as our washers, dryers, door locks, refrigerators, cars, stoves, and many more things are now equipped with Wi-Fi capabilities. Wi-Fi is short for wireless fidelity and is known as WLAN (wireless LAN). Wi-Fi is based on IEEE 802.11 standards. Wi-Fi operates using unlicensed wireless frequencies 2.4 and 5 GHz. Generally, devices can connect to frequencies or both, which would be considered dual-band capability.

There are different revisions of the IEEE 802.11 standard. In consumer Wi-Fi, the first to show its excellence to the world was 802.11a. 802.11a operated using the 5-GHz frequency and was introduced in 1999. 802.11a offers speeds up to 54 Mbps and was not generally used in a consumer-based configuration. 802.11b was the next Wi-Fi technology brought to the attention of home users. 802.11b operated using the 2.4-GHz frequency and was introduced the same year as 802.11a. 802.11b was popular with home users, accomplishing speeds of 11 Mbps.

When Wi-Fi was created, ownership of personal devices such as laptops and personal data assistants (PDAs) exploded. Consumers could have the wireless capability in their homes, and they loved it. Although in 1999, when consumers were rocking the high speeds of 11 Mbps, a new 802.11 technology was introduced. 802.11g was brought into the force of Wi-Fi, and consumers could now enjoy speeds of up to 54 Mbps using the 2.4 GHz frequency. To look at the frequencies used by Wi-Fi, there is a reason consumers prefer 2.4 GHz over 5 GHz. 2.4 GHz is half as narrow of a frequency as 5 GHz, which has a greater potential to avoid obstacles such as walls, trees, or floors.

802.11g was introduced sometime in the year 2003. Until 802.11g, consumers chose the range of 802.11b over the speeds of 802.11a. Not to mention, the prices of 5 GHz equipment were drastically more expensive than 2.4 GHz equipment. Wi-Fi continued to grow increasingly popular, and devices continued to support the growing need for this new way of communicating data. Just as in 2003, when we thought we had more than we would ever need, in 2009, 802.11n was introduced. 802.11n introduced the new 802.11 technology capability of MIMO (multiple-in, multiple-out). MIMO gave Wi-Fi and 802.11n the capability to transmit and receive using multiple transmitters and antennas to increase the throughput of a Wi-Fi access point and devices. 802.11n can have up to eight antennas but most only use four antennas. Using the four-antenna system, two antennas would be used to transmit simultaneously, as the other two antennas would be used to receive data packets simultaneously. Vendors claim that 802.11n can deploy speeds up to 250 Mbps. The most recently deployed consumer 802.11 technology is that of 802.11ac, which is supposed to be able to support actually up to 7 Gbps of throughput with wider channels and quadrature amplitude modulation (QAM) increase to 256 from 802.11n's 64 QAM. QAM is an abbreviation of quadrature amplitude modulation. QAM is an adjustment of amplitude and phase that allows transmission of data wirelessly in which each path is represented as a symbol. The greater the QAM, the more paths a wireless signal can transmit, increasing the throughput. Table 2.2 is a chart of consumer Wi-Fi mentioned thus far and their corresponding speeds and other facts.

Wi-Fi not only has grown as technology itself, but it has also helped the world economically. Many businesses take advantage of Wi-Fi and its greatness. When passing through an area with businesses or looking for a hotel, we are attracted to the "Free Wi-Fi" notice associated with the services offered. Wi-Fi has had a tremendous effect on the economy of Africa. As mentioned in an article by Montegray [10], there is a projection that by 2025, Wi-Fi will be responsible for an increase of up to $300 billion in its GDP. Africa started their projects to implement Wi-Fi for its citizens to improve the education and economics of the country. Free Wi-Fi does several things for your business [11]. Businesses found that customers spend more time at their business after offering free Wi-Fi. During a study, it was determined that 62% of businesses stated that people spend more time at their businesses after offering free Wi-Fi. While only a few businesses believe that offering free Wi-Fi just crowds their business with freeloaders looking for Wi-Fi, the study showed that 50% of the customers spent

Table 2.2 Maximum speed comparison between the Wi-Fi generations [9].

WI-FI Generation	IEEE Standard	Bands	Top Speed	Year
Wi-Fi 7	802.11be	1-7.25 GHz	46 Gbps	2024
Wi-Fi 6, Wi-Fi 6E	802.11ax	2.4, 5, 6 Ghz	9.6 Gbps	2019–2021
Wi-Fi 5	80.11ac	5 Ghz	3.5 Gbps	2014
Wi-Fi 4	802.11n	2.4, 5 Ghz	0.6 Gbps	2008
Wi-Fi 3	802.11g	2.4 Ghz	0.05 Gbps	2003
Wi-Fi 2	802.11a	2.4 Ghz	0.05 Gbps	1999
Wi-FI 1	802.11b	5 Ghz	0.01 Gbps	1997

more money. The article also mentions that customers are more likely to sit alone at restaurants or cafés when free Wi-Fi is offered. A survey showed that 53% of people are happy to sit alone when free Wi-Fi is offered.

With the growing trend of free Wi-Fi, businesses must embrace the free service because a study showed that 1 in 10 people would leave a business if free Wi-Fi was not available. This case is more effective in hotels when people are staying for a long period. Free Wi-Fi at businesses can also give a business the capability to get customers to view ads. When a customer connects to the free Wi-Fi at the business, the customer could be initially forced to a webpage displaying daily deals or ads about discounts.

In my opinion, Wi-Fi has changed the world of technology to levels we would never have imagined. The thing with consumers is that more is never enough, and as long as the speeds and technology progress, we will want it. This has a great effect on our society, meeting our needs technically and economically. As Wi-Fi progresses, I believe there will be a day when ethernet data cabling is completely replaced with Wi-Fi. Although data cabling will be replaced, infrastructure needs will continue to increase. Now with 802.11ac wireless, there are not many devices that can support these wireless speeds. The effects, however, will soon start to change with 802.11ac and infrastructure. I will use an education facility as an example for my reasoning. Each classroom has approximately 30 students, and in many situations, each student has a wireless device. The infrastructure to the wireless access point could currently have a 1 Gbps to the access point. As more and more devices begin to embrace 802.11ac, this could eventually cause network congestion at that access point. Access points, such as extreme networks, are manufactured with two 1 Gbps ethernet ports. This enables network administrators to aggregate two 1 Gbps ports to an access point creating a 2 Gbps trunk to the access point. While this helps today, there will be a need in the future for a 10 Gbps connection to each access point. The point being made is that technology will never stop progressing, and infrastructure will always need to be upgraded to support faster connections to access points.

Wi-Fi has grown and will continue to grow in function and potential since its creation. Wi-Fi 6, based on 802.11ax, added compatibility with 6 GHz band operation, has best-in-class WPA3 security, and a max data rate of 9.6 Gbps. Wi-Fi 6E, the improved version of Wi-Fi version 6 that launched earlier this year, extended Wi-Fi 6 with 6 GHz band operation. Wi-Fi 6/6E exponentially increased scalability, reduced interference, improved security, tripled performance speed, and reduced latency by up to ~75%.

The next version of Wi-Fi will be Wi-Fi 7. Wi-Fi 7 will be based on P802.11be, offer maximum throughput of at least 30 Gbps, and offer a frequency range between 1 and 7.25 GHz. Some other physical enhancements coming with Wi-Fi 7 technology are changes in packet formats, increases in single-channel bandwidth, multi-RU, and an increase in data rate from 9.6 to 46.1 Gbps. Wi-Fi 7 will also offer many improved reliability features. First, it will bring multi-link operation, allowing link aggregation at the MAC layer. This will improve latency while running multiple links in parallel, increase reliability by duplicating packets over multiple links, and offer dynamic data flows that adjust to devices and application needs. Wi-Fi version 7 is designed to work optimally through multi-access-point configuration, so it can use MIMO on small and large scales to improve network performance and reliability. Multi-link functions can reduce latency even further than Wi-Fi version 6 because it uses several links and access points instead of a single one to optimize routes and reception.

Wi-Fi has grown exponentially since its inception and will only continue to grow in the future. Still, with Wi-Fi version 7, Wi-Fi and likely all other forms of wireless communication will grow in quality and reliability to levels that will change wireless communication forever.

2.3.2 LTE

In association with Wi-Fi, technology mobility is our desire as consumers in our everyday life. When we cannot the capability to attach to Wi-Fi, we still want internet connectivity. The phenomenon of cellular data has changed our lives forever. Through the generations of cellular data starting with 1G in the 1970s, things have come a long way for cellular data technology. 2G was later released in the 1990s and is known as GSM. 2G introduced better coverage along with texting, voicemail, faxes, and paging. 3G was four times faster than 2G with speeds up to 7.2 Mbps. 4G technology was introduced to the cellular world, although we have never truly seen the full capabilities of 4G. Figure 2.2 is a depiction of cellular data speeds and technologies [12]

We are now at 4G LTE, which stands for long-term evolution, and is the first generation implementation of 4G. NTT DoCoMo of Japan first introduced LTE in 2004. Although studies and actual work began in 2005, the first deployment of LTE was in Oslo and Stockholm in 2009 in the form of a USB modem. LTE is offered by all leading cellular carriers, and most advertise using "LTE." There are companies that use the terms "LTE" and "4G" interchangeably in advertising as having "4G," which is not exactly true. 4G was a cellular data standard created for the next level of cellular technologies. The standard was intended for speeds to reach up to 100 Mbps in high-movement environments such as riding in a car or a train. The 4G standard was also intended to reach speeds up to 1 Gbps for mobile users in a slow-motion state, such as sitting still or walking.

The average speed produced by Verizon Networks LTE service is 31.1 Mbps download and 17.1 Mbps upload [13]. LTE's predecessors, GSM and CDMA, moved data in smaller amounts, a holdback for the technology and its speeds. LTE moved data in larger packets. LTE's major benefit over the third generation is that it reduces latency by using a technology called time-delayed duplex (TDD). Our beloved cellular devices are operational using frequency and spectrum. LTE operates on a frequency spectrum between 700 MHz and 2.6 GHz.

Many of us remember the days of using cordless phones that seemed to work for very long distances. Those cordless phones that worked so well were interconnected using a base in your house that was 900 MHz. In basic principles of frequency wavelengths, it is known that the lower the frequency, the better the object penetration it can accomplish. While this gives a better range, the data transmission can be less. Similarly, with LTE, places such as rural areas will most likely use the lower frequency spectrum of around 700 MHz. This will allow the signal to travel further and reach rural households and roadways. Are the transmission speeds lower than those of higher frequencies? Yes, but consumers that live in rural areas will tell you that they will take what they can get considering their location. In areas such as a city full of businesses and homes that are close by, frequencies closer to 2.6 GHz will be used. Greater speeds can be reached at this

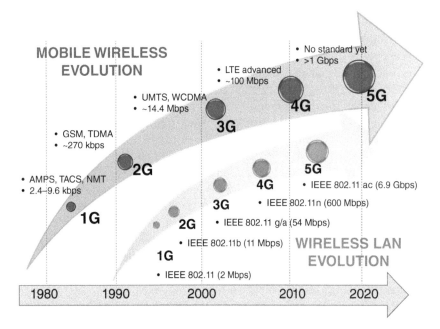

Figure 2.2 Characteristics of deployed wireless cellular technologies.

frequency with a smaller area to cover. Economics in the cellular data business is very hard to follow. Consumers are at the mercy of the technology and the carriers that choose to implement it. When consumers were getting 3G, they were buying cellular devices as quickly as they were put on the shelf. With LTE, another economic explosion took place. With the speeds LTE can produce, cellular carriers are now able to compete with home internet providers.

The first mobile phone that was introduced with LTE capabilities was Verizon's HTC Corporation Thunderbolt brand, also known as the T-Mobile G1 in the United States and parts of Europe. It is a CDMA/LTE variant of the HTC Corporation Desire HD. The Thunderbolt increased HTC's sales, and it was recorded that the HTC thunderbolt sold 28% better than iPhones during the quarter when the HTC Thunderbolt was released. LTE was showing its effect on economical sales data. Deloitte LLP did a projection on the US economy effects of 4G LTE implementations during the years 2012–2016. The studies showed that 4G helped contribute to a 41% annual growth rate in the use of cellular data. According to an article in [14], 4G could be accounted for $73 Billion in GDP growth and 371 000 new jobs. The data again proves that LTE has grown the economies around the world. The US is the leading economical country in cellular technology, as they have developed three of the five top cellular device operating systems.

LTE has truly changed the world of cellular technology. Most of us are enjoying the benefits of fast speeds and low battery usage costs. While LTE is deployed by most carriers in the US, other countries are beginning to deploy if they haven't already. I also believe that LTE will stay around in cellular devices for a while. While some people will always desire faster speeds, I believe most people are satisfied with the average speeds offered by LTE. New technologies are not expected to be deployed until the end of the decade. 5G is expected to be 40 times faster than its predecessor, 4G. For example, a 3D video on 4G technologies would take approximately six minutes to download a 3D movie. With 5G implementation, it would only take approximately six seconds. I believe the timing for major 5G deployment will work out, being closer to the expected year of 2025.

2.3.3 Heterogeneous Networks

Heterogeneous networks, also referred to as HetNet, are a very important aspect of the movement of cellular technology. A HetNet is made up of different types of wireless stations that complement each other to provide high-quality services to end users [15]. With the growing number of users and the growing amount of required data in cellular services, something has to be done for the future. A heterogeneous network consists of macro base stations, small cells, distributed antenna systems (DAS), and even Wi-Fi access points. There are three types of cell base stations in a mobile network that can be used in a HetNet. A macro base station is the backbone, and can typically be found on rooftops or towers covering large areas and many users. Another type of cell base used is the micro base station, which can be used in indoor and outdoor crowded areas. The pico base station can also be used in a heterogeneous network, and generally takes over when a user moves indoors. An even smaller portion of a HetNet is known as a femtocell. The femtocell is generally a small access point that improves cellular service in a home or small office via the internet. The pico base station and femtocell are a critical portion of HetNet considering that 70% of cellular traffic is generated indoors [15]. Figure 2.3 shows a femtocell [16].

2.3.4 Femtocell Repeater

Also involved in a heterogeneous network is wireless transmission between base stations and the rest of the network. In other words, there has to be a relay device involved for connections between the pico, micro, and macrocell bases in order for the different technologies to work together. Heterogeneous networks are used today, whether we realize it or not, although it is mostly deployed in larger cities. The reason for the need for heterogeneous networks is due to the growing numbers of devices and data needed to operate a mobile network. In 2016, there will be more than 8 billion mobile users in the world. Overall, mobile data is expected to grow with an envisions that there will be more than 50 billion connected smart devices by the year 2020 [15]. Figure 2.3 is a depiction of a basic topology of a heterogeneous network [17].

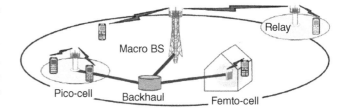

Figure 2.3 Heterogeneous network topology.

I believe that heterogeneous networks are an important part of the future of wireless mobility. Without it, we will eventually exceed the limits of networks that we have today and the data demands.

2.3.5 The Dawn of 5G Wireless Systems

In the mid-1970s, wireless communication became popular. Over the next 40 years, cell innovation progressed from the original to the fifth generation. The fifth generation's invention provides an unprecedented level of throughput [18]. The fifth-century innovation introduced a slew of new and complicated capabilities, making it the most notable and well-known arrangement later on. Wi-Fi, WIMAX, and other networks like service or private networks, e.g. Bluetooth, are all available. For the exchange of services, many cell terminals have interfaces like GSM. IP principles apply to all the networks, like wireless and circular networks. The web convention level sends all indicators and information. Clients that utilize Bluetooth and Picante technology are entirely unaware of fifth-generation innovations such as music players, storage, graphics, video, etc. [19]

The universe of unlimited, unrestricted data, entertainment, and communication will create a new metric for modifying peoples' perspectives toward their way of life. The fifth-generation wireless sight and online wireless network capabilities are entirely wireless, culminating in a completely wireless world [20]. As a whole, worldwide wireless network (WWWW). For the fifth age, 4G innovation is critical. Large code division multiple access (LASCDNA), orthogonal frequency multiplexing (OFDM), multi-carrier multi-access codes (MC-CDMA), and local multipoint service (LMDS) are the inventions of the wireless network (fifth generation). It is a Wi-Fi environment. Fifth-age innovation delivers broad information transfer capabilities, unlimited calls, and endless information transferring abilities in the most recent wireless network [19].

Mostly, the worldwide 5G industry is yet in the beginning phases of network development. For the most part, the business accepts that "the future 6G innovation" will not be applied until 2030. Along these lines, regardless of whether as far as business situations, network innovation, modern advancement, arrangement pace, and so on, the following three to five years will, in any case, be basic for 5G turn of events. Thus, 3GPP is not set in stone 5G-Advanced as the idea of 5G organization advancement at the PCG (Packet core Gateway) gathering held in April. Later on, all parts of the media communications industry will continuously work on the structure and enhance the substance for 5G-Advanced beginning from R18. In the course of start to finish 5G-Advanced network evolution, the development of the central organization assumes a crucial part. From one viewpoint, the central network is associated with different administrations and applications, the combination point of the whole network business, and the driving force of future business advancement. Then again, the central network is associated with different standard terminals and access organizations, the entire network geography [21]. The middle moves the whole body. Accordingly, advancing 5G center network innovation and design advancement dependent on genuine business needs will assist administrators with further developing profit from speculation and help industry clients better utilize 5G organizations to accomplish computerized change.

Unlike past ages of communication networks, 5G is viewed as the foundation of the business's advanced change. The world's significant economies have mentioned 5G as a modern long-haul, modern long-haul turn of events. For instance, the European Union proposed the 2030 Digital Compass plan, which plans to computerize business change and public help digitalization. It took on 5G as the reason for industry 4.0 as the main nation to convey 5G; South Korea has additionally reinforced the development of a 5G+ combined system and advanced 5G joined services. Japan keeps on advancing the worth of B5G (Beyond 5G) to individuals' work and society. China has likewise advanced a drawn-out objective for 2035 driven by demanding logical and technological development and extending the "5G + Industrial Internet" as its current objective [22].

Subsequently, 5G-Advanced necessities to thoroughly consider the development of the engineering and upgrade capacities, from the current consumer concentric broadband (MBB) network to the genuine modern Internet. Notwithstanding, it is conceivable to utilize network cutting, MEC (multi-access edge computing), and NPN (nonpublic network) to serve the business. Regardless of whether it is network arrangement status, business SLA (service-level agreement) ensures capacities, simple activity, upkeep abilities, and some assistant capacities required by the business, the current capacities of the 5G network are as yet lacking. In this manner, it needs to keep on being improved in 3GPP R18 and ensuing adaptations [23].

Above all else, later on, XR (extended reality) will turn into the principal assortment of business conveyed by the network. Not exclusively will the meaning of XR be redesigned from 8K to 16K/32K, or much higher. AR (augmented reality) business situations for industry applications will likewise advance from single-terminal correspondence to multi-XR cooperative connection, and it will grow quickly past 2025. Because of the effect of business traffic and business attributes, XR administrations will advance higher prerequisites for SLA assurances, for example, network limit, postponement, and transfer speed. Simultaneously, fundamental correspondence benefits have a great deal of space for improvement. Multiparty video chats and virtual conferences cut down on the need to travel to connect. Working from home or working a hybrid arrangement of going into an office some days and working from home on other days will be commonplace. The current meeting method of fixed admittance and video and call will change into distant multiparty cooperation of mobile access and rich media and continuous communication in business. For instance, corporate representatives can get to the corporate office climate with virtual pictures whenever at home and speak with them. Associates impart effectively. Consequently, 5G-Advanced necessities to give an overhauled network engineering and improved intuitive communication abilities to meet the business advancement needs of the current intelligible voice-based specialized strategies developing into full-mindful, intelligent, and vivid technical methods. It should likewise empower customer experience updates.

Second, industry digitization has achieved a more complex business climate than the consumer network. Organizations in various enterprises – for example, the industrial IoT, energy IoT, mines, ports, and medical health – need the organization to give them a separate business encounter as well as give deterministic SLA certifications to business results. For instance, the Industrial IoT requires deterministic correspondence transmission defers that are limited here and there, and insightful lattices need high-accuracy clock synchronization, high confinement, and high security. Mines need to give exact situating deep down, ports need far-off gantry crane control, and clinical wellbeing needs real-time determination and treatment data, synchronization, and backing of far-off finding with super low dormancy. Subsequently, 5G-Advanced necessities to thoroughly consider the deterministic experience ensure for industry services, including ongoing assistance insight, estimation, booking, and lastly, framing a generally closed control circle. For various ventures, 5G requirements to take on open networks, nearby private networks, and different crossbreed organizing modes to meet the business segregation and information security prerequisites. In this manner, 5G-Advanced should focus on the organization's engineering, organizing plan, hardware structure, and services supportability that coordinates with the various and complex business climates.

There is no one-size-fits-all cybersecurity solution for any field of digital technology, but there are many common and inherent flaws that can and should be fixed. As networking technologies continue to change and grow, so does the laxness in security culture and faith in established standards. In reality, the policies and best practices for security planning should be continually evolving. In enterprise network management, organizational structure and culture often severely affect risk mitigation and can make it either very effective or detrimental to a security system. It is very important to approach security planning with a proactive mindset because "should haves" after a breach occurs do not prevent breaches. Many security teams do not focus on the correct aspect of preparedness. Oftentimes, when evaluating policy and department effectiveness, questions such as "Am I compliant?" or "Am I secure?" are asked, but these are too narrow and only part of a responsive mindset. Organizational culture with these objectives is going to only be reactive to an attack and base changes on what has happened. More broad-ranging and nuanced questions such as "Are we effective?" or "How can we improve the security of this system?" are much more effective approaches to security planning. These approaches represent a team proactively looking for weaknesses in a system and ways to improve efficiency before attackers can exploit these systems. Especially with the new developments coming with 5G networks, proactive risk-mitigation techniques are the only ones that will hold up in a landscape of changing threats and tech.

5G technology allows for previously unimaginable speeds and efficiency of data systems, but this tech is also very dangerous for careless integration. For one, the low latency can allow attackers to complete their attacks much faster, and before response, teams have the chance to save anything. The capabilities are also causing a rapid expansion of low-power devices and IoT networks that introduce many more security threats to a network. 5G functions allow for much more efficient use of these technologies. Still, these low-power and low-function devices have very limited security capabilities and provide endless gateways into otherwise secure networks. Every host is a threat, and when a fish tank sensor can be used as a backdoor to breach a network, the value of these network capabilities becomes much more questionable. While the evolution of 5G technology is inevitable, security specialists must have an active mindset when designing new policies and finding new technologies and methods to protect their networks. For an enterprise to run effectively in the new technology climate, digital security is an absolute must. In an ever-evolving landscape, there is no other way to handle it. Without a widespread change in threat mitigation and security culture, the growth of the technology will be cut short.

5G-Advanced development is innovatively introduced as far-reaching coordination of ICT technology, modern field network technology, and information technology. The communication network after 4G ultimately presents IT innovation, and the telecom cloud is largely utilized as the foundation. In the real telecom cloud landing measure, innovations like NFV (network functions virtualization), compartments, SDN (software defined network), and API (application programming interface)–based framework ability openness have all gotten real business confirmation. Then again, the network edge is the focal point of future business improvement. Overall, its plan of action, arrangement model, activity, and support model, particularly asset accessibility and asset productivity, is not the same as the incorporated network of distributed computing. The Linux Establishment suggested that after presenting the idea of Cloud Native to the edge, it additionally needs to join the different elements of the boundary to frame an edge local (Edge Native) application structure. In this way, the development of 5G-Advanced requirements to coordinate the qualities of cloud-local and degenerative, accomplish a relation between the two through a similar organization engineering, and lastly move toward the drawn-out development bearing of and cloud-network reconciliation.

For CT (core network and terminals) innovation itself, 5G-Advanced necessitates applying its network union abilities further. These reconciliations incorporate the coordination of various generations and various models of NSA/SA, just as the mix of individual customers, family access, and industry networks. Moreover, with the development of satellite communication, the 5G-Advanced center network will likewise get ready for a completely joined network design situated to the reconciliation of ground, ocean, air, and space. In the expansion of ICT innovation, there will be additional interest in creation and activity later on, and OT (operational technology) will carry new qualities to versatile organizations. For instance, the industrial IoT for modern assembling is unique to the conventional purchaser internet. It has tougher prerequisites for network quality. It is essential to think about the presentation of 5G while supporting negligible systems services. Quality review situations dependent on machine vision require the organization to keep huge data transfer capacity and low dormancy capabilities. Remote mechanical control requires the network to help deterministic transmission and ensure the number of associations and data transfer capacity that can be guaranteed. The intelligent creation line for adaptable assembling too should be given by the same network situating information assortment as well as different capacities. Therefore, remote access needs to have unwavering quality, accessibility, determinism, and continuous execution similar to wired admittance. The coordination of OT and CT will turn into a significant course for the advancement of mobile networks. The 5G-Advanced network will turn into the basic foundation for the exhaustive interconnection of individuals, machines, materials, techniques, and the climate in a modern environment, acknowledging modern plans, R&D, creation, and executives. Universal interconnection, everything being equal, is a significant main thrust for the advanced change of business.

Moreover, DT (data technology) innovation will infuse new catalysts into network advancement. The turn of events establishment of the computerized economy is a vast association, computerized extraction, information, demonstration, and examination and judgment. Joining a 5G organization with colossal information, AI, and different innovations can accomplish more exact advanced extraction and construct information models dependent on rich calculations and business highlights. It can likewise make the most proper investigation and decisions dependent on advanced twin innovation and give full play to the computerized sway, which will also advance the development of the organization. In rundown, the entire combination of these technologies will together drive network changes as well as limit updates and help the advanced improvement of the whole society in all fields.

2.3.6 Advancing from 5G to 6G Networks

Intent-based networking (IBN) is a form of network administration that employs machine learning and deep analytics to computerize network maintenance and management (Figure 2.4). IBN is gaining traction, but there is still a lack of clarity concerning its possible benefits and applications [25]. IBN systems adopt a business intent as the input, automate the identification of a needed state of the network to match the business intent, and then allot available network resources to generate the needed state and administer matching network guidelines.

There are three blocks of IBN [26]:

- *Assurance.* This block deals with end-to-end verification of the wide network. The assurance block also makes predictions on the modification regarding the primary intent and then offers correction recommendations. The recommended procedure is primarily implemented by AI/ML into the network. The performance and security factors of the network undergo regular study and need reconfigurations.
- *Translation.* This block can capture and decode business intents into guidelines and recommendations all through the system.

- *Activation.* This block comes after the specification of the intent and progress of policies. The block uses network-wide automation to authenticate the configuration of devices before they are deployed.
- The functioning of the three blocks can be understood with the use of the image below.

IBN systems dynamically keep track of and preserve the state of the network. Machine learning and optimization implements come in handy when determining the most viable means of attaining the needed network state and in automating the implementation of corrective courses if necessary. Consequentially, IBN can bring down the intricacy of network management and maintenance, improve network performance, enhance network agility, and reduce the risk of failures and errors compared to traditional manual and rigid techniques.

IBNs most exciting return is the ability to offer an operational abstraction of an enterprise's general network framework. IBN allows full visibility, making it easy to map and decipher high-

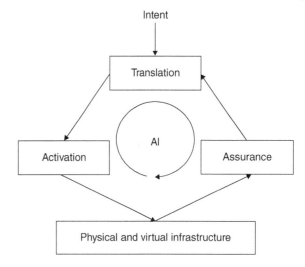

Figure 2.4   Internet-based network [24].

level intent into lower-level network constructs [25]. This quality works as a plus to permit seamless change from challenging management of CLI-based imperative technique to an additionally manageable intent-based declarative technique. Nonetheless, this does not mean that IBN does not come with hurdles. Even though IBN is a powerful and highly efficient technology, it is not a suitable match for all enterprises. IBN is most appropriate for companies that experience a fast-growing network scale and intricate matters which need high network security and performance.

IBN proves efficient for companies that need stronger network security, free network administrators to pay attention to other tasks, or enterprises that need to match up to network compliance guidelines. IBN can take care of tiresome manual errands like network verification and configuration and search and remediation [25]. The IBN technique can also pave the way for faster troubleshooting and remediation and improved predictions and analytics.

With such influential returns, IBN is a suitable means to ensure that organizations maintain proactivity concerning networking and enhance their internal procedures to bring down the rate and consequences of human mistakes. Several IBN specialists believe that choosing suitable tools is the most significant decision those new adopters of the same need to consider. However, a more crucial matter needs attention. The most challenging component of adopting IBN is considering a cultural change. To help with the cultural shift, DevOps and additional practices can bridge the gap between permitting organizations to partner and coming up with effective CI/CD workflows founded on IBN.

IBN can be employed through various applications. One of the applications is security, which offers improved security through backing from ML and AI algorithms [26]. Another application is performance testing, where IBN systems help with other applications. Assisting in the filtering of web traffic is also an IBN application. IBN systems offer firewalls to web applications to aid internet traffic and improve security measures.

2.4 Cognitive Networking

The most recent generation of cellular, 5G New Radio, comprises three essential features: backing for massive internet of things devices, multi-gigabit per second throughput, ultra-low latency, and exceptionally high reliability [27]. Consumer-facing 5G network deployments are already in place on a global scale. Nonetheless, the collaboration of the three mentioned features is still in the initial stages. The vision of a comprehensive 5G feature is on the way. However, this vision needs a new level of arrangement and automation [27].

IEEE publication Xplore in 2005 defines cognitive networks to comprise elements that pay attention to network conditions and, with the use of past knowledge obtained from past interactions with the network, go on to plan, make decisions, and act on the same [27]. The objective is for the embedded intelligence in cognitive networks to consider the end-to-end objectives of data flow, thereby effectively turning the consumers' end-to-end objectives into a comprehensible cognitive procedure [28].

In the setting of a progressed 5G system, an enterprise user can onboard thousands of environmental sensors. Instead of abiding by a manual procedure of linking, validating, and other functions, cognitive networks, instead of "see" devices, comprehend the objective of deploying the connected sensor and provision because of the same. Cognitive networks then use feedback information from all through the network to unceasingly optimize. Fundamentally, a type of data network, cognitive networks have grown significantly of late because of their widespread use in communication networks. This networking approach combines leading-edge technologies from various research areas like machine learning, computer network, knowledge representation, and network management to take care of matters prevailing networks encounter.

Cognitive networks use a cognitive process that can only recognize prevailing network conditions. However, the cognitive process can also make plans, decide, and assume actions on the same network conditions. Beyond that, this kind of network borrows from the penalties of its actions while abiding by the end-to-end objectives. The loop, also called the cognition loop, can sense the environment and plan actions founded on the inputs from network policies and sensors.

2.4.1 Zero-Touch Network and Service Management

There is expected intricacy in operating and managing 5G and future networks, and this probability propels trends concerning closed-loop automation of network and service management operations. The ETSI (European Telecommunication Standard Institute) Zero-touch network and service management structure are visualized as a next-generation management system to have all operational procedures and tasks carried out automatically with preferably 100% efficiency [29]. Artificial intelligence is an essential enabler of self-managing tasks, causing reduced operational costs, lower risk of human mistakes, and enhanced time to value. Nonetheless, the increasing acceptance of leveraging AI in a zero-touch network and service management structure should also consider the possible restrictions and risks of using AI approaches. Figure 2.5 represents the overall view of how a zero-touch network and service management structure functions [30].

The progress of mobile applications and services in the last years challenges the prevailing network infrastructures. As a result, the entrance of many management solutions to counter the explosion of mobile applications and an end-to-end network chain also increases the intricacy in the synchronization of various parts comprising the whole infrastructure. The zero-touch network and service management idea automatically organize and manage network resources while guaranteeing the quality of experience needed by most consumers. ML is part of the essential enabling technologies that numerous zero-touch network and service management structures take up to offer intellectual decisions to the network management system.

Diverse standardization entities have constructed reference structures to lay the basics of new automatic network management functions and resource organization. Different ML approaches are now used to contribute to improved zero-touch network and service management developments in various aspects like architecture coordination, multi-tenancy

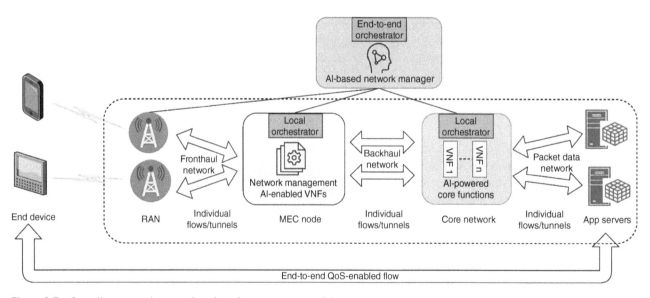

Figure 2.5 Overall zero-touch network and service management vision.

management, and traffic tracking [30]. Nonetheless, various challenges like scalability, complexity, and security of ML approaches are present. There should be future research guidelines to complete stable progress in the zero-touch network and service management framework.

2.4.2 Zero-Trust Networking

Zero-trust networking is founded on a security framework that institutes trust via unceasing authentication and keeping track of every network access effort [31]. This kind of networking differs from the traditional assumption that everything in a corporate network deserves trust. A zero-trust network comes with significant benefits like:

- *Improved security*. Attacks normally come far from the target, and attackers mostly piggyback on the access of sanctioned users before making lateral movements within the network to access their targeted assets. Zero-trust networking assists with such attacks.
- *Simple security technique*. Traditionally, corporations have layered security solutions to keep attackers at bay. Nonetheless, through time, these layered security solutions can generate security breaches for invaders to compromise. Zero-trust networking offers seamless security with additional integrated networks.
- *Management of dispersed infrastructure*. Network infrastructure keeps on getting additionally complex and dispersed with applications, data, and assets spread through various cloud and hybrid settings. Consumers work from diverse locations, making it additionally challenging to describe a secure limit. The simple solution of securing a perimeter is outdated to a complex hurdle changing from one organization to another.

The generation of a zero-trust network encompasses a series of well-defined steps. The first thing to do is to identify the assets. At this stage, the company can take an inventory of assets and evaluate the value and susceptibility of corporate assets like intellectual property and proprietary information. The second step is to verify the users and devices. In most cases, invasions come from spoofed devices. To attain zero trust, users and devices need verification of what and who they claim to be. The verification process can be backed using user multifactor authentication, behavior analytics for connected IoT devices, and embedded chips in devices. The third step is to map the workflows. This involves defining who has access to what assets, when they ought to gain access, and how and why access is granted as part of daily business. The fourth step is to define and automate policies. At this stage, the organization can use the evaluation outcomes to describe authentication policies, including metadata on the device, time, location, origin, and contextual information like recent activity and multi-factor authentication. Process automation can be attained with firewalls capable of screening for mentioned attributes. The fifth and final step is testing, keeping track, and preserving. With a zero-trust technique, like threat modeling, testing needs to ensure the effect on productivity is insignificant and hypothetical security threats are deactivated. Following implementation, security teams can observe device trends over time to recognize any irregularities that point to new invasions and proactively become accustomed to policies to keep attackers at bay [31].

2.4.3 Information-Centric Networking

Information-centric networking (ICN) is a novel technique in networking content as opposed to devices that have the content. ICN has of late warranted significance in network research and communities in charge of standardization. National and multinational supported research projects progress on a global scale. For instance, the ITU in the Telecommunication Standardization Sector began the ICN standardization process in 2012. Additionally, the ICN Research Group of the Internet Research Task Force offers cooperation on standards-oriented research. Such global efforts collectively progress the new network structure of ICN. Nonetheless, there are limited research and debate on the comprehensive ICN standardization position.

Currently, consumers use networks more for content recovery instead of direct communication between people or between a server and a device [6]. This trend in consumer needs demands a novel network structure suitable for content recovery, which results in the discovery of novel network structures referred to as ICN [32]. The ICN idea emerges in an era with more consumers moving their interests to the content instead of the server or the location where the content is in storage. This content-specific tendency of user applications makes the point-to-point communication model of IP networks ineffective. ICN natively backs in-network caching approaches to efficiently use network resources and apt delivery of contents to the end consumers.

ICN primarily offered to change the internet structure with a novel network framework for the native support of access to named data objects. Because of direct access to data objects, affordable and pervasive in-network caching and replication are attainable. There is enhanced efficiency and improved scalability concerning data distribution and network bandwidth use. To attain maximum benefits, there are technical research challenges of ICN as locating named data objects and deliver to consumers, how to back mobility in an information-centric and location-independent way, how to name data objects to specifically identify, and how to secure named data objects widely disseminated in the network [33].

2.4.3.1 Basic Concepts of ICN

- *Naming.* Naming data objects is an essential idea for ICN. Naming allows the identification of data objects minus their location. There are two principal techniques to name a data object in ICN: a flat naming scheme and a hierarchical naming scheme [33]. Hierarchical naming scheme functions based on the prefix of a publisher, making it possible to aggregate routing data and improve the scalability of a routing scheme [34]. The flat naming scheme employs a hashing value of the data object like the name or the content. Because of using the hashing value, the flat naming scheme generates a fixed-length name for every data object, which improves the speed of name lookup time.
- *Routing.* ICN comprises three stages: name resolution, discovery, and delivery. During name resolution, the name of a data object is translated into the locator. What follows is the discovery step courses a consumer request to the data object. The requested data object is routed to the consumer in the third stage. Relying on the combination of these three stages, ICN routing is classified principally into two: lookup by name routing and route by name routing.
- *Security.* Data objects can be recovered from whichever location in ICN because of the in-network cache; hence, their origins cannot at all times be a trusted unit [33]. A security approach ought to be offered to ensure that the recovered data object lacks any malicious modification from the time of publication from the genuine publisher. Additionally, data origin validation, which involves verification of integrity between data objects and their publishers also functions as another axis of ICN security. Failure of such security techniques results in susceptibility to various attacks by inserting adulterated content into the network [33].
- *Mobility.* ICN mobility aims to allow consumers unceasingly get content minus foreseeable interruptions in ICN applications. Because of a receiver-driven content recovery in ICN, a modification in the physical location does not disturb incessant content reception. Then again, server-side mobility presents a challenge in offering support. Specifically, ICN mobility needs close work with ICN routing to uncover a duplicate data object similar to the one passed on.

2.4.4 In-Network Computing

One of the most promising techniques for increasing application performance is in-network computing [35]. In-network computing denotes a specific type of hardware acceleration with network traffic interrupted by the accelerating network device before getting to the host, and where computations traditionally carried out in software are implemented by a network device, like networked FPGA (field-programmable gate array), smart network interface card, or a programmable ASIC [36]. Programmable network hardware can operate services traditionally set up on servers, ending with orders-of-magnitude enhancements in performance. Regardless of these performance enhancements, network operators are still doubtful of in-network computing. The conventional wisdom is that the operational expenses from higher power consumption supersede any performance returns. Without in-network computing justifying its costs, it will be dismissed as an additional academic exercise.

One of the principal reproaches of in-network computing is its power-hungry nature [37]. Every application's power consumption is assessed for hardware and software-based implementations, including the overheads like the power supply units. Different applications possess different power consumption abilities, and different servers tend to implement different power efficiency optimizations with different cores to attain different peak throughputs. In the same way, different smart NICs, FPGA cards, and programmable network devices tend to end up in various performance and power consumption outcomes [38].

2.4.5 Active Networking

Current research in active networking displays the prospects of this new kind of technology. From a service perspective, active networking makes room for customized packet processing within the network on a per-packet, per-flow, or per-service routine. It is possible to use customized packet processing like packet filtering, application-aware routing, multi-party communications, and information caching. From a management point of view, active networking technology makes it possible to attain fast service deployment and flexible service management [39]. The model of active networking permits an organization to set up and operate a service on a network in an identical manner to a program that can be set up and run on a computer. In this comparison, the network assumes the computer's responsibility, which is simultaneously shared between numerous parties like consumers.

End systems attached to a network remain open so that it is possible to program them in suitable languages and implements. Contrastingly, the nodes of a traditional network like ATM switches are IP routers, closed and integrated frameworks whose functions and interfaces are determined through the standardization procedure. The nodes transport packets exist between end systems. The processing of such packets inside the network is restricted to operations on the packet headers, initially for routing. Particularly, network nodes do not interpret nor change the packet payloads. In addition, the node's functionality is permanent and can only change with a principal attempt. Active networks break this tradition by letting the network carry out customized computations on whole packets, including payloads. Resultantly, the active network technique paves the way for the prospects of packet processing functions in network nodes concerning service-particular and user-particular needs and computation of regulation, user, and management data packets within the network [40].

2.5 Mobile Edge Computing

The significance of cloud computing to mobile networks is on a mounting course. Social network services like Twitter, Facebook, navigation implements from Google Maps, and content from Netflix and YouTube are all located on clouds. Moreover, consumers' added dependence on mobile devices to implement computing and storage-specific operations, whether they are business-oriented or personal, needs offloading to the cloud. This offloading helps attain improved performance, thereby prolonging battery life [41]. These objectives would be challenging and expensive to attain, minus bringing the cloud nearer the network's edge and the consumers. In reaction to this need, mobile operators improve mobile edge computing, where the storage, computing, and networking resources fit in with the base station. Figure 2.6 captures a representation of mobile edge computing architecture [42].

Mobile-edge computing offers an increasingly distributed computing setting that can set up applications and services, and store and process content near mobile consumers. Applications can be split into small tasks, with some of them carried out at the regional and local clouds, provided the latency and precision are maintained [41]. Several challenging matters come up in disseminating sub-tasks of an application with edge and additional clouds. If an application is split, the mobile edge cloud addresses the low latency, locally pertinent jobs, and high bandwidth.

2.6 Quantum Communications

After a long time of thinking about the prospects, researchers now build and test the tools set to form the basis of a powerful quantum internet where quantum communications begin incorporating the global hyperconnected economy. This process of integration takes place faster than most people imagine. The initial commercial quantum links are almost present, and the

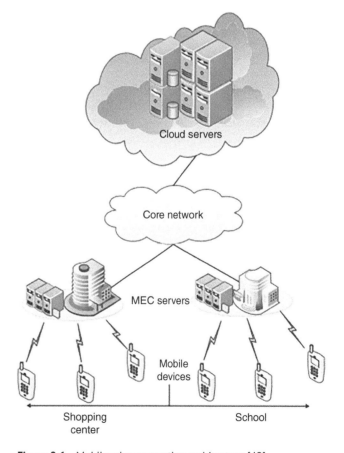

Figure 2.6 Mobile edge computing architecture [42].

quantum internet will link distant networks in approximately three years. Each new sector starts with initial movers. The movers come up with products that set standards; initial entrance consumers, with their brands, are always linked to a novel innovation. The same thing is happening with quantum communications [43]. The initial corporations are already positioning themselves and tapping into the knowledge, talent, and resources to make sure they go into the quantum era before anyone else gets there.

Quantum communications are not similar to powerful quantum computers capable of processing data exponentially quicker than conventional ones. In place of that, quantum communications encompass the sending of data. To comprehend why quantum communications will revolutionize cybersecurity and change the face of the internet, it is important to understand the science supporting them. The quantum particles are excessively, so it is important to define the quantum properties to better understand quantum communication.

The first property is the no-cloning principle. This is the most exciting property, especially for the cybersecurity sector. The principle denotes the concept that the act of observing quantum data results in its modification. The second property is entanglement. Albert Einstein said that quantum entanglement was "spooky" [44]. Some quantum particles connect so that a modification in one triggers a modification in another, even if they are not physically in contact. Even if one of the connected articles exists in a separate room, it will respond to a modification in its twin. The third property is teleportation. Researchers established a new means to get the status of one of the entangled particles and send it as a quantum unit of data referred to as a qubit to the entangled twin. The efficient transfer of quantum data is referred to as teleportation. However, teleportation gives rise to a challenge. The qubits of data sent from one entangled twin have to travel through conventional communications like satellites and fiberoptic cables [43]. The qubits come in the form of photons. Photons tend to bump into molecules in cables and in the air, which absorb them along the way.

2.6.1 Quantum Computing and 6G Wireless

In line with the progress of cellular frameworks from 5G to 6G, quantum information technology has likewise developed fast in the past few years. Quantum communications and quantum computing is a representation of this development. It is projected that quantum information technology will allow and enhance coming 6G systems from computing and communications standpoints [45]. Quantum computation is a fast-advancing sector with considerable breakthroughs attained in the past few years. Quantum computing has progressed from an emerging part of science into a developed research sector in engineering and science.

Quantum information processing (QIP) is meant to serve and function with quantum computing. Because of some of the amazing properties essential to quantum computation, most specifically parallelism and entanglement, it is forecasted that QIP technology will provide capabilities and performances yet to be rivaled by traditional counterparts. Such enhancements could be guaranteed security, computing speed, and fewer storage needs [46].

2.7 Cybersecurity of 6G

6G networks will affect significant coming developments. One of these developments is a network security and privacy, which ought to be fortified and developed for 6G technologies and applications [47]. Traffic evaluation, data processing, data encryption, and threat recognition are the most important matters in 6G networks. The security challenges resulting from massive traffic processing can be countered with decentralized security systems where the traffic can be monitored locally and dynamically. 6G use cases enforce more stringent security needs than 5G [48]. The Internet of Everything (IoE) offers a wide assortment of capabilities and services. However, it is challenging to operate and install distributed AI, privacy, and security solutions. The high mobility conditions of the new connected devices result in their changed interconnected networks and need services from additional networks, thereby having security hitches and privacy challenges.

Figure 2.7 depicts the changing security challenges through a progression from 1G to 6G networks. Because the 6G system security framework is designed to attain openness compared to 5G, the boundary separating the outside from the inside will keep blurring [49]. Consequently, prevailing network security measures like firewalls and IPsec will not be adequate to safeguard the network from outside attacks. The 6G security architecture ought to support the fundamental idea of zero trust in the mobile communication network to lessen the inadequacy. Zero trust is a security framework that

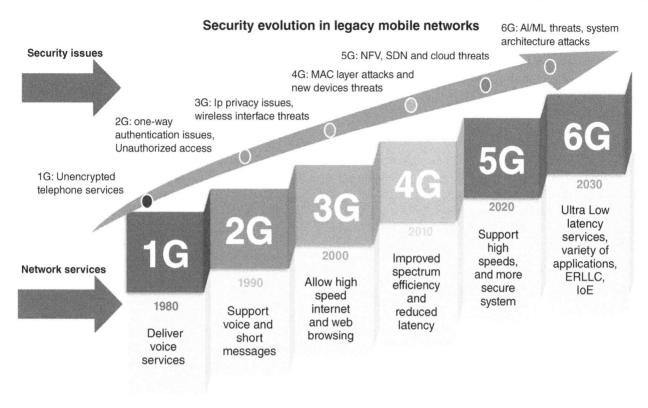

Figure 2.7 Security evolution in legacy mobile network [49].

focuses on safeguarding system resources over any other thing. Zero trust assumes that an intruder can exist in the confines of a network and that the network framework is untrustworthy and accessible to external players.

Assessments should be carried out regularly, with actions taken to lessen the risk of internal loss of assets. Zero trust frameworks that use the zero trust idea encompass relationships between access rules, protocol procedures, and network entities. Zero trust ought to be the basis of 6G security frameworks, and some security needs can be managed to back secure 6G networks using the zero-trust approach.

2.8 Massive Machine-Type Communications (MTC)

The description of machine-type communications (MTC) is intangible. The definition has to encompass a large assortment of emerging ideas like the IoE, the IoT, Smart X, and Industry 4.0, among many other things. These emerging concepts contribute to a new scenario with diverse assumptions and needs, from wireless factories with stringent requirements on reliabilities and latencies of wireless communication to long-term environmental observations encompassing restricted energy consumption in smart cities with numerous sensors. Now, a 5G design ought to regard all these scenarios with assumptions to satisfy the role of a universal enabler for emerging and prospective industries [50].

The FP7 (The Seventh Framework Program for research and development) Project METIS (Mobile and Wireless Communications Enablers for the Twenty-Twenty Information Society) took up the term MTC from 3GPP to account for all applications molding the requirements of a possible 5G solution. Additionally, MTC are differentiated concerning two principal challenges: ultra-reliable MTC and massive MTC. Massive MTC involves massive access by many devices offering wireless connectivity to billions of low-complexity low-power machine-type devices. A classic example of massive MTC is the assortment of measurements from a massive number of sensors like smart metering. In contrast, ultra-reliable MTC concerns the provision of enough wireless links for network services with somewhat strict requirements on reliability, availability, and latency [50].

2.9 Edge-Intelligence and Pervasive Artificial Intelligence in 6G

The promising AI applications motivate additional progress of wireless networks into 6G. It is forecasted that 6G will transform wireless space from connected things to connected intelligence. Nonetheless, state-of-the-art deep learning and big data analytics founded on AI systems need significant computation and communication assets, resulting in considerable latency, network jamming, energy consumption, and privacy leakage in the interference and training procedures.

Edge intelligence is outstanding as a disruptive technology for 6G networks to effortlessly incorporate sensing, computation, communication, and intelligence. There needs to be an embedded model training and inference ability into the network edge to attain this. There will be enhanced effectiveness, efficiency, security, and privacy of 6G networks [51]. The vision for trustworthy and scalable edge intelligent systems encompasses the integrated design of wireless communication techniques and decentralized machine learning models. Novel design principles of wireless networks. Include holistic end-to-end system architecture and service-oriented resource distribution optimization approaches to back edge intelligence.

2.10 Blockchain: Foundations and Role in 6G

Of late, blockchain technology and generally distributed ledger technology have gained traction and are adopted by research communities and industries worldwide. Some of the prospects of blockchain technology include transparency with anonymity, getting rid of one point of failure, thereby enhancing resistance and resilience to intrusions and decentralization. This can be done by doing away with the requirement of central trusted third parties and mediators, immutability and tamper-proofing for the disseminated ledger content, provenance and nonrepudiation of the transactions carried out, and relatively fewer delays in processing and its fees [52]. Blockchain counts as an indispensable technology that institutes incoming trust networks like 6G.

Because blockchain is projected as part of the essential enabling technologies for 6G networks, it is important to look into the different returns, challenges, and prospects expected with its utilization. Blockchain in 6G networks promises various returns. For instance, the embedded and ubiquitous integration of blockchains in prospective networks can progress prevailing healthcare systems and enhance performance regarding improved security, decentralization, and confidentiality. Another return is environmental monitoring and protection. Blockchains allow decentralized cooperative environmental sensing applications to be attained worldwide using the 6G network [53].

2.11 Role of Open-Source Platforms in 6G

Telecom domains encounter more than one challenge concerning enhancing flexible and agile network deployments. Telecom service providers are unceasingly seeking innovative ways to bring down network deployment. In contrast, the need for 6G networks will be additionally strict compared to 5G networks. It has the objective to back terabytes for every second throughput and latency of 0.1 ms or mere microseconds. Collaborative and innovative solutions will be needed to attain the needs of 6G networks. Open source projects like OpenDaylight, M-CORD, OPNFV, ONAP, and OpenStack will play an essential role [54].

There are several additional numerous initiatives assumed by the operators like disaggregation of radio network nodes and introduction of AI in the telecom networks, among many others. This inventiveness offers a collaborative setting to advance open-source platforms that offer optimized solutions concerning innovations and costs. In the 5G standardization course, ideas like service-based architecture are described. With that in mind, programmable secure plug-and-play open architecture intends to play an important role in standardizing 6G networks.

2.11.1 PHY Technologies for 6G Wireless

Machine-learning approaches were primarily used for upper layers like resource management in the wireless networking sector. Nonetheless, the focus now shifts to applying ML at the physical (PHY) layer because it primarily eliminates the necessity for traditional signal processing to be used a priori [55]. Such applications at the PHY layer include physical layer security, obstacle localization and recognition, channel equalization and estimation, modulation recognition, and channel coding.

6G networks look to accommodate extensive spectrum investigation in the mmWave sub-6 GHz, optical bands, endogenous network security from the network to physical layers, on-terrestrial/maritime/terrestrial communications for cost-aware/ubiquitous/seamless services, and AI/ML integration for improved network automation/management/dynamic network arrangement [56]. The use and change of promising physical layer technologies and important topics like channel modeling, transceiver design, complex signal processing, coding approaches, and hardware implementations need to be put in place in the 6G era.

2.11.2 Reconfigurable Intelligent Surface for 6G Wireless Networks

Reconfigurable intelligent surfaces (RIS) or intelligent reflecting surfaces count as one of the most hopeful and revolutionizing approaches for improving wireless frameworks' spectrum and energy efficiency [57]. These devices can reconfigure the wireless circulation setting by cautiously tuning the phase modifications of many low-cost passive reflecting components.

Incoming cellular systems, sensing, and localization intend to be internal with particular applications. Localization and sensing will also back seamless and flexible connectivity. Going by this trend, there is a necessity for fine-resolution sensing solutions and cm-degree localization precision. As luck would have it, with the recent progress of new materials, RIS offer a chance to remold and regulate the electromagnetic features of the setting, which can be used to enhance the output of localization and sensing. Aside from realizing low-cost MIMO systems, RIS display the possibility to enhance the performance of current wireless approaches. In the study of PHY technologies for 6G wireless networks, there is a consideration for physical limitations of RIS, where the number of phase changes for every element is restricted. There is also a consideration for the number of reflective elements of the RIS with an ability to give a suitable data rate. A hybrid beamforming framework gets into the picture in the RIS-based MIMO communications [58].

2.11.3 Millimeter-Wave and Terahertz Spectrum for 6G Wireless

5G is currently undergoing standardization meaning commercial millimeter-wave (mmWave) communications are now a reality. This reality is attained regardless of the concerns about the unfavorable propagation features of the frequencies involved. Even though the 5G systems are still being set up, there is a debate that their gigabits per second rates may not be efficient in backing several emerging applications like extended reality and 3D gaming. Such applications need several hundreds of gigabits per second to several terabits per second data rates with increased reliability and reduced latency [59]. However, such is expected to motivate the design objectives of the coming generation; 6G communication systems. With the possible terahertz communications systems to offer such data rates in short areas, they are widely considered the next cutting edge for wireless communications studies.

2.11.4 Challenges in Transport Layer for Terabit Communications

The introduction of 3GPP fifth-generation (5G) commercial cellular networks all over the globe signaled the research community's efforts to start paying attention to the configuration of the 6G system. One of the considerations is utilizing terabit communications to offer 1 Tbps and air latency of no more than 100 µs. Additionally, 6G networks look to offer more rigorous quality of service and mobility needs. Even though radio technologies and the physical layer can attain such objectives to a considerable degree, the end-to-end applications still encounter challenges in making use of the network capacity because of restrictions of the prevailing transport layer protocols. Some of the challenges are a result of the following:

- High-speed and high-bitrate communications.
- User mobility.
- Transport protocols are intricate in utilization The use of their service is attained by an API, which needs efficient know-how in several fields like system programming.
- Transport protocols are intricate by nature due to their operating principles, the approaches on which their mechanisms are founded, and their algorithms [60].

2.11.5 High-Capacity Backhaul Connectivity for 6G Wireless

A backhaul is a network that links the core network to the access network. Prevailing backhaul networks have dedicated fiber, microwave, satellite links at times, copper, and mmWave [61]. Backhaul connectivity with satellite links relies on the availability of additional options. A high-capacity backhaul network is an important part of 5G systems. It gives the 5G systems capability to exchange a considerable volume of data traffic between core and access networks. Minus a high-capacity backhaul network, the communication system fails to complete, even with access networks backing Gbps communication connections between the user equipment and access networks. With a low-capacity backhaul network, there will be a bottleneck in the system [62]. The entrance of 6G into the scene promises almost zero bottlenecks.

2.11.6 Cloud-Native Approach for 6G Wireless Networks

Cloud-native frameworks operate at scale, utilizing additional insights and data-founded approaches from ML algorithms. Cloud-native approaches also allow service providers to control open-source monitoring toolkits. Good examples of operators at the top of the market in their stance on cloud-native techniques are DISH Wireless and Orange. To put in place scalable, secure, and agile 6G core frameworks with the ability to adapt and progress to match emerging enterprise and industry use cases, operators need to take full advantage of Oracle's suite of cloud-native structure functions [63].

"Cloudification" and "software nation" of network elements need a considerable change in the attitude and abilities of operators. Sectors like software lifecycle management, data management, and infrastructure automation need adoption in a wholesale manner from cloud service providers. Most times, carriers who reinvent such sectors or invent them from the ground up can fail to acknowledge hard-won practices. Most of these practices have been codified in open-source tools and toolchains that corporations like Oracle now utilize to attain carriers' deployment.

2.11.7 Machine Type Communications in 6G

6G mobile networks look forward to deployment in the initial 2030s. At that time, the density of self-directed internet-linked machines will progress to close to 100 seconds of devices for every cubic meter. Such devices gain access to sophisticated AI founded services with increased frequency, generate voluminous multisensory information, and possess widely different limitations regarding energy, latency, computation power, and bandwidth. These devices are not under human operations. The devices can communicate using remote servers found on the network edge or core. Wireless communications from one machine to another are referred to as machine-type communications. MTC can exist between many machines that gather and process multidimensional data or between machines that relate to services on servers. With the coming of 6G, it will be possible to converge computing, communication, and energy for device-and-application-aware communications [64].

2.11.8 Impact of 5G and 6G on Health and Environment

The capacity, speed, and connectivity of 5G offer varied prospects to safeguard and maintain the environment. 5G technology coupled with the IoT increases energy efficiency, allows additional utilization of renewable energy, and reduces greenhouse gas emissions [65]. 5G helps to protect wildlife and reduce water and air pollution as well as food and water waste. Progress in cellular networks means expanding understanding and enhancing decision-making processes concerning pests, waste reduction, weather, industry, and agriculture. A good example can be cited from the case at St. Joseph River, located in South Bend, Indiana. There is smart sewer technology installed in the utility holes. These smart technologies in the sewers reduce overflows by 70%, saving the city over $500 million and keeping over 1 billion gallons of sewer material from polluting the environment [65].

Data from the United Nations indicates that 68% of the global population will reside in cities in 2050. City administrations look to 5G and coming 6G, IoT, and AI to generate smart cities where cameras, sensors, and smartphones operate connected. The speed and connectivity between such networks will allow cities to attain improved management and additional efficiency and sustainability.

6G promises efficiency in human activities, which will, in turn, contribute to the sustainability of the environment. There are various areas where 6G networks will offer improved efficiency. An example is agriculture. Come 2050; prevailing agricultural production will go up by 70% to address the needs of the whole population [66]. To attain this, the pervasive utilization of high-precision wireless technology plays a significant role. Some use cases include AR for training objectives, autonomous vehicles, sensors for tracking variables located on the farm, and information processing. Precision agriculture can also be realized through automated irrigation control [67]. Precision agriculture, also called smart farming, will enable all these by using wireless sensor networks to keep track of the farm variables and enable intelligent controls. There are specific stages to precision agriculture:

- Gathering of data
- Diagnosis
- Analysis of data
- Precision field operation and assessment
- Farm operations attain additional efficiency at the end of the day, thereby conserving environmental resources

2.12 Integration of 5G with AI and IoT and Roadmap to 6G

It is hard to believe that the internet was just created in the 1950s, and we are already on the fifth generation of cellular internet. Each generation has added something special to our lives and changed them dramatically. 1G was the beginning of everything, although extremely limited and nothing compared to what we have now. It truly got the ball rolling for what the internet would become today. 2G laid the groundwork for what was to come. It brought us texting and the ability to send pictures and other media to each other. 3G increased the speeds of the internet and allowed people to begin using video streaming. It gave customers the ability to better use the mobile internet from other places, allowing them to use the internet while roaming. 4G increased speeds even more and focused on high-quality video streaming for its customers. This helped gaming services and HD recordings immensely.

5G gives us much more, from higher speeds and better access across the globe. It also opens even more automation to devices. This could help with many things, from drones to AI. It even aims to bring us a way to seamlessly share data and connect our devices. This generation ushered in IoT, connecting objects to the internet. This allows us to do something as mundane as turning on a lightbulb with our phone to something as complex as tracking a patient's heart rhythm through a pacemaker connected online. IoT is a way to connect anything to the internet. The goal of IoT is to connect whatever you want to connect. Sadly, the capabilities of 4G just are not enough, which is why only some devices are connected to the IoT. With 5G and its more powerful capabilities, we will finally be able to make a huge transition. The greater coverage will also be a great boon. One of the big reasons IoT can be what it will be with 5G is because the coverage simply is not enough for all devices to work properly.

When people think of AI, they usually think of the integration of AI becoming more common, it will also help 5G immensely. AI has already been implemented into some networks with the intent to optimize performance. The two will make a fantastic duo – 5G's gigabit speeds, increased bandwidth, and low latency with the processing power that the AI can provide.

This will allow machines and devices to function with the intelligence that can rival or surpass a human. With the help of 5G, AI will be able to analyze data faster and learn faster, and with the increased capability of AI, it will help make massive IoT networks more feasible. Although it will take time for 5G to have an impact on AI processing, it is possible, and already many AI applications are being integrated into devices.

As you can see, 5G will do many great things in the future and help new developments, giving IoT the backbone it needs to stand and allowing it to have the connectivity it needs to sustain the number of devices we are going to have. It will also give AI the speed, connection, and lower latency that it needs to learn faster and evolve. Of course, these new developments also return the favor and help 5G, in addition, especially with the processing power that AI provides.

Now we can all see the great things that 5G will bring us, but how exactly are we going to get it? Well, the transition itself is already happening. Although 5G is not quite here, many devices are still being created with the capabilities to handle the change. Especially due to the continued data consumption we are experiencing. Carrier aggregation, small cells, and multiple inputs and multiple outputs are many advancements that have been created to help with the transition into 5G. They will increase data rates and network capacity while also helping to offload the strain networks put on the infrastructure and even help to better utilize the already existing bandwidth to lessen the strain on antennas.

The exact incorporation is expected to be gradual, though, with 4G still being used in situations where 5G is not accessible. This period of deployment will have both G's co-existing. This will be due to 5G having certain limitations, such as it needs high bandwidth usage and its range limitations. These, of course, will get better as time passes and more advancements are made but will be noticeable in the implantation phase. These implementations will begin in areas with high demand density. 5G coverage will be limited mostly to outdoors and areas where frequencies can easily reach users in the beginning. During the initial phase, it will give creators time to monitor how 5G really interacts with the environment, and then they will be able to upgrade their technology to better handle these new circumstances, and as the coverage is expanded, more implementation issues will arise. 5G still has questionable capabilities regarding range and penetration; this is due to its dependency on high-frequency ranges. Moving vehicles will also present difficulties in coverage.

These problems will be better addressed as technology advances and 5G is better understood, but ideally, technologies will work together in 5G deployment to form a better network architecture, allowing 5G to expand and cover most areas, but due to technical barriers, widescale implementation is not yet feasible; for the most part, 5G – like the generations before – will first be concentrated in dense urban areas until improved technology and reduced costs allow it to be more widely implemented.

Of course, there are other issues with the implementation of 5G. Devices themselves – especially older ones – might have issues connecting to this new generation of mobile internet. This could also carry over to different services; many might choose not to enable 5G. It may take industry time to transition products into 5G. Although the transition would certainly be a boon to them, it might not be cost-effective to make the change currently. Even if everyone decided to jump onto 5G as soon as it came out, it would still take time to transition everything over. The actual process of implementing it and an industry's implementation into their business plan may take time and add up. There very well could be a grace period that could last years in which 5G is gradually implemented into businesses and corporations. However, this could be a benefit because it would give players time to monitor and better learn the capabilities of 5G. The bottom line is that the implementation is going to take time from those who are creating it and the facilities that wish to use it. It will not be a flip of a switch but a coordinated effort that takes time.

You also must realize that financially it could be quite expensive to implement 5G. Building a network is expensive. Providers will have to buy and install new equipment such as antennas, base stations, and repeaters. Again, 5G is going to be more powerful and going to be able to cover more of the world. So, of course, it is going to need even better equipment than it does now. These devices might even need to be installed on more buildings and homes to keep up with consistent speeds and help in denser places. Your phone bill is going to go up too. Because of how expensive it is to build these better networks, carriers are most likely going to be increasing their costs, and consumers might find it worthwhile to have the top of the line.

There are certainly more challenges, too, one I find extremely important and which is a challenge for 5G is security and privacy. The very boons that 5G provides for the people can be used against them. People will have to be more careful with how they use their data, how authentication and key agreements work, and a system that is established to allow trust between different networks. It would be possible for those who can get through the security systems to track people and even eavesdrop on them. With how 5G is, supporting IoT could also lead to even more dangerous and terrifying acts of cybercrime. These criminals can hack into your car and maybe even mess with a medical device that is located inside of your body. Security should be a number 1 priority with how our everyday lives are beginning to use technology even more. Most carriers and network consortiums will be responsible for the safety of their customers through their security measures, but you can never be too careful. It is possible for cybercriminals to take advantage of new development such as 5G and to figure out a means by which to use it negatively before people are aware. One reason why the development of this technology may go slow is to make sure no vulnerabilities are left open.

Although those issues might have a bitter taste in your mouth, there are still some amazing applications and reasons to look forward to 5G. I have already touched on the IoT and AI, but there are still many more interesting pieces of technology that will benefit from 5G. One of the most impactful is autonomous vehicles. Smart cars are something many people have already seen today, with many of the newer models having an autopilot feature, but that is only the beginning. With the decreased latency that 5G provides, vehicles will be able to respond much faster. Imagine being able to use the autopilot feature in an actual city street instead of just on the highway. With these benefits, 5G cars will be able to respond more accurately to the objects in front of them and respond to road signs and even to people crossing the street. That is not the end of it either; with the help of IoT, cars will be able to communicate with each other. They will know where all other cars in the area are and be able to stay out of their path and correctly maneuver themselves efficiently. They would also be able to communicate with sensors that are being installed in most cities to detect other movements. Just imagine how much the crash rate and other vehicle dangers will go down with these new technologies.

They could even use them for drones; it would greatly help their uses if they could fly around without accidentally damaging themselves or a person. They would be able to perform many of their jobs even better, such as delivering packages to the correct houses or even being able to find the correct recipient while they are out on the town. With the new upgrades, they could even perform other jobs that could be dangerous to a normal person. They are already beginning to be used in surveillance, border security, and search and rescue missions. With the new developments of AI, they could potentially perform these tasks with the intelligence of a human. This could change how military operations are performed and save many lives in the process. Of course, many of these developments are still a way off in the future, but they are within sight.

Another application that I am thrilled about is augmented reality and virtual reality. With the ability to enjoy entertainment like you are right there in person and with almost no latency, it would be an enjoyable experience. This would also greatly help with different types of training, such as for a pilot and even for medical experts. They would have the ability to better train their skills to help during their jobs. If one of them were to make a mistake during a part of their job, it could be fatal, but with the help of VR, they could gain more experience with no disadvantages. I also must admit I am excited about the new types of video games that would be developed with the new technology. Current VR is lagging, and you can see the current latency problem, but with 5G, it is going to be so much smoother, and I cannot wait to see what kinds of new E-sports that will be developed with the new uprising of VR.

It was mentioned earlier that 5G would greatly benefit IoT with the new advancements, but now I will go into greater detail about how it will help. One of the biggest advantages that most people are aware of is the speed of 5G. The speed in transmission can approach 15–20 Gbps, having higher speeds like this can help with accessing files, applications, and other programs quickly and easily. Speeds like this will even help internal memory problems since we can add more data to the cloud now that we can access it more quickly. With the number of devices we are going to have connected to IoT, we must have the internet power to sustain them.

I have already talked about it before, but one of the biggest advantages that 5G is bringing us is lower latency. With it being 10 times less than 4G, it will give us the precision to perform remote actions in real time. This entails controlling machinery at a factor, controlling vehicles, and even controlling planes without an actual driver, but instead by someone from the safety of their own home. It opens up so many different avenues of thought; being able to perform actions from across the globe with reliability is simply amazing. As IoT grows, so many of our normal devices are going to be connected to the internet, especially our vehicles. With the low latency provided by 5G, we will be able to control all our devices connected to the IoT and have them react as we want them to. It does not necessarily mean you are controlling something, but it would mean your devices respond to your commands on time – like waking up in the morning and having your coffee machine prepare you a "cup of joe." It would not be helpful if it were still processing your command by the time you got downstairs. That is where the low latency comes into play. As soon as you click and make your decision, your devices should begin working with no delay. Figure 2.8 shows some of the 5G network capabilities.

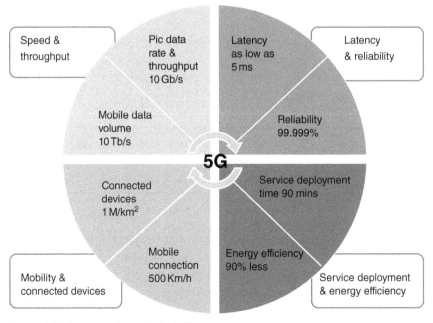

Figure 2.8 5G network capability [68].

Imagine a city, with millions of connected devices that can communicate with each other and share information, encompassing thousands of homes that are full of connected devices, and hundreds of business run with connected devices. Now imagine all these devices sharing information. The potential is that the city will be much safer and more efficient since devices will be able to tell people the best way to navigate through the city, and if there are any car accidents that weren't prevented by AI. It will even be able to tell us where available parking spots are.

With the help of these new technologies and the development they are bringing, we are entering an age of intelligent connectivity. What exactly does that mean, you may ask yourself? For some, it is a frightening thought – a *Minority Report* world where we are bombarded with information tailored to our fingerprint or facial image. For others, it is a bold and exciting combination of AI, 5G, and IoT.

I have talked about what 5G brings to each of them, but what about what they do when they are combined? More specifically, what does AI bring to the table with the fusion of these new technologies? With the advanced learning of AI, it will be able to analyze and present data to users in a more meaningful and useful way. The improved decision-making would help to deliver a completely personalized experience to the users. This would lead to more fulfilling interactions with devices that are connected to IoT. To better explain what I'm talking about, I'll bring up an AI we might have all had experiences with, Apple's virtual assistant, Siri. I, for one, have had situations when I've asked Siri to do something, and she's just completely missed the mark or not understood the lingo I've used. With the advanced AI that is coming, which will have the intelligence of a human, virtual assistants will be able to understand their users even better and will be able to adjust to us personally and know what to expect of a specific individual. Imagine with the help of an AI that knows you wake up early for work and can set the alarm without you asking and have the coffee maker prepare you a cup of coffee just when you awake. Of course, there will be many other uses, but it comes down to an AI that can personally adjust to a specific person to meet their needs.

AI will also process information much better than we will. With all the data gathered in the devices we use, we are going to need an AI to handle that information. When AI brings advanced algorithms to us, it will be able to manage the IoT and make it even more useful to us. AI will be the brains that help us to make all this possible. It will administrate the information being brought to it, and with the help of machines, learning will even grow better at understanding and putting the information to use. Of course, this will not be an overnight change. It will take time and financial effort to make a reality, and even then, the AIs will have to learn and adjust to meet the needs they are going to fulfill. Without a doubt, though, AI will be one of the greatest boons to helping IoT become what we envision for it. It will help IoT especially with internet connectivity, and AI will be used to help manage the newer, more advanced networks of 5G. Especially with network slicing and all the new subnets being created, AI will be such a tremendous help in managing these and making sure the information being gathered is correct. Data just continues to grow, and we humans are not perfect, we will need all the help we can get, and luckily, we are going to have AI to handle the processes that we cannot. In intelligent connectivity, I see AI as the brain, 5G as the heart, or the raw power keeping it going, and IoT as the body, which is controlled by the brain and powered by the heart, but which handles all the other functions.

As you can see, the future is quite exciting. Although it will take plenty of time for the vision to come to fruition, it is still an exciting process that will dramatically change our entire lives in the coming years. 5G is what we have been waiting for to help finally power all these amazing technologies that are being invented. As it becomes integrated with AI and the IoT, the true power and benefit of 5G will be seen. It has many applications besides just helping AI and IoT, though, but it is certainly not perfect. There will be many issues and challenges associated with bringing 5G into the world and fully implementing it into our daily lives and workplaces.

Just as 5G is really taking hold and revealing its weaknesses, 6G will step in, somewhere around 2030, based on the usual amount of time it takes for a new wireless technology to develop. With 5G barely on its way to being used commercially and research of 6G already beginning, it makes you wonder if we are truly solving a human need problem or simply inventing new technology for the sake of it.

The "Vision Thing" is a popular idea among 6G visionaries – the increased interaction between physical, digital, and human worlds, which is aided by AI/ML advancements, as already discussed. This has begun happening in 5G with the use of digital twins. These are digital replicas of the physical world that are used in factories to predict possible machine failures. Samsung wants to take digital twins a step further, and it envisions that 6G-enabled users will be able to interact with digital twins by using VR devices or holographic displays. There is also the possibility that multiple devices in the 6G world will amount to an "intuitive human-machine interface" that has the ability to recognize the user's mood by using facial and

Figure 2.9 Quantitatively comparison between the 5G and future 6G in terms of requirements.

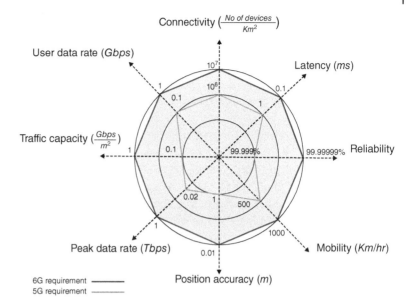

gesture recognition and body sensors that monitor stress levels. If these things are going to happen, there must be trusted artificial and ML technologies.

That is one vision of 6G; however, there is another one. The other vision of 6G is that the focus should be on ensuring greater capacity for apps that consume a lot of bandwidth. In 2020, IMEC (Illinois manufacturing Excellent Center) launched the Advanced RF program, and its goal was to have single-link data rates of 100 Gbps, microsecond latency, and significantly higher energy efficiency of less than one nano-joule per bit. Figure 2.9 [49] shows a vision of 6G compared to a 5G system capability. Advanced RF program plans to develop a new kind of semiconductor using a different material. Traditionally, complementary metal-oxide semiconductor (CMOS) has been used; however, it will not be able to handle the new 6G environment, so they are planning to use indium phosphide.

5G has a lot of promise; however, that promise may be overlooked and disregarded with the focus being pointed toward 6G. Karri Kuoppamaki, vice president of technology development and strategy at T-Mobile USA, advises that we should not fall for shiny object syndrome in regard to 6G and rush to its development when so much has been invested into 5G [69]. He also advises that we should monitor the 5G ecosystem and use that as a guide to 6G R&D. He recommends that a patient and evolutionary approach toward 6G can lower the risk of making suboptimal technology choices. This has also been an overview of how 5G and 6G should be handled now and in the future. Figure 2.9 shows some comparisons between 5G and 6G.

MTC will play an important driving force in regard to anything that moves autonomously and with intelligence. An example is ITS (Intelligent Transportation Systems), which demands the provision of reliable connectivity between multiple actors. It also improves the level of driving automation, and it is expected to increase the number of sensors and edge processing systems integrated within the vehicles. ITS will also use systems that have real-time wireless connectivity to the sensor and actuator nodes and external processing units.

2.13 3GPP

Without the combined intelligence of the 3GPP organization, cellular services would not be what they are today. The 3GPP was established in 1998 and is an organizational guidelines group comprising of seven telecommunication standard development organizations. 3GPP is not to be confused with 3GPP2, which is an organization for standards on the competing technology of CDMA. The seven group members of 3GPP are ARIB (The Association of Radio Industries and Businesses, Japan), ATIS (The Alliance for Telecommunications Industry Solutions, USA), CCSA (China Communications Standards Association), ETSI (The European Telecommunications Standards Institute),

TSDSI (Telecommunications Standards Development Society, India), TTA (Telecommunications Technology Association, Korea), and TTC (Telecommunication Technology Committee, Japan). These group partners are responsible for the maintenance of the partnership agreement, approval of applications for additional 3GPP partnerships, and making decisions related to the dissolution of 3GPP. There are four technical specification groups (TSG) in 3GPP, and those are radio access networks (RAN), service and systems aspects (SA), core network and terminals (CT), and GSM Edge radio access networks (GERAN). According to 3GPP.org, RAN is responsible for the functions, requirements, and interfaces of the UTRA/E-UTRA (universal terrestrial radio access/evolved-UTRA) in its two modes, FDD (frequency division duplex) and TDD (time division duplex). The SA is responsible for the overall architecture and service capabilities. Also, according to 3GPP.org, CT is responsible for specifying logical and physical terminal interfaces to the core network portion of 3GPP systems. The 3GPP site specifies that GERAN is responsible for specifying the radio access portion of the GSM/EDGE network. These groups meet regularly and come together quarterly. The quarterly meetings are used for discussion and approval of work done by the groups and inclusions of any specifications. The 3GPP organization also has what they call "Observers." An Observer of 3GPP is an organization that does not meet the qualifications to become a future partner of 3GPP. 3GPP currently has four Observers as a part of their organization. These partners are the Telecommunications Industries Alliance (TIA), ICT Standards Advisory Council of Canada (ISACC), and Communications Alliance – Former Australian Forum (ACIF). Having these observers keeps the group from getting lashings over being partial or having motives that may benefit partners individually. 3GPP is ultimately responsible for developing technical specifications for 3G systems. For example, 3GPP created a triple mode design giving a cellular radio the ability to be accessed by second-generation general packet radio service (GPRS), third-generation technology high-speed packet access (HSPA), and next-gen long-term evolution (LTE) radios. In the cellular data world, 3GPP is the heartbeat that allows different generations of cellular communications to operate and communicate. In the operational aspect of 3GPP, there are two types of federated access. Those two types of access or either 3GPP access or Non-3GPP access. 3GPP access is communications between an LTE network and another UMTS (Universal Mobile Telecommunications System) or GPRS network. For clarification, a UMTS network is considered the technology of third-generation (3G) cellular access, and GPRS is between 2G and 3G technologies, sometimes called 2.5G. Non-3GPP access would be networks such as CDMA, WiMAX, and Wi-Fi. These networks do not use the same authentication method as 3GPP access systems. Non-3GPP networks do not know how to deal with AKA authentication considering that they do not use the same authentication methods. AKA is Authentication and Key Agreement and is a security protocol used in mobile networks. The fix to this issue is the added authentication piece of Extensible Authentication Protocol (EAP). EAP is an extensible authentication protocol, which is a protocol for wireless networks and authentication methods used by point-to-point protocol (PPP). These two authentication methods combined created EAP-AKA. 3GPP is responsible for the EAP-AKA authentication method, which they created for non-like authentication methods to work together involving third-gen technologies and non-third-gen technologies. This is just a small example of how important and influential the 3GPP organization has made an impact on the GSM cellular movement in the world. For clarification, CDMA and GSM are ways in which cellular data networks communicate wirelessly with cellular networks. LTE (a 3GPP specification) involves the movement of data on a cellular network in conjunction with CDMA/GSM. Considering that GSM networks are used by 80% of the mobile network market share, it is clear that 3GPP continually improves the economy globally. According to an article on Nasdaq.com [15], the number of LTE technology users exceeded 1.29 billion in the first quarter of 2016. 3GPP is the group responsible for standardizing and improving LTE. Also, in the first quarter of 2016, LTE subscribers grew by approximately 182 million. The numbers are astronomical, and 4G LTE has already surpassed 3G subscribers globally. Also mentioned in the Nasdaq.com article is the growing number of LTE-Advanced technologies. LTE Advanced (LTE-A) is a technology standardized by the 3GPP organization. LTE-A is the latest version of LTE that is supposed to be more closely aligned with actual 4G speeds of around 300 Mbps. LTE networks are expected to reach 550 by the end of 2016. Currently, 168 of those LTE networks are LTE-Advanced. None of these economical feats would be possible without the help of 3GPP and its knowledge pool. In my opinion, 3GPP is an organization that has truly made a change in the world today. Although sometimes misused by cellular providers, 3GPP's LTE has been one of the greatest impacts on mobile data and its speeds. Sure, 3GPP may not have been every aspect behind the creation of the technologies that are involved in the high speeds offered by LTE. However, I believe without the group's standardization practices and guidance, the technology would be behind today. I am certain that 3GPP will continue to strive as a great organization and keep providing the economy with great mobile standardizations that keep us mobile-hungry as citizens.

2.14 Conclusion

To conclude this chapter, it is clear that these technological advancement has paved the road for better and more efficient communication. Of course, Wi-Fi is something that we all know – at least what it is – and have used it. The interesting part of Wi-Fi with its evolution to Wi-Fi 7 is how it affects businesses when free Wi-Fi is offered. LTE has changed the way we use mobile internet today. With more demand speed and the applications that cellular devices bring to us, LTE and other 4G technologies have been great, offering us speeds that we never dreamed of having a mobile. With providers moving to LTE Advanced (version 17), we are only going to get faster connections in the future. 3GPP has a great contribution to technology and where we are toward 5G and 6G. Without their guidance in standardization and implementation, mobile internet would not be where it is today. A heterogeneous network is a look at the future. Although currently used in larger cities, HetNet is what we need to guarantee availability and coverage for our mobile wireless needs. Using HetNet correctly in the world will create little to no "dead spots" while we are moving from point A to Z. All of the topics in the chapter are an influential part of our lives.

References

1 Khan, S.U., Sherali, Z., and Chilamkurti, N. (2011). *Green Networks*. Springer Science+Business Media, LLC https://doi.org/10.1007/s11227-011-0640-2.

2 Moon, S., Yi, Y., and Kim, H. (2015). Energy-efficient user association in cellular networks: A population game approach. https://www.youtube.com/watch?v=Jc2HXP5qwsY.

3 Jading, Y. (2015). *Network Energy Performance of 5G Systems*. Ericsson https://www.eit.lth.se/fileadmin/eit/group/71/2015/ylva_jading%7D.pdf.

4 Lancaster University (2021). Emissions from computing and ICT could be worse than previously thought. https://www.sciencedaily.com/releases/2021/09/210910121715.htm.

5 Tariq, F., Khandaker, M.R.A., Wong, K. et al. (2020). A speculative study on 6G. *IEEE Wireless Communications* 27 (4): 118–125.

6 Cisco, V.N. (2018). *Cisco Visual Networking Index: Forecast and Trends, 2017–2022*. White Paper.

7 Maier, G. and Reisslein, M. (2019). Transport SDN at the dawn of the 5G era. *Optical Switching and Networking* 33: 34–40.

8 Singh, R.K., Bisht, D., and Prasad, R.C. (2017). Development of 5G mobile network technology and its architecture. *International Journal of Recent Trends in Engineering & Research (IJRTER)* 3: 196–110, 201.

9 NewsTeam (2021). Wi-Fi 7 (802.11be) About Four Times Faster Than Wi-Fi 6. https://www.ktm2day.com/everything-about-wifi-7/.

10 Jackson, T. (2016). How the growth of free Wi-Fi is transforming life in Africa. https://thenextweb.com/news/growth-free-wi-fi-transforming-life-africa.

11 Chaney, P. (2015). Why offering free Wi-Fi to your customers is wise. https://smallbiztrends.com/2015/11/small-businesses-need-offer-free-wifi-customers.html.

12 Sutisna, N., Nagao, Y., and Ochi, H. (2017). SoC design with HW/SW co-design methodology for wireless communication system, https://doi.org/10.1109/ISCIT.2017.8261177. In: *2017 17th International Symposium on Communications and Information Technologies*. IEEE.

13 Gompa, N. (2016). ExtremeTech explains: What is LTE. ExtremeTech (April 1) http://www.extremetech.com/mobile/110711-what-is-lte.

14 Deloitte (2012). The impact of 4G on the U.S. economy, Part 1. *CIO Journal* (October 1). http://deloitte.wsj.com/cio/2012/10/01/the-impact-of-4g-technology-on-the-u-s-economy-part-1.

15 Uyehara, L. (2015). What's the difference between SON, C-RAN, and HetNet? http://electronicdesign.com/communications/what-s-difference-between-son-c-ran-and-hetnet.

16 Auer, G., Gódor, I., Hévizi, L. et al. (2010). Enablers for energy efficient wireless networks. https://www.researchgate.net/publication/221640959_Enablers_for_Energy_Efficient_Wireless_Networks#fullTextFileContent.

17 Auer, G., Godor, I., Hevizi, L. et al. (2010). Enablers for energy efficient wireless networks. In: *38th Vehicular Technology Conference*. IEEE.

18 Alhilal, A., Braud, T., and Hui, P. (2020). Distributed vehicular computing at the dawn of 5G: A survey. arXiv preprint arXiv:2001.07077.

19 Kumar, M.S. (2019). Revolution of 5G wireless technology – Future direction. *Journal of Computer Engineering (IOSR-JCE)* 21 (4): 36–42.

20 Bor-Yaliniz, I., Salem, M., Senerath, G., and Yanikomeroglu, H. (2019). Is 5G ready for drones: a look into contemporary and prospective wireless networks from a standardization perspective. *IEEE Wireless Communications* 18–27.

21 Wong, V.W., Schober, R., Ng, D.W., and Wang, L.-C. (2017). *Key Technologies for 5G Wireless Systems*. Cambridge University Press.

22 Jamthe, D.V. and Bhande, S.A. (2017). Nanotechnology in 5G wireless communication network – an approach. *International Reasearch Journal of Engineering and Technology* 4 (6): 58–61.

23 Lin, S. (2020). Research on the development of 5G wireless communication in the industrial field. *Journal of Electronic Research and Application* 4 (4): 9–13.

24 GeeksforGeeks (2020). Intent-based networking. https://www.geeksforgeeks.org/intent-based-networking-ibn.

25 Edwards, J. (2021). getting started with intent-based networking. https://www.networkcomputing.com/networking/getting-started-intent-based-networking.

26 Deep, M. (2020). Intent-based networking (IBN). https://www.geeksforgeeks.org/intent-based-networking-ibn

27 Kinney, S. (2021, August 17). What are cognitive networks? https://www.rcrwireless.com/20210817/5g/what-are-cognitive-networks.

28 Thomas, R.W., Dasilva, L.A., and Mackenzie, A.B. (2007). Cognitive networks. *IEEE Xplore* 1–196. https://www.researchgate.net/profile/Ryan-Thomas-2/publication/4194092_Cognitive_networks/links/004635304818662c7a000000/Cognitive-networks.pdf?origin=publication_detail.

29 Benzaid, C. and Taleb, T. (2020). AI-driven zero touch network and service management in 5G and beyond: challenges and research directions. *IEEE Network* 34 (2): 186–194. https://doi.org/10.1109/MNET.001.1900252.

30 Gallego-Madrida, J., Sanchez-Iborra, R., Ruiza, P.M., and Skarmeta, F.A. (2021). Machine learning-based zero-touch network and service management: a survey. *Digital Communications and Networks* 30 (40): 1–19. https://doi.org/10.1016/j.dcan.2021.09.001.

31 Cisco (2021). What is zero-trust networking? https://www.cisco.com/c/en/us/solutions/automation/what-is-zero-trust-networking.html#~terms.

32 Xylomenos, G., Ververidis, C.N., Siris, V.A. et al. (2014). A survey of information-centric networking research. *IEEE Communications Surveys & Tutorials* 16 (2): 1024–1049.

33 Yu, K., Eum, S., Kurita, T. et al. (2017). Information-centric networking: research and standardization status. *IEEE Access* 1–12. https://doi.org/10.1109/ACCESS.2019.2938586.

34 Raicu, I., Schwiebert, L., Fowler, S., and Gupta, S.K.S. (2004). E3D: An energy-efficient routing algorithm for wireless sensor networks. In: *Proceedings of the 2004 Intelligent Sensors, Sensor Networks and Information Processing Conference*. Department of Computer Science, Purdue University http://datasys.cs.iit.edu/reports/2002_Purdue_CS590n_2003.pdf.

35 Jin, X., Li, X., Zhang, H. et al. (2017). NetCache: Balancing key-value stores with fast in-network caching. *SOSP* 121–136.

36 Firestone, D., Putnam, A., Mundkur, S. et al. (2018). Azure accelerated networking: SmartNICs in the public cloud. *NSDI* 18 (1): 51–66.

37 Mogul, J. and Padhye, J. (2017). In-network computation is a dumb idea whose time has come. HotNets-XVI Dialogue. https://conferences.sigcomm.org/hotnets/2017/dialogues/dialogue140.pdf.

38 Tokusashi, Y., Dang, H.T., Pedone, F. et al. (2019). The case for in-network computing on demand. In: *Proceedings of Fourteenth EuroSys Conference 2019*, 1–16. Dresden: ACM https://doi.org/10.1145/3302424.3303979.

39 Tennenhouse, D., Smith, J., Sincoskie, W. et al. (1997). Survey of active network research. *IEEE Communications Magazine* 35 (1): 80–86.

40 Brunner, M. (2000). Active networks and their management. *IEEE Xplore* 1–20. https://doi.org/10.1109/ECUMN.2000.880793.

41 Gupta, L., Jain, R., and Chan, A.H. (2016, March). Mobile edge computing – An important ingredient of 5G networks. IEEE Software Defined Networks. https://sdn.ieee.org/newsletter/march-2016/mobile-edge-computing-an-important-ingredient-of-5g-networks.

42 Ahmed, A. and Ahmed, E. (2016). A survey on mobile edge computing. In: *10th IEEE International Conference on Intelligent Systems and Control*, 1–9. ISCO https://doi.org/10.1109/ISCO.2016.7727082.

43 Poloni, A., Sandilya, M., Ricotta, J. et al. (2021). *Untangling the Future of Quantum Communications*. New York: Accenture https://www.accenture.com/_acnmedia/PDF-167/Accenture-Future-Quantum-Communications.pdf.

44 Lawler, D. (2022). Quantum entanglement: the 'spooky' science behind physics Nobel. https://phys.org/news/2022-10-quantum-entanglement-spooky-science-physics.html.

45 Rahman, A. and Wang, C. (2021, June 18). Quantum-enabled 6G wireless networks: Opportunities and challenges. https://www.techrxiv.org/articles/preprint/Quantum-Enabled_6G_Wireless_Networks_Opportunities_and_Challenges/14785737/1

46 El-Latif, A.A., Maleh, Y., Banerjee, A. et al. (2022). Quantum technology for 5G/6G wireless communication systems: applications, opportunities, and challenges. *IEEE Wireless Communications* (October). https://www.comsoc.org/publications/magazines/ieee-wireless-communications/cfp/quantum-technology-5g6g-wireless.

47 Yang, H., Alphones, A., Xiong, Z. et al. (2020). Artificial-intelligence-enabled intelligent 6G networks. *IEEE Networks* 34 (1): 272–280.

48 Huang, T., Yang, W., Wu, J. et al. (2019). A survey on green 6G network: architecture and technologies. *IEEE Access* 7 (1): 175758–175768.

49 Hakeem, S.A., Hussein, H.H., and Kim, H. (2022). Security requirements and challenges of 6G technologies and applications. *Sensors* 22 (1): 1–43. https://doi.org/10.3390/s22051969.

50 Bockelmann, C., Pratas, N., Nikopour, H. et al. (2016). Massive machine-type communications in 5G: Physical and MAC-layer solutions. *IEEE Communications Magazine* 54 (9): 59–65. https://www.researchgate.net/publication/305881263_Massive_Machine-type_Communications_in_5G_Physical_and_MAC-layer_solutions.

51 Letaief, K.B., Shi, Y., Lu, J., and Lu, J. (2021). Edge artificial intelligence for 6G: Vision, enabling technologies, and applications. https://arxiv.org/pdf/2111.12444.pdf.

52 Hewa, T., Gur, G., Kalla, A. et al. (2020). The role of blockchain in 6G: Challenges, opportunities, and research directions. https://digitalcollection.zhaw.ch/bitstream/11475/20116/3/2020_Hewa-G%C3%BCl-etal_The-Role-of-Blockchain-in-6G.pdf.

53 Zhang, Z., Xiao, Y., Ma, Z. et al. (2019). 6G wireless networks: vision, requirements, architecture, and key technologies. *IEEE Vehicular Technology Magazine* 14 (3): 28–41.

54 Bhat, A., Gupta, N., Thaliath, J. et al. (2021). Role of open-source in 6G wireless networks. In: *6G Mobile Wireless Networks* (ed. A. Bhat, N. Gupta, J. Thaliath, et al.), 6. Springer International Publishing.

55 Sattiraju, R., Weinand, A., and Schotten, H.D. (2019). AI-assisted PHY technologies for 6G and beyond wireless networks. In: *The 1st 6G WIRELESS SUMMIT*, 1–2. https://arxiv.org/pdf/1908.09523.pdf.

56 Khalid, D.W., Yu, D.H., Noh, D.S., and Ali, D.R. (2022). Future of Physical Layer Technologies for 6G Wireless Communication Networks. *Sensors* (Special Issue): https://www.mdpi.com/journal/sensors/special_issues/PLT_6G.

57 Pan, C., Ren, H., Wang, K., and Kolb, J.F. (2021). Reconfigurable intelligent surfaces for 6G systems: principles, applications, and research directions. *IEEE Communications Magazine* 59 (6): https://doi.org/10.1109/MCOM.001.2001076.

58 Zhang, H., Di, B., Song, L., and Han, Z. (2021). *Reconfigurable Intelligent Surface-Empowered 6G*. Switzerland: Springer Cham https://doi.org/10.1007/978-3-030-73499-2.

59 Tripathi, S., Sabu, N., Gupta, A., and Dhillon, H. (2021). Millimeter-wave and terahertz spectrum for 6G wireless. In: *6G Mobile Wireless Networks* (ed. Y. Wu, S. Singh, T. Taleb, et al.), 83–121. Cham: Springer https://doi.org/10.1007/978-3-030-72777-2_6.

60 Oulmahdi, M., Chassot, C., and van Wambeke, N. (2015). Transport protocols: limitations, evolution obstacles, and solutions for actual deployment on the internet. *International Journal of Parallel, Emergent, and Distributed Systems* 30 (6): 515–535. https://doi.org/10.1080/17445760.2015.1053807.

61 Jaber, M., Imran, M., Tafazolli, R., and Tukmanov, A. (2016). 5G backhaul challenges and emerging research directions: a survey. *IEEE Access* 4 (1): 1143–1166.

62 Chowdhury, M.Z., Shahjalal, M., Hasan, K., and Jang, Y.M. (2019). The role of optical wireless communication technologies in 5G/6G and IoT solutions: prospects, directions, and challenges. *Applied Science* 9 (4367): 1–20. https://doi.org/10.3390/app9204367.

63 Hill, K. (2022, March 22). Three crucial changes in cloud-native networks: A Q&A with Oracle. RCR Wireless. https://www.rcrwireless.com/20220314/5g/three-crucial-changes-in-cloud-native-networks-a-qa-with-oracle.

64 Braud, T., Chatzopoulos, D., and Hui, P. (2021). Machine type communications in 6G. In: *6G Mobile Wireless Networks* (ed. Y. Wu, S. Singh, A. Roy, et al.), 207–231. Springer Cham https://doi.org/10.1007/978-3-030-72777-2_11.

65 Cho, R. (2020). the coming 5g revolution: How will it affect the environment? https://news.climate.columbia.edu/2020/08/13/coming-5g-revolution-will-affect-environment.

66 Jawad, H., Nordin, R., Gharghan, S. et al. (2017). Energy-efficient wireless sensor networks for precision agriculture: A review. *Sensors* 17 (1): 1781.

67 Sahota, H., Kumar, R., Kamal, A., and Huang, J. (2010). An energy-efficient wireless sensor network for precision agriculture. In: *Proc. IEEE Symp. Comput. Commun*, 347–350. IEEE Xplore.

68 Attaran, M. (2021). The impact of 5G on the evolution of intelligent automation and industry digitization. *Journal of Ambient Intelligence and Humanized Computing* https://doi.org/10.1007/s12652-020-02521-x.

69 Kuoppamaki, K. (2018). Evolution to 5G: The uncarrier view. In: *Brooklyn 5G Summit*. IEEE.

3

The Future of Wireless Communication with 6G

3.1 Introduction

The rollout of the fifth generation of wireless communications technology, or 5G, is still ongoing. Even though the current generation is not even fully available yet, the industry is already looking to the future. 6G technologies, which are likely still a decade away from being available, will require considerably more power than even our current standards. By the 2030s, our devices and technologies will require higher bandwidth, lower latencies, and tighter security than ever before. These innovations will surely revolutionize the possibilities of what can be accomplished in several worldwide industries and in our daily lives on our mobile devices and through the Internet of Things (IoT). This chapter will discuss the impact of 6G on industries such as healthcare and aerospace, security challenges related to the 6G implementation, artificial intelligence (AI) in 6G networks, and more, including the role of 6G in the future.

3.2 Recent Trends Leading to 6G Technology Evolution

5G systems are currently experiencing several challenges due to the rapid growth of data-centric intelligent systems. Even though it is considerably faster than 4G, the latency of 5G technology is not fast enough at 1 ms to satisfy the needs of several applications such as telemedicine [1]. The Covid-19 pandemic exacerbated these issues because many people worked from home for two years or longer, and some are still working from a home office or splitting time between corporate and home offices. Healthcare checkups over meeting software such as Zoom can potentially cause issues when 5G bandwidth, computation power, and latency cannot keep up to satisfy the users' needs. As our world moves toward being more connected than at any previous point in history, it will strain the capabilities of 5G. It will become apparent that 6G will be needed, likely within the next few years. Figure 3.1 compares the capabilities of 5G systems to the expected capabilities of 6G technologies.

As we continue moving forward in our 5G world, there will be even more instances where 5G does not meet the needs of society regarding data rates, latency, and more. Diseases may go undiagnosed, industries may falter, and we may even fall behind economically to other countries. 6G will revolutionize the communications industry with higher data rates, density, latency, efficiency, and reliability than ever before seen with previous generations of mobile technology.

3.3 Security and Privacy Challenges in 6G Wireless Communications

Whenever a new technology is developed, there are concerns regarding bugs and oversights that could allow malicious actors to steal data and cause mayhem. This will be an issue with 6G, especially since many of the technologies driving 6G could even be experimental initially. Figures 3.2 and 3.3 illustrate some of the potential security and privacy issues within 6G.

As we move into 6G, cells will shrink to "tiny" cells instead of small cells, plus a denser cell deployment, mesh networks, multiconnectivity, and D2D communications will become the norm [4]. Since the network will be more distributed than ever before, and devices will be mesh connected, the threat surface for bad actors will considerably increase. Since there will be many devices connected within each subnetwork of the WAN, it will be impractical to apply security provisions to

From 5G to 6G: Technologies, Architecture, AI, and Security, First Edition. Abdulrahman Yarali.
© 2023 The Institute of Electrical and Electronics Engineers, Inc. Published 2023 by John Wiley & Sons, Inc.

Major factors	6G	5G
Peak data rate	> 100 Gb/s	10 [20] Gb/s
User experience data rate	> 10 Gb/s	1 Gb/s
Traffic density	> 100 Tb/s/km^2	10 Tb/s/km^2
Connection density	> 10 million/km^2	1 million/km^2
Delay	< 1 ms	ms level
Mobility	> 1000 km/h	350 km/h
Spectrum efficiency	> 3x relative to 5G	3~5x relative to 4G
Energy efficiency	> 10x relative to 5G	1000x relative to 4G
Coverage percent	> 99%	About 70%
Reliability	> 99.999%	About 99.9%
Positioning precision	Centimeter level	Meter level
Receiver sensitivity	< −130 dBm	About −120 dBm

Figure 3.1 Comparison of 5G and 6G characteristics [2].

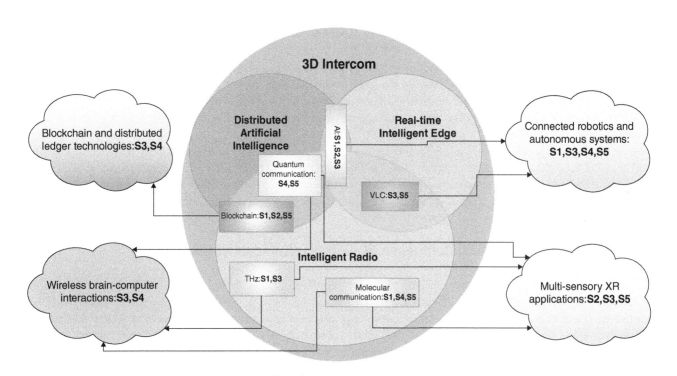

Security and Privacy Issues:
S1: Authentication, **S2**: Access Control, **S3**: Malicious behaviors, **S4**: Encryption, **S5**: Communication

Figure 3.2 Potential security and privacy issues for 6G [3].

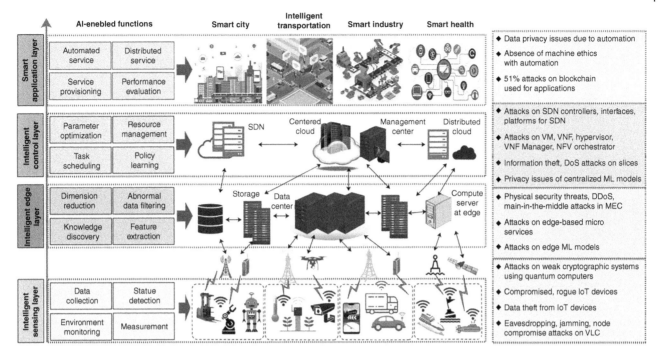

Figure 3.3 AI-enabled functions for 6G [4].

individual devices. Instead, a hierarchical approach that could distinguish between inter-subnet and subnet-to-WAN communications could be a better approach when dealing with 6G [4].

Since 6G will rely on AI and machine learning (ML) to run autonomous networks, that will be another avenue by which malicious actors can attack. AI and ML will allow 6G systems to perform more efficiently. Still, attacks such as AI poisoning and data manipulation, also known as adversarial ML attacks, could cripple AI systems [5, 6]. For example, a malicious actor could inject false data into an ML model that could cause it to believe security breaches are normal and not worth reporting. There are even ways to trick AI without "hacking" it – such as causing a self-driving car to go the wrong way by placing stickers on the road or placing a piece of tape onto a stop sign to make it think it is a speed limit sign [6].

A blockchain is a network of devices connected using software. Blockchain involves storing digital information (blocks) on a public database in which all blocks have cryptographically secured links between them [7]. Most public blockchains allow anyone to operate and create a node, making blockchain a decentralized, open computerized record that monitors cryptographic money exchanges in a subsequent request. The weakest link for hacker access is a node in the blockchain that approves a fraudulent transaction. Blockchains are viewed by many as problematic in the currency-related world because no individual or regulatory body has oversight over transactions. Others consider it a good substitute for physical monetary establishments like banks. Transparency, scalability, regulation, security, and energy consumption are some of the areas of concern that must be addressed to make blockchain a credible distributed ledger technology.

The advantages of blockchain – disintermediation, immutability, nonrepudiation and proof of provenance, integrity, and pseudonymity – are particularly important as 6G networks grow and look to enable many services in a trusted and secure manner. This leads to a focus on distributed ledger technology to offer some layer of trust for the developing and potential uses of 6G. This could be in the form of protecting the integrity of AI data, for example.

Blockchain will be able to help eliminate many of the challenges presented in 6G, such as scalability, throughput, confidentiality, and the threat of DDoS (distributed denial of service) attacks. Blockchain can:

- Ensure resource integrity.
- Improve spectrum sharing security.
- Ensure trust between requesters and providers.
- Eliminate intermediaries.
- Enable massive connectivity [8].

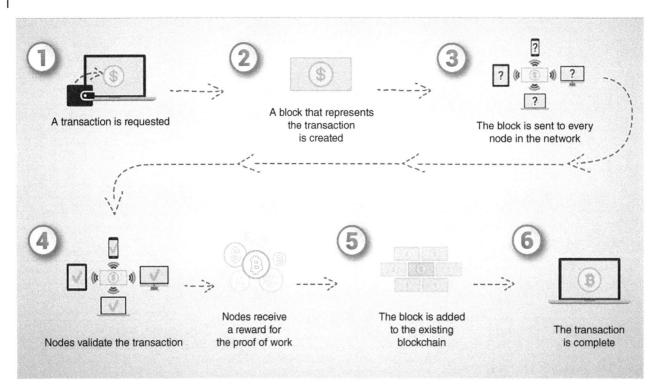

Figure 3.4 Blockchain and security issues [9].

Blockchain is not without its potential problems, though. Several potential security and privacy issues in using blockchain with 6G could be concerning. There have been several instances where private keys or personal digital signatures have been stolen, resulting in the theft of millions of dollars in cryptocurrencies. A few other types of attacks on blockchains include 51% attacks, routing attacks, phishing attacks, blockchain endpoint vulnerabilities, and transaction privacy leakage. With the advent of quantum computing, *51% attacks* may be the most troubling of all these attack types. In a 51% attack, also known as a majority attack, a malicious actor attacks a blockchain by gaining control of at least 51% of its computing power. A successful attack can destabilize the blockchain by allowing the attacker to block transactions, reverse their transactions, and change the ordering of transactions. Figures 3.4 and 3.5 illustrate how AI can be used by enablers, offenders, defenders, and targets in 6G security.

3.4 The Impact of 6G on Healthcare Systems

6G technology will dramatically change the healthcare industry by creating an intelligent healthcare system. There are a plethora of applications in the medical field in which the advent of 6G technologies can vastly improve them or make them possible at all. Telehealth checkups, telesurgery, telerehab, and more will expand once 6G is available. It will greatly improve both the quality of service and the quality of patient experience. Figure 3.6 shows some of the possibilities of an intelligent healthcare system.

One major benefit of 6G technology will be development of intelligent devices that patients can wear. The devices would monitor body functions such as heartbeat, blood pressure, nutrition, and other health conditions [10]. Basic tests could quickly be sent to a doctor or nurse from home without going to a clinic, and AI and ML could be utilized to advise the patient on what to do during minor health and body issues/events. Devices such as these will be especially helpful for elderly patients and those who are otherwise immunocompromised. It will lessen their need to be around large amounts of sick people in a hospital or clinic setting. Even outside of the immunocompromised, intelligent wearable devices will allow everyone to have fewer overall hospital/clinic visits and lower their yearly healthcare costs.

Remote health services such as telehealth checkups and telerehab, usually conducted through meeting software such as Zoom, exploded in popularity during the COVID-19 pandemic. The increase in the use of these types of services

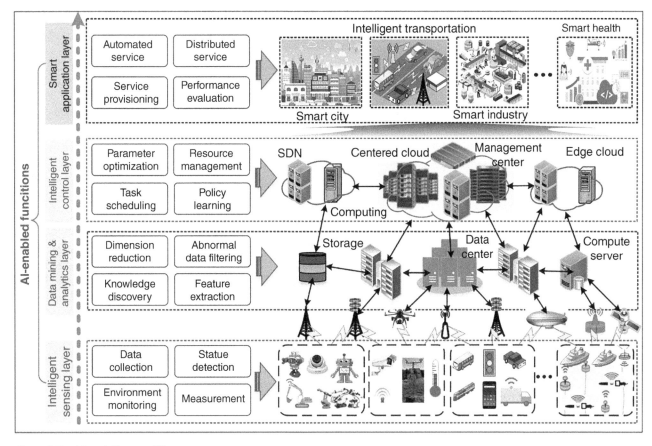

Figure 3.5 6G and AI usage [4].

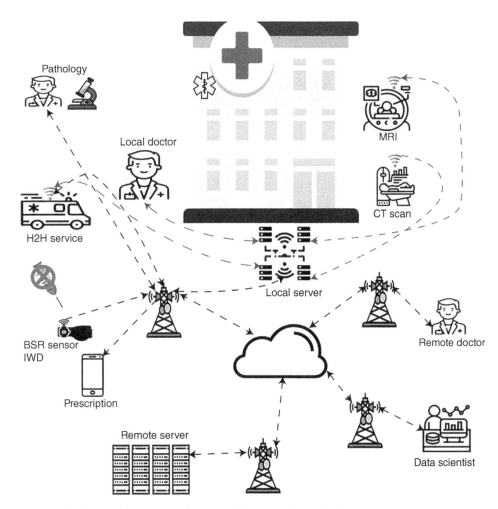

Figure 3.6 Potential 6G connectivity for intelligent healthcare [10].

highlights the deficiencies of 5G, but one of the most exciting remote services that will be made possible through 6G technologies is telesurgery. Telesurgery, which will require robots, nurses, and at least one remote doctor, is not currently possible with the constraints of 5G because of the constant real-time verbal and visual communication required. However, holographic communications that will be made possible with 6G will be a game-changer for telesurgery since the remote doctor will visually provide directions and "physically" move around the area to have a better view if needed. Doctors have even begun operating remotely, controlling robotic operating equipment from hundreds of miles away, but internet speed and reliability are current hindrances to widespread adoption. Virtual reality and augmented reality devices, as well as tactile and haptic technology, will also assist the doctors and other healthcare workers who are taking part in telesurgery.

Another aspect of healthcare that 6G can improve upon is epidemic and pandemic control. This is yet another need that was highlighted by the COVID-19 pandemic, as healthcare workers are often exposed to and even killed by deadly viruses. With an intelligent 6G medical system, this number could be drastically reduced by allowing tests to be taken at home, which would then be sent to a testing center. The data could be tracked to keep a real-time outlook on the outbreak [10]. Allowing for all this to be done without physical contact could drastically reduce the number of new infections and potentially curb a new epidemic before it can get started.

Finally, AI and ML with 6G can improve intelligent diagnosis, patient monitoring, and smart disaster response. Patients may be able to input their symptoms into a device that would use AI and ML to learn based on past patients to send the doctor a list of potential diagnoses, potentially lessening the workload of the doctors and nurses. 6G will allow for even more efficient monitoring of patients staying overnight in hospitals, allowing nursing staff to concentrate on urgent situations instead of constantly checking up on patients whose condition has not changed since their last check. Smart disaster response would allow emergency crews to be automatically notified of events like car accidents and building fires through sensors throughout a city. This could increase emergency response time exponentially and potentially save many lives.

3.5 The Impact of 6G on Space Technology and Satellite Communication

By 2030, we expect to see a new generation of mobile communication, 6G. We can expect to see antennas, arrays, and wearable devices that all communicate with each other. These are all expected to machine high download rates and lightning-fast uploaded rates compared to what we have accessible today. New, compatible applications will also have to be co-developed. People use the internet more than ever, and communication systems are higher. To meet this, we are challenged to develop a higher-capacity architecture infrastructure for mobile communication systems worldwide. If we do not do this, old technology will become obsolete and unusable. GEO (geostationary), MEO (medium Earth orbit), and LEO (low Earth orbit) satellites are all parts of communications for extension and integration to terrestrial networks. Satellites are the backbone of all communications bridging services across the globe and are efficient communication means. Satellite communications comment to unmanned aerial vehicle (UAV) communications through mmWave/THz band and to maritime broadband, remote location automation, satellites on buildings, professional mobile network, aeronautical broadband, and massive-MIM using sub 6 GHz and mmWave.

Many people have concerns about the safety and privacy of 5G communication system, such as health concerns and spy technology. However, 5G communication technology already exists, such as the mmWave band. They are in security systems, radar guns for police, satellites, and remote sensors. Research has been done on all these technologies, and there is no evidence that these waves create adverse health conditions. With 6G, we can expect to see connectivity across the globe in more rural areas. There will also be more security efforts, by necessity.

Since the launch of the first satellite into space in 1957, humanity has been making strides in space technology. From the Apollo Guidance Computer with 4 KB of RAM to the supercomputers used by NASA today, we have certainly come a long way technologically. 6G might be the next step to further our conquering of space. Figures 3.7 and 3.8 show the layout of a space-integrated computing network that could be made possible. 6G will be the first generation to incorporate space, terrestrial, and IoT in a massive network of connectivity. The key features of 6G include connectivity, mobility, security, broadcasting, ubiquity, and data rates.

A space-air-ground integrated network could be made possible with the help of 6G by utilizing a combination of satellites orbiting Earth at various levels. With the growing network-related demands of humanity as a whole, terrestrial networks are likely to become even less efficient as we move into the future [13]. This is where the Space-Air-Ground Integrated

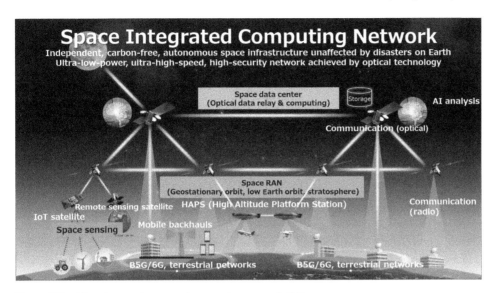

Figure 3.7 6G and space-integrated computing network [11].

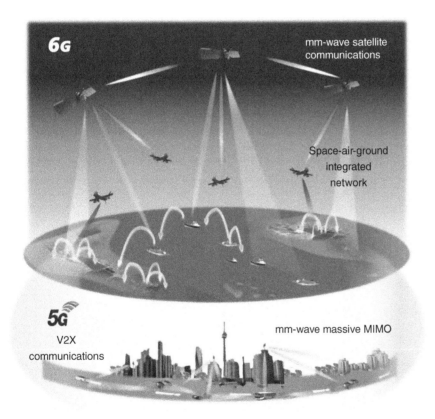

Figure 3.8 3D coverage of 6G [12].

Network (SAGIN) comes into play. With the expected speed and reliability of 6G, a SAGIN network could potentially offer internet connections to vast swaths of the area and people who otherwise would have a hard time being connected at all and fill the digital gap in underserved areas – not to mention allowing for higher network traffic in urban areas due to being able to handle a much higher load than terrestrial networks.

6G will also be beneficial for human-crewed space missions. Astronauts performing research on the International Space Station should be able to communicate with their countries more easily and make use of more IoT and IoE devices on the station itself. There is also the possibility of future moon missions, human-crewed missions to Mars, or even ventures further into space. 6G technologies will aid all of these ventures and make them more likely to happen at all.

3.6 The Impact of 6G on Other Industries

The healthcare and space industries will not be the only areas of our lives affected by 6G when it is implemented. From the manufacturing industry to the automotive industry and even the agriculture industry, many aspects of our life will be touched by 6G. Likely, we cannot even perceive at this point how many things will be changed with the advent of 6G since it is still at least a decade away from commercial availability. Figure 3.9 illustrates potential use cases for 6G systems, though that list is not exhaustive.

The agriculture industry is generally slow to adopt and accept technological change. Society often looks down on it as a "lesser" trade, but our world could not exist without it. With the rapid growth of the world's population, the agriculture industry will likely need to increase output by up to 70% in the coming decades [15]. 6G and the technologies it allows for will take a major role in achieving this growth. Wireless sensor networks will allow for remote farm monitoring, and AI will be able to intelligently make decisions based on the data read by the sensors. Farm vehicles such as tractors and combines may be remotely or autonomously controlled, augmented reality may be used to train new farmers and farmhands, and livestock can be tracked and monitored using remote sensors.

During the COVID-19 pandemic, many students from kindergarten to university were forced to learn remotely, often with poor outcomes. 6G will allow for better options when remotely learning through online classwork. Online learning can be enhanced by utilizing augmented reality devices and holographs to improve the experience greatly and even simulate a physical classroom [15]. In nonremote education, 6G can also benefit by allowing for rapid communication and response times in an emergency such as earthquakes or an active shooter situation.

In the manufacturing and transportation/logistics industries, 6G should have a massive impact. 6G will allow for extremely low latency factory automation, and sensors will be able to provide precise, real-time updates on parts as they move through the manufacturing process. Upon completion and shipment, 6G will aid the logistics process by improving inventory management, providing opportunities for logistical feasibility analysis, optimizing communications in warehouses, automizing forklifts, and potentially even allowing for nearly autonomous delivery truck driving [15]. Other functions such as shipment tracking and real-time delivery updates will likely be improved with 6G.

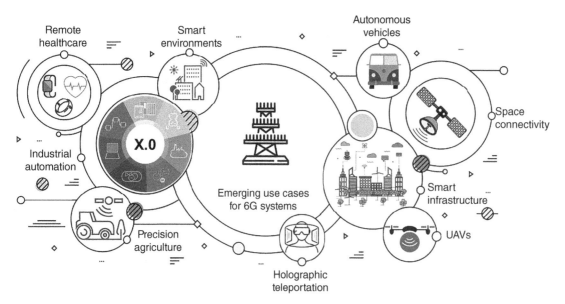

Figure 3.9 Various potential use cases for 6G [14].

3.7 Terahertz Wireless Systems and Networks with 6G

Over the last several years, global mobile data traffic has been dramatically increasing due to increasing demand for secure, fast, and large data transmission rates in broadcasting, IoT, automobiles, smart cities, energy, communication, wearable devices, and more. This of course puts pressure on the existing mobile communications systems to meet the new and evolving data rates and demands. There are several potential ways to improve capacity and data rates, including increasing the bandwidth, which is directly proportional to the data rate. High-frequency bands, or mmWave bands, can provide significantly higher bandwidth than those used by our current systems. Some of the main challenges to massive MIMO (multi-input, multi-output) and mmWave bands include limited available space, significant path loss, and coverage area.

6G communication, which is in the conceptual and development phase, is being envisioned to have high data rate capabilities as well as ultra-low latency, extreme capacity, high security, and wider coverage. 6G will likely be a combination of sub-6 GHz, mmWave, and THz bands. Some of the key features of the 5G to 6G transition are massive connectivity, ultra-low latency allowing for high mobility, extreme data privacy and protection, high-quality video streaming and live broadcasting, ubiquitous coverage area, and massive data rates. Possibilities for antennas include beam steerable antennas, switchable phased arrays, and dual-polarized antennas. THz frequency bands from 100 to 300 GHz are being considered for 6G, which will require a higher complexity of antenna design, creating even more challenges in terms of materials, design, fabrication, and experimental validation.

In the last decade, wearable electronics have progressed hugely, and it is expected to continue growing at a high rate. The worldwide market size of wearables technologies in 2022 was USD 61.30 billion with a revenue of CAGR of 14.4% [16]. Rapid worldwide internet connectivity expansion with the availability of smartphones and IoT smart devices connectivity has paved the way to connect any wired or wireless network. Demands in all sectors have seen growth, but the demand from sectors such as remote real-time medical monitoring with body-worn sensors is important due to the rapidly aging population and this sector will continue to drive the market forward.

An essential element of wearable electronics is a wearable antenna for devices such as medical for monitoring patient health, consumer smartwatches, wearable navigation systems, body cameras, wearable athletic devices, and more. Conventional modified microstrip antennas are designed on a thin layer of PET substrate for 5G applications, operating at around 5 GHz with a gain of 5dBi and 38 percent rad. efficiency. The utilization of higher frequencies in wearable antennas will result in more propagation loss, so 5G wearable antennas need to be highly directive with greater gain to ensure

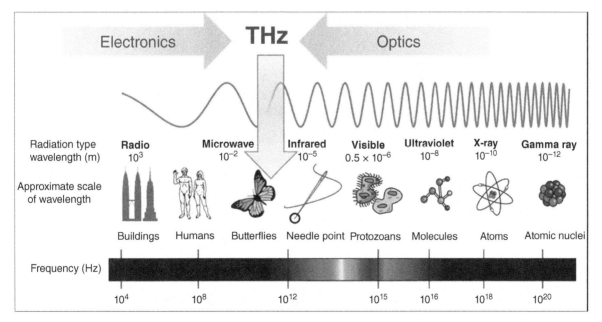

Figure 3.10 THz band and several pertinent applications [19].

mmWave antennas operate at acceptable levels. A test was run on a 60 GHz mmWave band applied to body-mounted antennas and concluded that a metal ground plane reduced electromagnetic exposure significantly, as well as helped to reduce the sensitivity of the reflection coefficient.

Since 6G aims to turn our current wireless world into an AI-based intelligent network connecting as much of the world as possible, networks of the future will need massive amounts of bandwidth – which will create other problems compared to our current networks. To combat this, we will need to operate on higher frequency bands like terahertz (THz). THz will be a requirement because contiguous radio spectrums allowing massive amounts of bandwidth are not possible below 90 GHz [17]. The high transmission frequency of the THz band will allow us to design high-directional antennas with extremely low beamwidths. This procedure will be used to counterbalance the high channel attenuation by establishing high-directional links [18]. Figure 3.10 illustrates the THz band and several applications requiring THz.

3.8 The Future of 6G and Its Role in IT

6G is being researched, and several countries are taking the first steps to create it. Countless hours of work will need to be done before it is rolled out to the public – probably not for another decade. Once it does come out, there will undoubtedly be countless bugs and glitches that will need addressing. IT professionals will likely have their hands full, whether helping consumers with troubleshooting, debugging applications or hardware, installing and maintaining new infrastructure, or doing further research into the future, perhaps even into the seventh generation of wireless technologies. Regardless of what specific job each IT professional holds, the 6G revolution will likely play a large part in it in the not-so-distant future.

References

1 Zhao, Y., Zhai, W., Zhao, J., Zhang, T., Sun, S., Niyato, D., and Lam, K.-Y. (2021). A comprehensive survey of 6G wireless communications. https://arxiv.org/pdf/2101.03889.pdf.

2 Chen, S. (2020, February). Possible capabilities of 6G in comparison with 5G.

3 Wang, M. (2020). Security and privacy issues in the 6G network. *ScienceDirect* https://ars.els-cdn.com/content/image/1-s2.0-S2352864820302431-gr1.jpg.

4 Siriwardhana, Y., Porambage, P., Liyanage, M., and Ylianttila, M. (2021). *AI and 6G Security: Opportunities and Challenges*. Dublin: Madhusanka Liyange.

5 Wiggins, K. (2021, May 29). Adversarial attacks in machine learning: What they are and how to stop them. *Venture Beat* https://venturebeat.com/2021/05/29/adversarial-attacks-in-machine-learning-what-they-are-and-how-to-stop-them.

6 Wiggers, K. (2021, May 29). Adversarial attacks in machine learning: What they are and how to stop them. https://venturebeat.com/security/adversarial-attacks-in-machine-learning-what-they-are-and-how-to-stop-them/.

7 Geroni, D. (2021). Top 5 blockchain security issues in 2021. *101 Blockchains* (June 13): https://101blockchains.com/blockchain-security-issues.

8 Central Blockchain Council of America (2020, November 12). Blockchain technology benefits in 6G networks: The way forward. https://www.cbcamerica.org/blockchain-insights/blockchain-technology-benefits-in-6g-networks-the-way-forward.

9 Bitcoin Warrior (2021). Top 6 blockchain security issues in 2021. https://bitcoinwarrior.net/2021/09/top-6-blockchain-security-issues-in-2021.

10 Nayak, S. and Patgiri, R. (2020, April). 6G communication technology: A vision on intelligent. https://www.researchgate.net/publication/341451844_6G_Communication_Technology_A_Vision_on_Intelligent_Healthcare.

11 Converge Network Digest (2021, May 20). NTT envisions a space integrated computing network. https://www.convergedigest.com/2021/05/ntt-envisions-space-integrated.html.

12 Bhat, J.R. and AlQahtani, S.A. (2020). 6G ecosystem: Current status and future perspective. *IEEE Access* 9: 43134–43167. https://doaj.org/article/61f82bc8195445a4a1774264ce9aa7c2.

13 Ray, P.P. (2022, October). A review on 6G for space-air-ground integrated network: Key enablers, open challenges, and future direction. *Journal of King Saud University – Computer and Information Sciences* 34 (9): 6949–6976. https://www.sciencedirect.com/science/article/pii/S1319157821002172.

14 Akyildiz, I.F., Kak, A., and Nie, S. (2020). 6G and beyond: The future of wireless communications systems. *IEEE Access* 8: 133995–134030. https://ieeexplore.ieee.org/document/9145564.

15 Imoize, A., Adedeji, O., Tandiya, N., and Shetty, S. (2021). 6G Enabled Smart Infrastructure for Sustainable Society: Opportunities, Challenges, and Research Roadmap. *Sensors* 21: 1–57. https://doi.org/10.3390/s21051709.

16 Emergen Research (2023). Wearable technology market, by product (wristwear, headwear, footwear, bodywear, and others), by technology, by application, and by region forecast to 2032, https://www.emergenresearch.com/industry-report/wearable-technology-market.

17 Alsharif, M. (2021, January). Toward 6G communication networks: Terahertz frequency challenges and open research issues. https://www.researchgate.net/publication/348204464_Toward_6G_Communication_Networks_Terahertz_Frequency_Challenges_and_Open_Research_Issues.

18 Boulogeorgos, A.-A.A., Yaqub, E., di Renzo, M. et al. (2021, September 9). Machine learning: A catalyst for the wireless networks. *Frontiers in Communications and Networks* 2: https://www.frontiersin.org/articles/10.3389/frcmn.2021.704546/full.

19 Weissberger, A. (2021, September 28). ITU-R Report in Progress: Use of IMT (likely 5G and 6G) above 100 GHz (even >800 GHz). https://techblog.comsoc.org/2021/09/28/itu-r-report-in-progress-use-of-imt-likely-5g-and-6g-above-100-ghz-even-800-ghz.

4

Artificial Intelligence and Machine Learning in the Era of 5G and 6G Technology

Artificial intelligence (AI) is a system or machine that simulates human thought processes. Most AI systems are not sentient, but instead use algorithms to think like humans and make calculations. While these might seem underwhelming at first, they can actually be used with another breakthrough known as ML (machine learning). Machine learning is the process of using pattern recognition to let computers learn without being programmed to do the task at hand. This is what can make them so much like humans. With the ability to learn from things that happen, AI can complete tasks many thought machines would never be able to do.

The real magic happens when we combine AI with Internet of Things (IoT), so we give machines that have the ability to learn the ability to communicate with each other. This is what can create some of the most amazing breakthroughs in technology. For example, when Tesla Motors Inc. decides to upgrade a part of its cars, it is able to do it remotely. When Tesla experienced a sudden power-steering failure in its Model S electric sedan, technicians at the carmaker's California head-quarters quickly pinpointed the likely cause – an electrical connector that had been damaged three days earlier by an overzealous employee using a standard air gun to remove adhesive from an assembly line. This is just one example of how AI and IoT combine to make something amazing. With these two technologies, we can create products and features that weren't originally thought of before.

So how has this new technology affected us? Well in order to figure that out we have to look at the past. Before the integration of AI and IoT, factories had to use manual machines. This means each machine had to be operated by a person, which was incredibly tedious for each worker. With the integration of AI and IoT, machines can now be operated by the machines themselves, cutting down on labor costs. Citing the example of Amazon's warehouses, "Those warehouses are filled with robots and sensors that need to be maintained and improved. If something goes wrong, Amazon does not have time for its workers to fix a machine or recalibrate it manually. Adding in AI to automate these processes saves money and labor". Using these technologies instead of humans for repetitive tasks is saving companies time (and money) by reducing their costs for labor.

Not only has this combo changed the way factories work, but it also changes standard companies as well. Using the machine learning aspect, normal companies can virtually minimize security risks. By analyzing the data, companies can predict potential security risks, such as the likelihood of theft or a cyberattack. Responding to such threats can also be automated and accelerated, further enhancing security.

AI is a core component of 6G as well. Created from the melding of machine learning and deep neural networks. The AI is comprised of three subsections of intelligence. The first is operational intelligence, which deals with allocating resources for efficient operation in extremely complex and dynamic environments. Second is environmental intelligence, which would allow for self-organizing and self-healing properties in everything from networks to complex autonomous devices. The final is service intelligence, which extend human focused applications in ways that increase user satisfaction by using deep learning techniques to adjust things for optimal efficiency. Then come the concerns of data privacy and security features. The advent of quantum computing technologies like quantum key circulation will be crucial in improving the security of 6G far beyond that of 5G and 4G.

From 5G to 6G: Technologies, Architecture, AI, and Security, First Edition. Abdulrahman Yarali.
© 2023 The Institute of Electrical and Electronics Engineers, Inc. Published 2023 by John Wiley & Sons, Inc.

4.1 Artificial Intelligence and Machine Learning: Definitions, Applications, and Challenges

Machine learning analyzes data structures that use algorithms and statistical models to learn by inference and patterns without being explicitly programmed. ML has proven to be one of the most game-changing technological advancements. Machine learning is helping organizations expedite their digital transformation and move into the age of automation in today's more competitive business climate. Some argue that AI/ML is required to stay relevant in some businesses, such as digital transactions, financial fraud detection, and product recommendations. The adoption and pervasiveness of ML algorithms in enterprises have been extensively documented by multiple companies. ML is currently employed in practically every other application and program on the internet in some way or another. ML has grown in popularity to the point that it is now the go-to tool for organizations to solve various problems [1].

To comprehend machine learning, we must first grasp the fundamental notions of AI. AI is described as software with cognitive abilities comparable to a person's. One of the major concepts of artificial intelligence is to have computers think like people and solve problems in the same manner we do. AI is a broad phrase that refers to all computer programs that can think similarly to humans. AI is defined as any computer program that exhibits self-improvement, inference learning, or even fundamental human abilities like image recognition and language processing. In AI, the categorizations of ML techniques are covered [1].

4.1.1 Application of Machine Learning and Artificial Intelligence

However, there is a distinction between machine learning and artificial intelligence. Deep learning is deterministic, whereas machine learning is probabilistic. Machine learning is fundamentally different from AI in that it can evolve. ML algorithms can analyze vast volumes of data and extract meaningful information using a variety of programming approaches. In this approach, they may learn from their given data and improve on earlier versions [2].

Big Data, one of the most significant components of deep learning algorithms, cannot be discussed without mentioning machine learning. Because AI mainly relies on statistical approaches, any sort of AI typically relies on the quality of its dataset for successful results. A solid flow of structured, diverse data is essential for a successful ML solution, and deep learning is no exception. Companies have access to a tremendous quantity of data on their clients in today's online-first world, typically millions [2]. The sheer volume of information included in this data, which is huge in the number of data points and the number of fields, is referred to as Big Data. It is time-consuming and difficult to interpret by human standards, yet good data is the best feed for training a ML system. The more clean, usable, and machine-readable data there is in a huge dataset, the better the ML algorithm's training will be. Big Data, which refers to massive data that can be algorithmically analyzed for predictive modeling, trends, and connections is developing forms of insight and knowledge with significant benefit for countering humanity's greatest challenges when combined with the power of AI and high-performance computing. Agriculture production can be increased by digitizing and analyzing photographs taken by autonomous drones and spacecraft. Improving the collection, processing, and dissemination of health information or data might aid patients in receiving better diagnoses and treatments, particularly in rural and remote areas. Better data on meteorological and environmental conditions can also aid governments in better anticipating malaria prevalence, limiting disease transmission, and allocating medical resources [2].

4.1.2 Challenges for Machine Learning and Artificial Intelligence

While AI holds a lot of potential, it also has many drawbacks. Gender, racial, and ideological prejudices can be reflected or reinforced by datasets and algorithms. When AI relies on insufficient or biased datasets (supplied by humans), it may result in skewed AI conclusions. Humans increasingly use deep-learning technology to select who receives a loan or a job. However, the inner workings of deep-learning algorithms are opaque, and humans have no way of knowing why AI makes particular associations or draws certain conclusions, when mistakes may occur, or when and how AI may be replicating prejudice [3]. By automating ordinary chores and replacing occupations, AI might exacerbate disparities. Any software can have security issues, including the software that operates cell phones, surveillance cameras, and electricity grids. These can result in money and identity theft and internet and energy outages. Advances in AI technology may potentially create new dangers to world peace and security. ML, for example, may be used to create phony multimedia content to sway elections, policy making, and legislation.

4.2 Artificial Intelligence: Laws, Regulations, and Ethical Issues

AI is slowly infiltrating every aspect of our society, from the vital to the mundane, such as healthcare and humanitarian assistance. AI may enhance economic, social, and human rights results, such as embodied AI in robots and ML methodologies. These cutting-edge technologies can help a variety of sectors. At the same time, AI has the potential to be misinterpreted or to act in unanticipated and perhaps dangerous ways. As a result, it is more important than ever to understand the intersection of law, ethics, and technology when it comes to managing AI systems. AI systems are used in a variety of applications, the bulk of which use statistics-based learning techniques to find patterns in large amounts of data and make predictions based on those patterns [4]. Because AI is becoming increasingly common in high-risk sectors, there is a growing demand for responsible, fair, and transparent AI in its design and control. What frameworks may be utilized to do this, and how can this be done? This is one of the major themes covered by the writers of this special issue. They offer in-depth analyses of the ethical, legal-regulatory, and technological issues that arise when putting up AI governance frameworks.

Complex, high-risk tasks, including providing parole, detecting illnesses, and processing financial transactions, are increasingly assigned to AI systems. This raises new difficulties such as autonomous car liability, the limitations of present legal frameworks in dealing with "Big Data's uneven impact" or preventing algorithmic damages, social justice issues relating to automating law enforcement or social welfare, or online media consumption. Given AI's extensive reach, these significant challenges can only be fully addressed by a multidisciplinary approach [4].

4.2.1 Ethical Governance in Artificial Intelligence

Fresh ideas on building and supporting the ethical, legal, and technological governance of AI are required. Ethical governance is centered on the most pressing ethical issues highlighted by AI, including fairness, integrity, privacy, service, and product allocation. Clarity and interpretability are two principles that may be used to improve algorithmic fairness, transparency, and accountability. In Europe, for example, the concept of a "right to explanation" of algorithmic choices is being explored [5]. Individuals would be entitled to an explanation if an algorithm decided against them under this right. Accountability measures for opaque and sophisticated algorithmic systems cannot exclusively rely on interpretability. Rather than dissecting how the system works, auditing tools are presented as viable remedies that evaluate the inputs and outputs of algorithms for bias and damages.

The ownership of AI is a third ethical problem that is crucial for evaluating concerns like responsibility. Who is responsible for the activities, flaws, and liabilities of robots? For example, who is responsible if a robot causes injury to someone due to negligence? Is it more likely to be the robot's owner or itself? The fourth ethical question is whether robots are considered natural persons in the eyes of the law. This is still a gray area, and it's unclear if a robot would be considered a real person in the eyes of the law. AI thinking should be capable of evaluating societal values, moral and ethical dilemmas, balancing the relative significance of values held by many stakeholders in a variety of multicultural contexts, explaining its rationale, and ensuring transparency. The concept that robots can be ethical agents who are accountable for their actions, or "autonomous moral agents," is at the heart of some machine ethics discussions. Machine ethics is already making its way into practical robotics, where the assumption that these machines are artificial moral agents in any meaningful sense is rarely made. It has been highlighted that a robot that has been designed to follow ethical standards may easily be reprogrammed to follow immoral ones [5].

4.2.2 The Future of Regulation for AI

Individuals worldwide are concerned that AI and robots will eventually replace the human workforce. As a result, fewer individuals are prepared to reveal personal information, as their faith in the media, online social platforms, and search engines erode. Some argue that the divide between trust and technology has grown for a good reason: There hasn't been any regulation around AI for most of its life. The regulations may appear a little vague and unclear at times of how world-changing they might be. It is tempting to leap to conclusions without building a guardrail for AI to help us grasp its position in our daily lives and determine if society is going to be co-piloted by AI or directed by AI.

4.3 Potentials of Artificial Intelligence in Wireless 5G and 6G: Benefits and Challenges

4.3.1 Artificial Intelligence in Wireless 5G and 6G

The 5G cellular network was launched to enable Internet of Everything (IoE) applications by delivering high data rates, low latency, and ultra-reliable connectivity. To facilitate connections between people, machines, and things in 5G, network equipment must be software-driven, intelligent, virtualized, and energy-efficient as the number of connected edge devices grows. The networks should also support edge computing to decrease backhaul bandwidth consumption. It will be difficult for 5G to provide a single network architecture that can support various services, such as increased mobile broadband, 3D connectivity, and ultra-reliable, low-latency communication with heterogeneous connected devices. As a result, 6G networks must be intelligent networks that employ AI to improve network services and edge computing to enable low latency and high bandwidth.

At the network level, new technologies and applications such as the IoT, social networking, and crowd sourcing create massive volumes of data. Models based on AI are frequently created from acquired data to assist in detecting, categorizing, and predicting future occurrences. It is typically impractical to transport all data to a single place due to traffic, storage, and privacy considerations. In wireless communication, federated learning (FL) has been utilized to meet the needed goals. The rollout of 5G wireless communication networks has reached a major milestone in academia and industry worldwide, and the industry now aims to advance beyond 5G. The use of 6G in the AI platform to collect sensor data with the help of high-performance computer networks is gaining traction. The typical method entails collecting data in a centralized way, and the heterogeneous data from numerous sources is then aggregated in the server, resulting in communication challenges. FL based on AI is a bottleneck technology that improves the wireless paradigm's online privacy challenges.

4.3.2 Benefits and Challenges of AI in 5G and 6G

Current 4G networks transfer data through an inefficient internet protocol (IP) broadband connection. ML and AI allow 5G networks to be proactive and predictive, essential for 5G networks to go live. Thanks to the incorporation of ML into 5G technology, intelligent base stations will be able to make decisions for themselves, and smart gadgets will create dynamically adaptable clusters based on learned data. As a result, network applications will be more efficient, have reduced latency, and be more dependable. Machine learning will become increasingly critical in implementing the 5G goal as the network becomes more intricate and new applications such as self-driving cars, industrial automation, virtual reality, e-health, and others emerge. There are enormous potentials and limits to overcome with each new technology.

The following are some of the benefits of ML for 5G communications:

- *Enhanced mobile broadband (eMBB).* Enables the generation of new applications that require higher data rates in a consistent service region. Two examples are virtual reality and ultra-high-definition video streaming.
- *Massive machine-type communications (mMTC).* Scalable connection needed for extending the number of wireless devices with the effective transmission of tiny amounts of data across broad coverage areas is a major feature of 5G communication services. This sort of traffic will be generated by applications such as body-area networks, smart homes, IoT, and drone deliveries. mMTC must be able to enable new and previously unimagined applications.
- URLLC (ultra-reliable low-latency communications) will be prioritized above data rates in connected healthcare, remote surgery, mission-critical applications, autonomous driving, and smart industry applications.

4.3.3 How Can AI Be Used to Enhance 6G Wireless Security?

Even as 5G is recognized for its network cloudification and microservices-based design, the next technology of networks, or the 6G generation, is strongly linked to intelligent network orchestration and management. As a result, AI will play a huge role in the envisioned 6G paradigm. However, in many circumstances, the partnership of 6G and AI might be a double-edged sword due to AI's potential to safeguard or infringe on security and privacy. End-to-end automation of future networks, in particular, necessitates proactive threat detection, intelligent abatement approaches, and ensuring the realization of auto network systems in 6G [6]. Nonetheless, AI applications are rare, and support for AI-driven technologies in 5G networks is constrained by the limitations of the conventional architecture that existed at the network's inception. As a result, no assistance is provided for distributed AI or intelligent radio, as these two domains are entirely AI-based.

Furthermore, while 5G networks have already included real-time intelligent edges, such as car networks, emergencies cannot be addressed in "real time" owing to latency difficulties. 6G networks, on the other hand, can.

Several important research fields have risen to the top of 6G wireless summits and symposiums. The subjects explored were THz communications, quantum communications, Big Data analytics, cell-free networks, and omnipresent AI. An efficient network will be required to offer greater rates and lower latency performance benefits. The network must dynamically allocate resources, adjust traffic flow, and interpret signals in an interference-filled environment. For fulfilling these jobs, pervasive AI is a viable candidate. AI and ML will play a crucial role in enabling 6G technology by optimizing networks and developing new waveforms [6].

Innovative network slice administration and dynamic security approaches are also being investigated. All of these activities are aimed at making the environment more intelligent. Furthermore, 6G technology will allow for greater AI/ML breakthroughs, such as using data that is local to the 6G sensor and swiftly transferring the AI/ML algorithms. Despite its early stage of development, 6G shows evolutionary improvements that can broaden users' experiences and open up new applications.

4.3.4 The 6G Era's Edge Intelligence and Cloudification

Since there is a strong relationship between AI and edge computing, it is natural to combine the two. In some 6G wireless applications, it is critical to move processing to the network's edge. The deployment of AI/ML algorithms to acquire, store, or process data at the network edge is called edge intelligence (EI). In EI, an edge server collects data from many linked devices. Data is shared around numerous edge servers for training models, which are then evaluated and forecasted, allowing devices to benefit from quicker feedback, reduced latency, and lower costs while enhancing their operations. However, because data is collected from various sources, AI/ML algorithms' outcomes become extremely data-dependent [7].

According to Zhang et al., EI is subject to a range of security vulnerabilities since data is collected from some sources and the outcomes of AI/ML algorithms are mostly data-dependent [7]. In this environment, user identification and access control, model and data integrity, and mutual platform verification all need trust in EI services. It is demonstrated how blockchain may be used to protect distributed edge services and resource transactions from malicious nodes. The blockchain ensures the consistency of deconstructed tasks and bits of learning data required for AI implementation. Edge computing is distributed by nature, and hackers might use its accompanying dependencies to carry out different attacks such as data theft.

Hackers might use the dispersed nature of edge computing and its corresponding dependencies to execute various assaults such as data poisoning, data evasion, or a privacy attack, damaging the AI/ML applications' results and undercutting the benefits of EI. Furthermore, EI service delivery may necessitate unique secure routing algorithms and trust network architectures. Because edge devices may capture sensitive data like a user's location, health or activity records, or manufacturing information, among other things, security and privacy are tightly linked in EI.

In edge AI models, federated learning is one way to enable local ML models with privacy-friendly distributed data training. Experts are also investigating safe multiparty computation and homomorphic encryption to build privacy-preserving AI framework configuration techniques in EI solutions. 6G will be the major communication technology for linking intelligent healthcare services in the future. As a result, ensuring secure communication, device authentication, and access control for billions of IoT and wearable devices will be major security challenges in the 6G era. Privacy protection and the ethical components of user data or electronic health records will be critical challenges in the future healthcare system. To handle billions of IoT devices and evaluate health-related data, AI is essential. However, there are certain limitations [7].

4.3.5 Distributed Artificial Intelligence in 6G Security

The quick AI technology in the 6G vision is closely linked to ML, with privacy having a stronger influence in two ways. In certain ways, the proper use of AI/ML in 6G can preserve privacy. Privacy issues may also emerge as a result of AI/ML assaults. While 5G networks have always been IoT ready, 6G networks would need to be IoE ready. As a result, the 6G network will probably be a largely decentralized system capable of intelligent decision-making at several levels.

Furthermore, because everything in the Internet of Things is connected to the internet, distributed AI will need to meet various criteria. The training data set should be dispersed unevenly across most edge devices, with each edge device having access to and control over a portion of the data. In addition, the edge device should be able to process and handle data on its own.

4.4 Cybersecurity Issues in Advanced 5G and 6G

Our essential networks will undergo a physical transformation due to 5G, which will have long-term ramifications. Given that 5G is mostly a software-based network, future upgrades will be software updates, similar to how your smartphone is now upgraded. The more challenging part of the actual 5G "race" is rethinking how we safeguard the most important network of the twenty-first century and the ecosystem of devices and apps that sprout from it, from cyber software vulnerabilities. The cost of missing a proactive 5G cybersecurity opportunity after the fact will be far higher than the cost of cyber vigilance in the first place. Businesses lost $10 billion in 2017 due to the NotPetya ransomware attack. Of course, 5G networks did not exist at the time of the assault, but it does show the excessive cost of such incursions, which pales in comparison to an attack that causes human pain or death. In order to be a viable business for all 5G players, we must first establish the circumstances for risk-based cybersecurity investment to adopt and implement written cybersecurity policies and procedures reasonably designed to address cybersecurity risks [8].

5G networks are more vulnerable to assaults in various manners than previous generations:

- The network has changed from centralized, hardware-based switching to distributed, firmware-based digital routing. Prior networks used hub-and-spoke topologies, all traffic passing via physical chokepoints where cyber hygiene could be implemented. Meanwhile, in a 5G software-defined network, such activity is transferred outside of the network to a web of digital routers, removing the capacity to monitor and manage chokepoints.
- Even if the network's software flaws could be patched, the network is still run by vulnerable software – often, early-generation AI. If intruders gain control of the computer systems, they can also get control of the network.
- New attack vectors emerge due to the large increase in bandwidth that makes 5G practical. In terms of physical security, low-cost, short-range small-cell antennas placed across urban areas have become new hard targets. These cell sites will use 5G's Dynamic Spectrum Sharing features, which allow multiple streams of data to share bandwidth in "slices," each with a distinct level of security risk. When software allows network functions to change constantly, the cyber defense must be dynamic rather than relying on a consistent lowest-common-denominator strategy.
- Finally, linking tens of billions of hackable smart gadgets (actually, small computers) to the IoT poses a security concern. Plans are being developed for a wide range of IoT-enabled activities, including public safety, combat, medical, and transportation applications, which are amazing and especially vulnerable.

As a result, fifth-generation networks are especially vulnerable to a multidimensional cyberattack. This reimagined nature of networks – a new network "ecology of ecosystems" – also requires a reimagined cyber strategy. While many big network providers devote significant resources to cybersecurity, small- and medium-sized wireless internet service providers (ISPs) servicing rural areas with 5G networks have struggled to justify a sophisticated cybersecurity program. Some of these businesses have less than 10 workers. They cannot afford a dedicated cybersecurity officer or a cybersecurity operations center that is open 24 hours a day, 7 days a week [8].

4.5 Benefits and Challenges of Using AI in Cybersecurity: Help or Hurt?

AI is a method of simulating human intelligence. In the realm of cybersecurity, it has a lot of promise. If implemented correctly, AI systems may be trained to deliver risk alerts, detect new types of malware, and safeguard crucial data for enterprises. AI is intelligent, and it uses that intelligence to continuously improve network security. It uses ML and deep learning to learn the behavior of a business network over time. It searches the network for patterns and clusters them together. Before taking action, it looks for any deviations or security concerns from the norm. By learning trends over time, deep learning algorithms can help to improve security in the future. Potential risks with the same characteristics as those described are recognized and stopped as soon as possible [9].

There is a lot of activity on a company's network. A typical mid-sized company receives a lot of foot traffic. This means that customers and the company exchange a lot of data daily. This information must be protected from malicious people and software. On the other hand, cybersecurity experts are unable to scrutinize every communication for potential dangers. AI is the most effective method for detecting dangers disguised as normal behavior. It can filter through vast data and traffic due to its automated nature. The management of vulnerabilities is crucial to a company's network security. As previously said, a normal business encounters a variety of threats daily. It must recognize, identify, and restrict them to be safe. Vulnerability management might be enhanced by applying AI research to analyze and appraise existing security solutions. AI allows you to study systems faster than cybersecurity experts, dramatically improving your problem-solving abilities. It assists businesses in focusing on critical security responsibilities by detecting weak points in computer systems and corporate networks [9].

AI can provide an IT organization with unrivaled real-time monitoring, threat detection, and reaction times. Cybersecurity systems powered by AI can respond to threats faster and more precisely than people, and it may also allow cybersecurity specialists to focus on more critical internal tasks. Using these strategies, AI can quickly comprehend diverse IT trends and change its algorithms depending on the most up-to-date data or information. Similarly, AI in cybersecurity is familiar with complicated data networks and can quickly detect and eliminate security issues with minimum human intervention. In cybersecurity, AI does not replace cybersecurity expertise, and instead, it assists cybersecurity specialists in detecting and addressing hostile network behavior as quickly as possible.

4.6 How Can AI Be Used by Hackers Attacking Networks?

Instead of constantly monitoring harmful behavior, cybersecurity professionals may utilize AI to promote cybersecurity best practices and reduce the attack surface. On the other hand, cybercriminals can use the same AI systems for evil reasons. ML models with adversarial AI misread inputs entering the system and respond in a way that benefits the attacker. Simultaneously, attackers employ AI and its derivative, ML, to launch more automated, forceful, and coordinated strikes. It's not enough to use automation to combat machines with machines in today's world; it's also important to use the correct tools and technology to fight smart with insight. Hackers can be more efficient in understanding how corporations try to prevent them from breaching their surroundings by employing AI and ML [10].

Manipulating an AI system may be simple if you have the correct tools. AI systems are constructed on the data sets used to train them, and even modest, subtle alterations might cause AI to go in the incorrect direction over time. Modifying input data may easily cause system failures and disclose security flaws. Hackers can also use reverse engineering to gain access to private data sets that train AI systems directly. Cybercriminals may use AI to search and locate vulnerable applications, devices, and networks to scale their social engineering attacks. On a personal level, AI can detect behavioral patterns and weaknesses quickly, making it simpler for hackers to access sensitive information. With the aid of AI, fraudsters can enhance everything they do on social media platforms, emails, and even phone conversations. Creating deepfake content and sharing it on social media, for example, can spread misinformation and encourage people to click phishing links and travel down the wrong path that compromises their security [10].

Cybercriminals employ social engineering to trick and convince others into providing personal information or performing a specific action, such as making a wire transfer or opening a harmful file. ML capitalizes on criminals' behaviors by making it easier and faster to gather information on firms, workers, and partners. ML, in other words, amplifies social engineering assaults. Spoofing and impersonation are fraud strategies in which hackers attempt to imitate a corporation, brand, or well-known individual. Hackers can examine several characteristics of a target in great depth using various techniques [11].

Systems for ML, which are at the foundation of today's AI, have several shortcomings. Attack codes that exploit these holes have propagated quickly, while defensive solutions are limited and struggling to keep up. Hackers can alter ML systems' integrity, confidentiality, and availability by exploiting weaknesses. ML attacks are distinct from classic hacking vulnerabilities, necessitating novel defenses and remedies. ML flaws, for example, are difficult to fix in the same way that traditional software flaws are, creating persistent weaknesses for attackers to exploit. Worse, some of these flaws need little or no access to the victim's system or network, giving attackers more opportunities and defenders less capacity to identify and protect themselves from assaults [11].

As the world of cybersecurity is so dynamic, no one can anticipate a silver bullet that would keep them secure indefinitely. Looking back through history, cybersecurity technology and hackers have always played catch-up. As a result, if AI can be used to improve cybersecurity tools, it can also improve attack tools. Even though AI-powered hacks are currently uncommon, they have the potential to become commonplace soon. Phishing emails are traditionally created by hand and distributed to many people without particularly targeting each one. Anyone who pays careful attention to the content will be able to tell them apart. So, how might AI go about doing things differently?

To begin with, AI can comprehend and converse in written language via natural language processing. This allows AI bots to participate in email discussions and respond to the victim. Furthermore, AI prototypes are already capable of self-identifying an employee's role within a business based on their LinkedIn profiles or email signatures. In this way, high-profile targets may be quickly recognized [11].

4.7 Conclusion

Both physical and virtual communications helped mitigate the harmful impacts of the COVID-19 pandemic on several sectors, including health, education, transportation, and industry. The epidemic would have had a bigger impact if these communication technologies hadn't been available, and many industries would have come to a standstill. Because these industries increasingly rely on telecommunication networks to support their distant operations, existing networks have been subjected to unprecedented traffic demands. Therefore, understanding the role of AI in the era of 5G and 6G communication is key.

References

1 Venkateswaran, C., Manickam, R., Saravanan, V., and Vennila, T. (2021). A study on artificial intelligence with machine learning and deep learning techniques. In: *Data Analytics and Artificial Intelligence*, 32–37. REST Publisher.

2 Chen, Z. and Liu, B. (2018). Lifelong machine learning, second edition. *Synthesis Lectures on Artificial Intelligence and Machine Learning* 12 (3): 1–207. https://doi.org/10.2200/s00832ed1v01y201802aim037.

3 Agidew, A. (2020). Artificial intelligence and advances. *Advances in Machine Learning and Artificial Intelligence* 1 (1): 5. https://doi.org/10.33140/amlai.01.01.03.

4 Reier Forradellas, R. and Garay Gallastegui, L. (2021). Digital transformation and artificial intelligence applied to business: legal regulations, economic impact, and perspective. *Laws* 10 (3): 70. https://doi.org/10.3390/laws10030070.

5 Dadhich, A. (2018). A critical view of laws and regulations of artificial intelligence in India and China. *Kathmandu School of Law Review* 1–10. https://doi.org/10.46985/jms.v6i2.202.

6 Li, W., Su, Z., Li, R. et al. (2020). Blockchain-based data security for artificial intelligence applications in 6G networks. *IEEE Network* 34 (6): 31–37. https://doi.org/10.1109/mnet.021.1900629.

7 Zhang, S. and Zhu, D. (2020). Towards artificial intelligence-enabled 6G: state of the art, challenges, and opportunities. *Computer Networks* 183: 107556. https://doi.org/10.1016/j.comnet.2020.107556.

8 Raj, V. and Ancy, C.A. (2021). Understanding the future communication: 5G to 6G. *International Research Journal on Advanced Science Hub* 3 (Special Issue 6S): 17–23. https://doi.org/10.47392/irjash.2021.159.

9 Basnet, M. and Ali, M. (2021). Exploring cybersecurity issues in 5G enabled electric vehicle charging stations with deep learning. *IET Generation, Transmission & Distribution* 15 (24): 3435–3449. https://doi.org/10.1049/gtd2.12275.

10 Petrov, I. and Janevski, T. (2020). 5G mobile technologies and early 6G viewpoints. *European Journal of Engineering Research and Science* 5 (10): 1240–1246. https://doi.org/10.24018/ejers.2020.5.10.2169.

11 Ramachandran, K. (2019). Cybersecurity issues in the AI world. *Deloitte* (Sept. 11): https://www2.deloitte.com/us/en/pages/technology-media-and-telecommunications/articles/ai-and-cybersecurity-concerns.html.

5

6G Wireless Communication Systems: Emerging Technologies, Architectures, Challenges, and Opportunities

5.1 Introduction

The use of 5G in mobile devices worldwide is growing, and almost everyone can use it. But just as its possibilities are revealed, so too are limitations. Because of this, the idea that we can go further has arisen, leading us to conceptualize and pursue 6G. Emerging 6G technologies would help us further our horizons, as we have never seen before. Artificial intelligence (AI) and the Internet of Everything (IoE) will be deeply affected by this development. Considering how 5G has created smart homes, smart cars, and everything in between, 6G could further create much more. New architectures have been designed for 6G, and there are new challenges and opportunities within this technology. Research on 6G has only been recent, and there is a world of opportunities we could create as the years go by.

Cognitive service architecture is a new architecture designed specifically for the core network, meaning it can potentially enhance the core network to give the highest quality of service (QoS) and different scenarios. Cognitive service architecture would greatly enhance the performance of 6G. There will be frequencies up to terahertz, unlimited bandwidth, and microsecond latency. This new architecture would support all the needs and expectations that 5G was not able to satisfy.

The 6G network should eventually replace the 5G network in the coming years. While 6G makes sense as a successor to 5G, it could never be referred to as such. If it were not for anything like 5G Enhanced or 5G Advanced, we might just say "we're connected" instead of all the numbers and names. Every decade or so, a new mobile network standard becomes the center of attention. This indicates that 6G networks could start rolling out by about 2030, or at the very least when many service providers will be conducting 6G tests, and phone manufacturers will be teasing 6G-capable devices. On a 6G network, anything you do now that requires a network connection would be considerably better. Every advance that 5G delivers will be mirrored in a 6G network as an even superior, advanced version. Statistics demonstrate the importance of strengthening communication networks. We're on our way to a society dominated by fully autonomous remote monitoring systems.

Autonomous systems have gained traction in various fields – notably industry, agriculture, transportation, marine, and aerospace. To create a smart life and automated processes, tons of sensors are incorporated into cities, automobiles, residences, enterprises, nutrition, sports, and other settings. As a result, these applications will want a high-data-rate connection with a high level of reliability. 5G communication protocols have already been installed in several regions of the world and were expected to be completely operational globally by 2022.

5.2 Important Aspects of Sixth-Generation Communication Technology

The transition from analog to digital communication of 2G has paved the road for flexibility by means of reuse frequency options, multiple modulation schemes, adaptive equalization, and dynamic frequency allocation. The third-generation (3G) and fourth-generation (4G) cellular systems incorporated bundles of services such as voice, data, and video streaming. Access technologies such as code division multiplexing (CDMA) and orthogonal frequency division multiplexing (OFDM) have enhanced multiplexing methods, flexible and adaptive rate, and interference management via the utilization of frequency farming and multiple streaming [1].

From 5G to 6G: Technologies, Architecture, AI, and Security, First Edition. Abdulrahman Yarali.
© 2023 The Institute of Electrical and Electronics Engineers, Inc. Published 2023 by John Wiley & Sons, Inc.

5.2.1 A Much Higher Data Rate

Technology reports show that 5G is set to deliver 20 Gbps peak data rates and 100 Mbps user data rates. By contrast, the peak data rates for 6G are estimated to be approximately 1000 Gbps. Likewise, the user data rate experience is set to increase to approximately 1 Gbps. Hence, it is estimated that the spectral efficiency of the 6G network will be twice more than that of 6G. This advanced spectral improvement is set to change the face of multimedia services improvement both instantly and simultaneously. Therefore, the 6G network revamp is characterized by a much higher data rate.

5.2.2 A Much Lower Latency

The 5G infrastructure is set to reduce the existing latency by approximately one millisecond. The role of this ultra-low latency is to enhance the performance of various real-time applications. On the other hand, the 6G communication technology is projected to reduce the latency by less than 0.1 ms. Because of the huge decline in latency, real-time applications will realize improved functionality and performance. Besides, lower latency is known to improve emergency response calls, industrial automation, and remote surgeries. In retrospect, 6G networks are projected to make networks 100 times better than 5G networks.

5.2.3 Network Reliability and Accuracy

Without a doubt, the 6G network infrastructure will be more reliable than the 5G network. For instance, 5G is built to support the highest speeds in mobile devices by approximately 500 km/h, whereas 6G is set to serve over 10 devices per square kilometer. This means that many devices are able to interact with each other in real-time effectively. Furthermore, 6G is set to optimize M2M interaction by improving network reliability.

5.2.4 Energy Efficiency

6G is all about promoting real-time interactions seamlessly without any technical interruptions or disruptions. It offers end-users a great experience that is characterized by high-end services and powerful batteries for devices. It means that one of the goals of 6G is to improve the user experience while promoting environmental sustainability because its design uses less energy. This increases its optimization and energy efficiency, making it superior to 5G networks.

5.2.5 Focus on Machines as Primary Users

Both 5G and 6G aim at improving M2M communication, but they will do so at different levels and capacities. For instance, 5G is more focused on mobile and human experience, whereas 6G focuses on machine communication as its primary use. 6G network functionality is set to boost M2M interaction across different devices and machines such as robots, drones, home appliances, and smart sensors. Besides, these wireless networks enable the effective transition into next-generation devices that use technologies such as augmented reality (AR) glasses and holograms.

5.2.6 AI Wireless Communication Tools

Technologists believe that 6G will transform wireless communications, especially those using artificial intelligence. Unlike the 5G networks, the designs for 6G networks aim to leverage the most practical applications of artificial intelligence. Artificial intelligence is a technology that is beneficial in reducing infrastructural costs and improving operational efficiency. The real-time data collected in the connected devices and various communication devices are set to improve services.

5.2.7 Personalized Network Experience

OpenRAN characterizes 5G as an evolving technology, whereas 6G is set to utilize OpenRAN as a mature technology. According to technology engineers, an AI-driven RAN allows personalized network experiences for users dependent on real-time data from various sources.

Based on these characteristics of 6G networks, it is evident that reliable data connectivity is vital to a ubiquitous digital world. The development of networks from 4G to 6G over the years shows the vitality of communication and the roles technologies play in the social and physical world. In addition, technology research predicts that the current deployment of the 5G network will be unable to fulfill the current and future technological demands. Thus, telecommunication engineers are working toward developing and rolling out 6G networks because they are superior in their architecture, scalability, and reliability. Thus, understanding the evolution of 5G to 6G by assessing its technologies, architecture, and uses is critical for utilizing various communication paradigms [2].

Mobile technology generations are often designed to support an end to end-users and various network operators. Figure 5.1 shows how network generations have evolved over time due to physical and human demands. The current technological demands are more data-centric, dependent, and automated. Thus, it requires constant improvement from 4G and 5G technologies to future networks such as 6G. Autonomous systems characterize the world as many industries embrace sensors to operate their activities. For instance, machines embedded with many sensors characterize the world's roads, airspace, and waters and systems operated using artificial intelligence.

Even though the talks surrounding 6G are on the rise, credit should be given to its predecessor, 5G networks, because they have paved the way for further innovation. For instance, there are improvements in frequency bands, latency, spectrum usage, and management brought by the need to develop 6G networks.

Thus, it is rational to state that the development and rolling out of 6G networks are inevitable because they can optimize artificial intelligence and the internet of things systems. The hype to have an intelligent digital world puts pressure on engineers to focus on 6G network architectures. Table 5.1 shows the parameters and features of the latest wireless generations.

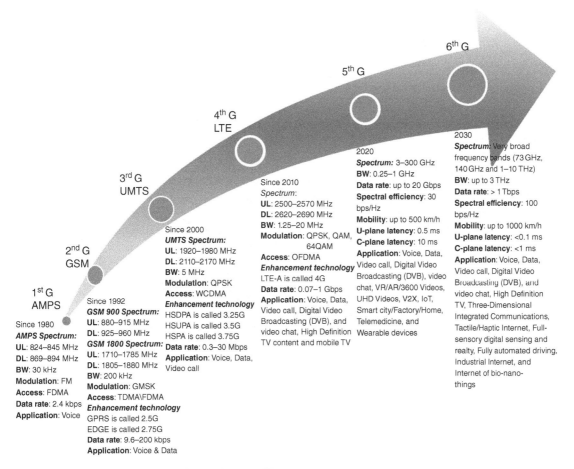

Figure 5.1 Network evolution graph over the years [3].

Table 5.1 5G to 6G parameters and features [4].

Issue	4G	5G	6G
Per device peak data rate	1 Gbps	10 Gbps	1 Tbps
End-to-end (E2E) latency	100 ms	10 ms	1 ms
Maximum spectral efficiency	15 bps/Hz	30 bps/Hz	100 bps/Hz
Mobility support	Up to 350 km/h	Up to 500 km/h	Up to 1000 km/h
Satellite integration	No	No	Fully
AI	No	Partial	Fully
Autonomous vehicle	No	Partial	Fully
XR	No	Partial	Fully
Haptic communication	No	Partial	Fully
THz communication	No	Very limited	Widely
Service level	Video	VR, AR	Tactile
Architecture	MIMO	Massive MIMO	Intelligent surface
Maximum frequency	6 GHz	90 GHz	10 THz

5.3 Enabling Technologies Behind the Drive for 6G

Originally, 6G networks were primarily built on the 5G framework, acquiring the gains acquired in 5G, given the past progress of mobile networks. Various technological (Figure 5.2) advances will be introduced, while others from 5G will be upgraded to 6G. As a result, a variety of technologies would power the 6G infrastructure.

5.3.1 Artificial Intelligence

The basic property of 6G autonomous networks is intelligence. As a result, AI is the most important and recently launched technology for 6G communications networks. AI was not used to develop 4G network technologies. Partially or very restricted AI would be supported by the next 5G communications networks. Nevertheless, AI will significantly assist 6G in automation [6]. The entire capability of radio transmissions will be realized with AI-powered 6G, allowing the transition from cognitive to smart radio. Increasingly smart systems for real-time connectivity in 6G are being created thanks to advances in machine learning (ML). The use of artificial intelligence in communication will make the transmission of real-time data easier and more efficient.

AI can determine how a difficult target job is accomplished using a variety of statistics. AI boosts productivity and minimizes the time it takes for communication stages to be processed.

AI would be used to quickly complete time-consuming processes like transfer and connectivity selection. M2M, machine-to-human, and human-to-machine (H2M) communications will all benefit from AI. It also encourages BCI interaction. Nanostructures, smart structures, intuitive networks, telecommunication technologies, adaptive cognitive radio, conscious wireless systems, and deep learning will all be used to enable ML communication systems. Integrating photonics and AI will accelerate the advancement of AI within 6G, allowing for the development of digital logic cognitive radio networks. AI-based encoder-decoders, deep neural networks for channel state prediction, and automatic modulation categorization are used in the physical layer. Deep learning-based allocation of resources and smart traffic modeling have been intensively investigated to meet the 6G needs for the data link layer and the transport layer [6].

The combination of big data analytics to discover the optimum route to convey data to end-users by giving data modeling would drastically minimize latency. Table 5.2 shows 6G emerging technologies for different services.

Enabling Technology	Potential	Challenges	Use cases
New Spectrum			
Terahertz	High bandwidth, small antenna size, focused beams	Circuit design, high propagation loss	Pervasive connectivity, industry 4.0, holographic telepresence
VLC	Low-cost hardware, low interference, unlicensed spectrum	Limited coverage, need for RF uplink	Pervasive connectivity, eHealth
Novel PHY techniques			
Full duplex	Continuous TX/RX and relaying	Management of interference, scheduling	Pervasive connectivity, industry 4.0
Out-of-band channel estimation	Flexible multi-spectrum communications	Need for reliable frequency mapping	Pervasive connectivity, holographic telepresence
Sensing and localization	Novel services and context-based control	Efficient multiplexing of communication and localization	eHealth, unmanned mobility, industry 4.0
Innovative Network Architectures			
Multi-connectivity and cell-less architecture	Seamless mobility and integration of different kinds of links	Scheduling, need for new network design	Pervasive connectivity, unmanned mobility, holographic telepresence, eHealth
3D network architecture	Ubiquitous 3D coverage, seamless service	Modeling, topology optimization and energy efficiency	Pervasive connectivity, eHealth, unmanned mobility
Disaggregation and virtualization	Lower costs for operators for massively dense deployments	High performance for PHY and MAC processing	Pervasive connectivity, holographic telepresence, industry 4.0, unmanned mobility
Advanced access-backhaul integration	Flexible deployment options, outdoor-to-indoor relaying	Scalability, scheduling, and interference	Pervasive connectivity, eHealth
Energy-harvesting and low-power operations	Energy-efficient network operations, resiliency	Need to integrate energy source characteristics in protocols	Pervasive connectivity, eHealth
Intelligence in the network			
Learning for value of information assessment	Intelligent and autonomous selection of the information to transmit	Complexity, unsupervised learning	Pervasive connectivity, eHealth, holographic telepresence, industry 4.0, unmanned mobility
Knowledge sharing	Speed up learning in new scenarios	Need to design novel sharing mechanisms	Pervasive connectivity, unmanned mobility
User-centric network architecture	Distributed intelligence to the endpoints of the network	Real-time and energy-efficient processing	Pervasive connectivity, eHealth industry 4.0
Not considered in 5G		With new features/capabilities in 6G	

Figure 5.2 Comparison of 6G enabling technologies and relevant use cases [5].

Table 5.2 Characterization of emerging technologies under different 6G services [4].

Technology	uMUB	uHSLLC	mMTC	uHDD
Artificial intelligence	√	√	√	√
Terahertz communications	√	√		
OWC	√	√	√	√
FSO fronthaul/backhaul	√	√		
Massive MIMO		√	√	√
Blockchain		√		
3D networking	√	√		√
Quantum communications		√	√	
Unmanned aerial vehicles	√	√	√	√
Cell-free communications	√	√	√	√
Integration of wireless information and energy transfer	√			√
Integration of sensing and communication	√			√
Integration of access-backhaul networks	√	√		
Dynamic network slicing	√	√		
Holographic beamforming		√		
Big Data analytics		√	√	√
Backscatter communication			√	
Intelligent reflective surface		√	√	√
Proactive caching		√	√	√
Mobile edge computing	√			

5.3.2 Terahertz Communications

The bandwidth can be increased to improve spectral efficiency. Widening bandwidths and employing advanced enormous multiple-input, multiple-output (MIMO) technology are two ways to do this.

5G uses mmWave frequencies to boost data throughput and open up new possibilities. 6G, on the other hand, wants to stretch the frequency band's bounds to THz to fulfill an even larger demand. The RF band is nearly full, but it is no longer able to ensure the high demands of 6G. For 6G communication, the THz band will be critical. The THz band is envisioned as the next step forward in high-data-rate transmission. Perspective 6th generation (6G) wireless networks will be made up of a slew of small radio cells coupled with high-speed communications cables. Wireless transmission at the Terahertz band is a particularly appealing and adaptable approach in this regard. The THz spectrum has been the electromagnetic family's real contender. Not only are they less well-known than radio or microwaves, but there are fewer technologies capable of generating and modulating these waves.

The terahertz realm is a technological dilemma, apart from submillimeter instruments in astronomy, and it is often alluded to as the "terahertz gap." Terahertz technologies are less well established and hence more expensive. However, there are some compelling reasons why THz is a better option. The advantages of THz wireless include aiming stability, responsiveness to atmospheric turbulence, and eye protection. They enhance each other's performance in inclement weather: In the rain or fog, THz performs better [3]. Tables 5.3 and 5.4 show different mid and high bands for wireless communications in 6G.

5.3.3 Optical Wireless Technology

OWC technologies are foreseen for 6G communications for any potential device-to-access connections; such networks also enable routing network access. Since 4G communication networks, OWC technologies have already been employed. However, to fulfill the needs of 6G communication networks, it is anticipated that they will have to be employed more

Table 5.3 mmWave, THz, and optical sub-bands for possible wireless communications in 6G [7].

						Optical
	mmWave part-1		30–275 GHz	10–1.1 mm	mmWave	
THz	mmWave part-2		275–300 GHz	1.1–1 mm		
	Far IR part-1		0.3–3 THz	1–0.1 mm	Infrared	
	Far IR part-2		3–20 THz	0.1–0.015 mm		
	Thermal IR	Long-wavelength IR	20–37.5 THz	0.015–0.008 mm		
		Mid-wavelength IR	37–100 THz	0.008–0.003 mm		
	Short-wavelength IR		100–214.3 THz	3 000 000–1400 nm		
	Near IR		214.3–394.7 THz	1400–760 nm		
	Red		394.7–491.8 THz	760–610 nm	Visible light	
	Orange		491.8–507.6 THz	610–591 nm		
	Yellow		507.6–526.3 THz	591–570 nm		
	Green		526.3–600 THz	570–500 nm		
	Blue		600–666.7 THz	500–450 nm		
	Violet		666.7–833.3 THz	450–360 nm		
	UVA		750–952.4 THz	400–315 nm	Ultraviolet	
	UVB		952.4–1071 THz	315–280 nm		
	UVC		1.071–3 PHz	280–100 nm		
	NUV		0.750–1 PHz	400–300 nm		
	Middle UV		1–1.5 PHz	300–200 nm		
	Far UV		1.5–2.459 PHz	200–122 nm		
	Hydrogen Lyman-alpha		2.459–2.479 PHz	122–121 nm		
	Extreme UV		2.479–30 PHz	121–10 nm		
	Vacuum UV		1.5–30 PHz	200–10 nm		

Table 5.4 THz channel features and impacts on 6G communication systems [8].

Parameter	Dependence on frequency	Impact on 6G THz systems	THz vs. Microwave and FSO
Spreading loss	Quadratic increase with decreasing area and constant gains; quadratic decrease with constant area and frequency-dependent gains	Distance limitation	Higher than microwave, lower than FSO
Atmospheric loss	Frequency-dependent path loss peaks appear	Frequency-dependent spectral windows with varying bandwidth	No clear effect at microwave frequencies, oxygen molecules at millimeter wave, water and oxygen molecules at THz, water, and carbon dioxide molecules at FSO
Diffuse scattering and specular reflection	Diffuse scattering increases with frequency; specular reflection loss is frequency-dependent	Limited multi-path and high sparsity	Stronger than microwave, weaker than FSO
Diffraction, shadowing and LoS probability	Negligible diffraction. Shadowing and penetration losses increase with frequency. Frequency-independent LoS probability	Limited multi-path, high, sparsity and dense spatial reuse	Stronger than microwave, weaker than FSO
Weather influences	Frequency-dependent airborne particulates scattering	Potential constraint in THz outdoor communications with heavy rain attenuation	Stronger than microwave, weaker than FSO
Scintillation effects	Increase with frequency	Constraint in THz space communications	No clear effects at microwave; THz less susceptible than FSO

broadly. Light fidelity, visible light communication (VLC), FSO communication optical, and camera communication based on the optical band are examples of OWC innovations that are well adapted. Pieces of research are focused on improving the performance of these technologies and solving their limitations. Wireless optical communications should deliver increased data speeds, low latencies, and secure connections. LiDAR is a revolutionary technology for very-high-resolution 3D mapping in 6G networking, as it is likewise based on the optical band. OWC is convinced that mMTC, uHSLLC, uMUB, and HDD technologies will be better supported in 6G communication networks. Advancements drive the OWC in 6G in light-emitting-diode (LED) technology and multiplexing methods. In 2026, both microLED technology and spatial multiplexing techniques are predicted to be robust and cost-effective.

5.4 Extreme Performance Technologies in 6G Connectivity

Substantial bandwidth and extremely directional antennas will be available to 6G handheld devices via the millimeter-wave (mmWave) and terahertz (THz) frequency bands, enabling software innovations and uninterrupted connectivity. The use of incredibly high computing power processors at both the infrastructure and smart objects will substantially reduce latency in 6G [9]. The upcoming platform's cell devices will be intelligent enough to sense the environment and take cautious and preventative steps. To enable the key characteristics of 6G, a combination of cutting-edge technologies should be used [10].

5.4.1 Quantum Communication and Quantum ML

Quantum technology makes ultra-accurate clocks, diagnostic imaging, and neural networks possible by utilizing quantum mechanics' features, such as the interaction of atoms, molecules, or even particles and electrons. The maximum potential, however, has yet to be realized. A quantum interconnection that connects quantum computers, models, and sensors to distribute resources and information around the world securely. The first service to use this infrastructure will be quantum key distribution. It offers an inherently secure random key to both the server and client of an encrypted message, preventing an adversary from eavesdropping or controlling the system. It will protect crucial sensitive communications from future intelligent computers breaking codes. This should offer a safe exchange of information, digital certificates, digital signatures, and time synchronization, among other things [11]. This infrastructure would benefit the economy and society while also ensuring important government data from the ground, sea, and orbit. QC and QML may handle complicated challenges like a multi-object exhaustive study in ad hoc routing protocols and Cloud IoT by delivering rapid and best-prop selection to information in the communication network backbone.

5.4.2 Blockchain

Blockchain is revolutionizing some of the world's most important businesses, including finance, operations management, accounting, and transnational remittances. The blockchain concept is bringing up new ways to do business. Blockchain establishes trust, openness, confidentiality, and autonomy among all participants in the network. The most crucial characteristic for successful enterprises in the telecommunication industry is innovation in a competitive environment with lower costs. The telecommunications business can profit from blockchain in several ways.

5.4.2.1 Internal Network Operations

Ethereum, the second generation of blockchain technologies, has transformed the automation system in various applications with smart contracts. When a specific event occurs, smart contracts allow mathematical equations to execute automatically. Because of this, blockchain has a lot of interest in the telecommunications business, where it may be used to automate things like billing, systems integration, and roaming. Blockchain can prohibit malicious traffic from entering the network system, saving a lot of bandwidth and expenses and, as a result, increasing the carriers' revenue. The telecommunications industry can save time by minimizing the time-consuming post-billing audit procedure by using smart contracts for bill validation and clearance. The telecommunications industry can automate auditing and accounting operations using this method [12].

5.4.2.2 Ecosystem for Productive Collaboration

The goal of next-generation wireless systems is to offer a wide range of new digital services. For the complicated transactions required for these services, blockchain is an appealing application. By efficiently exploiting user information, blockchain can potentially be employed in the advertising sector. Vast M2M interactions will emerge as a result of this. Telecommunications companies may lead the way in this sector by implementing blockchain technology and ushering in the next wave of digital operations. The desire for a huge connection in 6G has prompted system resource management difficulties such as power distribution, spatial multiplexing, and the deployment of computation power. Using smart contracts to manage the connection between operators and users, blockchain can bring solutions to the 6G network in several fields. Consequently, blockchain can help with spectrum power and risk management issues that aren't covered by licenses. In addition to flawless environmental protection and surveillance, blockchain can be employed in e-health and decrease cybercrime rates.

5.4.2.3 Tactile Internet

With the advancement of internet technology, data and social networking has become possible on mobile devices. The evolution of the Internet of Things (IoT), in which interconnection between smart devices is facilitated, is the next phase. Tactile internet is the next stage on the internet of networks, which incorporated real-time M2M and H2M communication by adding a new element of haptic sensations and tactile to the mix. The term *tactile internet* refers to touch transmission over large distances [10].

5.4.2.4 Spectrum Sharing (FDSS) and Free Duplexing

Previous wireless generations utilized fixed duplexing (TDD/FDD) or flexible duplexing (FDD) in the instance of 1G, 2G, 3G, and 4G, and flexible duplexing in the context of 5G. With the advancement of duplexing technology, 6G is expected to adopt a full free duplex, where all users are permitted to use all resources simultaneously. As a result, as AI and blockchain are projected to be major technologies in 6G, strong spectrum monitoring and spectrum management procedures will be established for the 6G rollout. AI-assisted 6G systems can dynamically control network resources. As a result, in 6G, open spectrum sharing will become a possibility.

Even though 6G is only a couple of years away, few service providers are currently considering it. However, as we pinpoint where 5G misses, 6G exploration is projected to ramp up. It won't be long until the forces that begin debating how to go about it, as 6G will improve on the unavoidable flaws and shortcomings of 5G. 6G network will mirror every advance that 5G delivers and, even better, enhanced version. As we've experienced with 3G, 4G, and 5G in the past, as a channel's capacity grows, so do its applications. This will have a huge impact because new goods and services will fully leverage 6G's throughput and other advanced capabilities.

5.5 6G Communications Using Intelligent Platforms

Society is becoming increasingly digitalized and will continue to do so into the future. The world is moving toward the deployment of 5G wireless networks. Wireless connectivity is now ubiquitous to connect people, and things allowing machines to communicate without the need for human intervention. 5G New Radio (NR) network should be able to handle high connection density and high user plane latency. 3GPP Release 15 defines the basic framework for NR, which supports ultra-low latency transmission through a very short mini-slot. The next release of 5G will feature enhanced URLLC (ultra-low latency connectivity), mMTC (massive Machine Type Communications), and NOMA capabilities, and NB-IoT and LTE-M will be integrated into 5G NR for MTC services. LPWANs such as SigFox, LoRA, and Ingenu have been developed for MTC applications. The 5G network can be used to provide mMTC and URLLC services, though the data rates are rather limited and the supported use cases are rather limited. To ensure robustness, connectivity solutions for MTC scenarios may require multiple radio access technologies. Although 6G research is still in the exploration phase, many publications have already been published on what 6G will be.

MTC is expected to be a key driver in the increasing trend toward making everything that moves autonomous and intelligent. 6G is enabling super smart cities with 30 billion connected devices by 2030. Industry 5.0 will involve more interactivity and novel human–machine interactions than Industry 4.0 and will be supported by MTC. MTC will play a fundamental role in facilitating the user experience of augmented/virtual/mixed reality, and in enabling novel and efficient man–machine interfaces. MTC will need to be energy-efficient, low-latency, and error-free to enable zero-energy, zero-touch systems. In 6G, data marketplaces will link data suppliers and customers and will add new KPIs like age of information, privacy, and localization accuracy. Industry 4.0 will require factories to be flexible, agile, and adaptable, supporting both automation and human-machine interface. Swarms of autonomous vehicles will be commonplace in 2030, requiring robust connectivity solutions to perform in such an environment. Wearable devices will be seamlessly integrated into our clothing in the 6G era. At the edge of the IoT, machines are low-power sensors that can keep going forever.

Wireless energy transfer and backscatter technology could make this possible. Human senses can interact with machines by using ultra-broadband and ultra-responsive network connectivity. Decentralized ledger technologies (DLT) will be used to transfer data and micropayments between IoT devices. As 6G becomes more prevalent, 5G will require more stringent requirements. In 6G, the ultimate goal is zero-energy MTDs, achieved through a combination of efficient hardware design and energy harvesting techniques. New KPIs are becoming increasingly relevant for 6G and relate to the Age of Information (AoI), interoperability, dependability, positioning, sustainability, and E2E EE. Seamless integration of heterogeneous network access technologies will help 6G technologies to provide reliable service for emerging applications. The EE vision for 6G will be wider and more stringent than the EE vision for 5G. In 5G, URLLC (critical) and mMTC (large/dense MTC) were established for small payloads, low-data rates, and sporadic traffic patterns.

In the coming decade, 6G will have to be scalable and multidimensional. To successfully meet 6G visions, novel approaches are needed to design enabling technology components. The MTC network architecture is meant to harmonize connectivity across the network and maximize performance. The MTC network architecture needs to accommodate a number of key trends affecting the MTC network architecture that including the evolution from cell-based networks to cell-free networks, the appearance of local micro-operators deploying new private or public networks, and the increasing number of traffic patterns and service classes.

The network architecture of the future should allow for the orchestration of applications across multiple heterogeneous networks and the use of technology-agnostic interfaces. The selection of the RAT(s) and the configuration parameters of the RAT(s) should be handled using machine-learning methods. Machine clients can change their traffic patterns in different places, and this postulates the need for centralized or distributed optimization mechanisms. Introduce manageable control and emergency traffic channels within the MTC architecture and ensure backward compatibility of new generations and releases within one generation. Networks require a new air interface to power zero-energy devices. The development of the underlying optimization mechanisms and technical solutions for the post-5G MTC network requires the development of common interfaces and procedures. Ultra-low-power (ULP) transceivers are essential for mMTC to operate in a sustainable way. Significant reductions in energy consumption can be achieved by considering the energy consumption of the MTD as a whole. The power consumption of the various building blocks of the transceiver can be reduced by better power schemes of the various building blocks.

5.5.1 Integrated Intelligence

Unlike previous generations of wireless communication technologies, 6G will be revolutionary, moving wireless technology from "devices connected" to "connected competence." Any step of communication, along with radio resource planning, will be automated with AI. AI's widespread implementation will usher in a new era of communication networks [11]. For ultra-dense complicated network scenarios of 6G, a complete AI system is required compared to 5G, empowering intelligent electronic devices to obtain and undertake the resource distribution operation [13].

5.5.2 Satellite-Based Integrated Network

To enable ubiquitous connection, satellite communication is required. It is almost unrestricted by geographical conditions. It may offer a seamless worldwide coverage of various geographic places such as land, sea, air, and sky to serve the user's seamless computing. To reach the goal of 6G, it is expected to link terrestrial and satellite technologies to deliver always-on broadband worldwide mobile connectivity. 6G will require integrating terrestrial, satellite, and aerial networks into a unified wireless network [13].

Each new generation of technology has its advantages, alongside its challenges that must be overcome before the technology can be fully realized. With technologies up to 4G focusing mostly on capacity and coverage, 5G wishes to push those boundaries, increasing connectivity and capacity while simultaneously further reducing power consumption as well as latency. 5G offers many advantages, but its challenges and restrictions are incredibly stringent and require existing infrastructure to be pushed to, or beyond, its limits.

4G functions on a 2D plane, where everything is flat and easy to work with. 5G and by extension 6G wish to function on a 3D plane, covering much more than the previous 4G-LTE generation was capable of covering. For this to become a reality, new frameworks and technologies will have to be created and put into place, starting with this so-called 3D architecture.

The 2D platform can be explained very simply as the networks that already exist on the ground here on Earth. This is everything from wireless access points in cities, cell and radio towers, and so on that provide internet and mobile coverage to the 2D plane of people on the ground. To introduce a 3D plane for internet connectivity, several technologies will need to be implemented to achieve the goal of lower power consumption, low latency, and high speed. These technologies include UAVs, LEO/GEO satellites, and more to bridge the gap between land, air, and space.

This new architecture is theorized to utilize a technology known as network slicing that is already being implemented for 5G networks. This technology works by splitting up existing networks and allowing for more customizability and use outside of just providing connectivity to those on this network. However, this slicing will be used not just across terrestrial nodes but over space and flying nodes as well.

Network systems have always had a standardization problem. For example, while most of the world utilizes their 4G networks and are standardized in that sense, the United States uses LTE technology instead, which is different in several ways from 4G and would potentially cause interoperability issues if the two were connected. For this new generation of technology to come to fruition, and its advantages be afforded to all, standardization across the board will have to occur. Without standardization, 6G and all tech connected to it will fail.

For the change from 5G to 6G to succeed, the 3D hierarchy and architecture must be implemented from the beginning. Enhancing existing systems and bolstering their capabilities while simultaneously creating new platforms of technology will be essential. The first obstacle that must be overcome is the ever-changing environment of connected devices. Stationary infrastructure will be incredibly inefficient at achieving the goal of 6G, so UAVs and satellites that can move to change their service area will be necessary. Of course, this has its challenges, too, such as attaining sufficient energy and overcoming interference from the atmosphere and other sources.

These UAVs will also be used to bring cloud services into the flying layer, as well as the space layer through the use of LEO/GEO satellites. By providing a UAV or satellite with sufficient computing power, field research and on-demand computations of datasets can be made easy by sending data to these UAVs, allowing built-in cloud capabilities to do the necessary processing and then retrieve the results, allowing for mobile research bases and other mobile data-collecting ventures to be much easier and more fruitful. This also brings along with it the idea of AI being distributed over this same sort of network. Having access to a pervasive and on-demand AI service could be beneficial for similar use cases.

The biggest obstacle alongside technological limits as well as funding for such a venture to create this new 3D network hierarchy will be interference. Satellites and other low-orbiting signal-producing apparatus have to contend with interference on a constant basis already, and with a more pervasive web of such devices being the future of 6G, that interference will only increase. To combat this, it is proposed to exploit the classification system of interference, as well as other technologies to detect interference before it happens to change transmission power at precise times to avoid interference such as this.

The internet of things, artificial intelligence, generations of various speeds, and satellites all began in their own little corner of the technology spectrum. Each advancing over time but when they started working together, coexisting through intelligent connected environments they became something greater than themselves. These combinations have revolutionized the technology industry and additional industries where technology is used – which is basically everywhere, by the way. Compacting these types of advancements into small satellites alone is leading to the creation of artificial constellations all managed and operated through the technologies through everything listed above. Satellites are relied on across the world for nations to operate successfully, generating information, regulating trade, and looking toward the stars' advanced satellites could potentially be the pinnacle of technology altogether.

5.5.3 Wireless Information and Energy Transfer Are Seamlessly Integrated

6G wireless networks will also send power to recharge battery devices such as smartphones and sensors. WIET (wireless information and energy transfer) will be incorporated as a result. Universal super 3D networking: Drones and very low Earth orbit satellites will access the network and key network functions, making super-3D compatibility in 6G ubiquitous.

5.6 Artificial Intelligence and a Data-Driven Approach to Networks

With 6G expected to bring humanity further toward autonomous applications, the question of protecting personal information (PI) comes into play. How is the data used, and who uses it? The challenge is quantifying what identified data means to develop proper measures. It becomes a tradeoff between managing privacy and building the required trust. The more trust that is given the more risks of privacy leaks grow. This means 6G will require new trust models to balance between maintaining customers' trust and privacy.

PI and personally identifiable information (PII) are used interchangeably in legislation frameworks. The date of birth is considered PI but not PII as it can't identify a single person. The question that needs to be answered is, "How many features must be linked before PI becomes PII for an individual known to be in a dataset?" With the definition of an individual's identity being very broad, it in principle covers any information that relates to an identifiable individual during their lifetime or decades after their death. As the more PI data is linked, the larger the personal information factor (PIF) becomes. Data above a specific threshold means that sufficient PI exists to identify an individual. Since there is no way to unambiguously determine when linked when the dataset crosses the threshold, it has become a major problem for many different technologies like AI and IoT. This has significant consequences for all smart services as a whole. This is still an issue unaddressed in 5G, and it is impossible to be ignored for the future of 6G

As 5G evolves, the reliance on AI smart applications increases. Due to the number of challenges for the future of wireless applications is the implementation of distributed ledger technology (DLT). DLT technologies such as blockchain provided security and privacy features such as immutability and anonymity. Privacy protection that uses differential privacy seems promising when addressing challenges that could arise in the future 6G wireless applications as it operates by perturbing the actual data using artificial designs.

Protection of individual privacy has been a challenge for standardization since nations have different perspectives in their area. International committees such as the joint technical committee 1 (JTC 1) are working on different privacy frameworks, but much more needs to be done. Due to the interconnectedness of the global markets, "Privacy by Design" will most likely prevail in newer products and communication techniques.

A future that sees us utilizing digital avatars will also have health and safety concerns. The future digital and physical worlds will become entangled with malicious cyberattacks, possibly leading to loss of property and life.

With 6G in 2030, AI has been proposed to become fully integrated with the communications backend, handling many time-consuming tasks like network selection, handoff or handover, and resource allocation. This will allow the network to dynamically adapt to changing usage conditions while tailoring an experience for each individual user. The improved specs

Table 5.5 Artificial intelligence data rates and latency features [13].

Peak data rates	100 Gbps to 1 Tbps
Extreme ultra-low latency	0.1 ms
Density	100 devices per m^3
Extreme ultra reliability	Max. 1 out of million outage
Extreme coverage	Sea, sky, space

and capabilities of 6G are also set out to change the industry in our world. The higher data rates and user densities possible prove promising for all sorts of technologies like extended reality (a new facet of augmented reality), production automation through robotics and other autonomous systems, wireless brain integrated computer interfaces, and high-performance real-time healthcare for anyone in the country. The new workings of 6G will also promote wireless power transfer. Think wireless charging, but without a charging pad. Now, your phone could very well charge in your pocket as you walk down the sidewalk.

Artificial intelligence technology is growing from simple tools used by average scientists to as far as to use within the professional development community for higher intelligence use. These various organizations can use AI technology to fill the current gap in the data science area, which could be a big game-changer for data science! This also always gives end-consumers the ability to take their business and personal data with them wherever they go. This, of course, is a big deal when it comes to making end-consumers happy while retaining their privacy with their online data.

When it comes to data processing with the help of 5G and 6G, AI, and IoT devices, modern technology and its data often demand that data management and processing capabilities have the data possessors brought to the data itself rather than sending the collected data off for processing elsewhere. This may also include the data processing and distributed data stores all to be included within the data management process to guarantee support is offered where the received critical data is being stored. 5G network technology will be revolutionary to end consumers, businesses, as well as data processing centers around the globe. Table 5.5 shows some features of AI.

5.6.1 Zero-Touch Network

Trust extends across all protocol layers, from IP to apps and data. In order to answer issues like: can this host interact with a remote party without being attacked or hacked, a trust system in network communication is needed. To sense, comprehend, and program the world, 6G will create a wide digital/physical world boundary. As a result, a breach of information security can put people's physical safety at risk and cause property loss, in addition to causing data loss, control loss, and financial loss. International wars could deploy foreign cyberwar warriors to wreak such damage in a country that traditional warfare is no longer necessary to persuade the victim to accept the attacker's demands. You can use trust networking to give remote access to specialized 6G networks or any other important infrastructure in local or national packet data networks (PDNs). The PDN service could be made available to customers. Last but not least, trust networking can be utilized across all platforms. Examples of use cases range from a simple, single administration handling the entire trusted network and all of its devices to cases where devices are owned by independent parties, and different layers, segments, or subsystems of the network are owned and managed by numerous different stakeholders with potential conflicts of interest. A mobile or IP network's adoption of trust networking calls for alterations to the existing network. Those changes should be able to be implemented one by one in a network. If device and network changes are required, they can really only be made by a single administrator. When a server can certify client software and vice versa, it's a good example of a single admin solution.

Networks are becoming increasingly complex as new technologies such as 5G, smart devices, and Cloud finds legitimacy. Humans are finding it difficult to keep up with increasing complexity and maintain networks running at the level required by evolving services. We have already been inundated with data, and it is only the beginning. Running a 5G network, incorporating data points from the IoT, while facing the challenge of mission-critical use situations would be only conceivable whenever AI, robotics, and Big Data are used to drive telecom networks' "Big Data operations" [14].

We are making progress toward our dream of "zero-touch networks" thanks to data-driven processes. The zero-touch network is an application network service that uses automation and AI to supplement the human experience in its operations. To realize the idea of zero-touch connectivity, systems and capabilities must operate together. Because a zero-touch network is built on intelligence, it necessitates the implementation of network elements and operational services that support AI technology. The system's products and services must be created from the ground up with intelligence in mind. Automation is a critical component of the overall strategy. Intelligent agents are software that can work alone and make intelligent decisions in a complex environment [10].

5.6.2 AI by Design

According to the technology strategy, AI will be used across the service or product portfolio in all components of the system architecture. The goal is to enable humans and automated systems to evolve an engineered and responsive system into a lifelong learning network that better meets the needs of our customers. AI can be used to solve specific challenges for telecommunication companies, creating value where everything matters most. It is not a broad-sense tool designed for every conceivable use case [15]. AI is being developed to address today and tomorrow's mobile communications challenges. Increasing the return on installed base without the need for site visits or the addition of new equipment. Providing consumers with always-on, high-performance networks. Providing mission-critical corporate infrastructure for national services. Adhering to elevated customer expectations and providing an impeccable QoS journey for enterprises. Acknowledge the importance of the Cloud by improving agility, lowering costs, and laying the groundwork for additional technologies. Managing growing demands while preserving operational expenses down [10]. Creating flexible infrastructures based on real usage and guaranteeing that it is kept to a minimum.

5.6.3 Technological Fundamentals for Zero-Touch Systems

6G components in communication systems need to embrace artificial intelligence from the onset to capitalize on the power that machine learning can infuse into technological advancements. An automation and strategic planning solution includes automated foresight operations, neurological risk mitigation, and data analytics for confidence, conscience networks, and innovative caseload positioning. The AI-based switchover, 5G-aware congestion management, evolutionary computation offloading at launch, uplink-traffic stimulated movement, transport network control, and amplified MIMO sleep are all features of cellular networks [16]. Deep learning paging on the cloud core is a premium feature. Automation is a critical component of the workaround. An intelligent agent is an application that can operate autonomously and make smart decisions in a dynamic environment.

Forward-thinking telecom companies and virtual network operators need an asynchronous network and commercial enterprise platform to allow them to switch to a new business reality characterized by 5G, the IoT, network virtualization functions, and application networks. Traditional ML systems offer numerous advantages to telecom service providers, especially in maintaining effective levels of quality. Notwithstanding, in addition to leaving a large network presence, the huge data transfer associated with traditional ML models can be problematic in the context of data protection [16]. Federated learning (FL) addresses these issues by bringing "each encoding to the data" rather than transmitting "the data to the encoding."

The idea of "intellectual purpose management," which uses innovative AI, is growing in popularity. Service providers and their end users will clarify "what" actions individuals need from the system. The framework will then determine "how" to achieve this query by employing closed-loop digitization and as little user intercession as feasible. We are investing heavily in robotic systems and enterprises to facilitate these product features in the value propositions.

5.7 Sensing for 6G

5.7.1 A Bandwidth as Well as Carrier Frequency Rise

The 6G would continue the trend started by 5G. NR networks continue to push for higher frequency ranges, bigger bandwidths, and larger antenna arrays. As a result, sensing solutions with the precise spectrum, harmonic and directional resolutions, and localization to cm-level precision will be possible. New materials, device kinds, and other innovations are also on the horizon [17].

Network operators will be able to meld and manipulate the electromagnetic field via customizable surfaces. Simultaneously, ML and AI will become more prevalent to take advantage of the extraordinary availability of data and computational resources to tackle the most difficult and complex problems. High carrier frequency RF-based sensing will enable more accurate approaches to measuring the surroundings, and detecting and recognizing things. It provides a larger spectral range chance to monitor and track new types of objectives and parameters that aren't detectable in conventional methods of radio spectrum currently in use [17].

Small cell deployments will begin to dominate as cellular systems evolve to mmWave bands in 5G and potentially sub-THz bands in 6G. Wide bandwidth systems deployed in cellular network architectures provide a practically unseen potential to use the data connection for sensing. 6G systems will indeed be highly smart wireless platforms that will enable extremely accurate localization and leading provider services in addition to widespread communication. They will be the driving force behind this popular revolt by introducing a completely new set of features and service capabilities. Localization and sensing would also intermingle with interactions. It is constantly able to share readily accessible resources in time, rate, and space [18].

We can identify four emerging technological enablers for 6G communication networks and localization and sensor technology:

1) The use of innovative wireless communications frequencies
2) Incorporation of an intelligent substrate surface
3) Smart beam-space handling, AI
4) Deep learning techniques

The transition to THz frequencies has several significant advantages in terms of localization and sensing. For starters, stimuli at these frequencies cannot permeate objects, resulting in a much more direct relationship between both the transmit power and the propagation environment. Furthermore, greater absolute data rates are obtainable at high frequency, resulting in far more correctable multi-path in the latency domain with more highly reflective components.

Finally, shorter wavelengths presuppose smaller antennas, allowing discrete components to be stocked with dozens or hundreds of transmitters, which is advantageous during angle estimation. This is useful across both passive and active sensing. To obtain the desired output, chip technology that supports cost advantages must be available. Furthermore, suitable channel designs that effectively classify the proliferation of 6G waves out over firmware and the air will be needed to facilitate innovative alternatives and methodologies.

5.7.2 Chip Technologies of the Future

There is now a concise need for further technological development to support the frequency mentioned above bands cost-effectively. One critical aspect is incorporating the necessary technology [19]. Presently, radio systems designed to operate throughout the numerous different 100 GHz normally entail antennas and signal processing facilities, which are outrageously massive to embed into conventional sensor nodes. While the front-end and wireless connection chips have taken time to develop, the processing method of preference is now sufficient to allow effective propagation, down-conversion, and further signal handling, even at huge frequency bands.

5.7.3 Models of Consistent Channels

Models of electromagnetic wave propagation are essential for the proper design, function, and efficiency of cellular networks. Indeed, when designing new heuristics or deploying new architectural design, modeling available radio complexities and process controls becomes critical for analyzing the overall effectiveness of mobile telecommunications frameworks. Even though it is difficult to build sonar, sensing, ranging, or direction approximation algorithms without audio suppositions on how microwaves perpetuate in the proximity of the sensor system, propagation models are equivalently vital to ensure reliable and efficient aligning. Frameworks are also effective in predicting and benchmarking the correctness of multiple sensing and localization techniques. If measured by standardized radio signals, greater systems are far more vulnerable to atmospheric conditions and do not continue to spread properly via the materials. These characteristics are also important for sensing.

Although we reach higher wavelengths and spectroscopic assessment, we should best understand and template the permeability of electromagnetic fields via various molecules and the reflection properties of various substances. Identifying the perfect band for the right frequency to maximize any use of prospective future applications is a major challenge. As we move closer to the data-rich 6G era, expert systems will become highly significant. A wide-ranging field of study establishes artificial intelligence and agents capable of achieving rational aims and objectives based on reasonable and predictive reasoning, prepping, and effective strategic thinking, possibly in dynamic situations. AI systems are typically built on machine learning, which further offers data-driven collaborative learning approaches frameworks that go beyond formal presuppositions [20].

5.7.4 X-Haul and Transport Network for 6G

Networks have difficulty providing extremely high data speeds (tens of gigabits per second) with very low latency (a few milliseconds). These demanding criteria occur when telecom operators' average income per user is declining, emphasizing the need for alternatives that can lower their operational costs to a competitive threshold. Shaul, dubbed "the option open transport method for future 6G networks," seeks to combine fronthaul and backhaul infrastructures, as well as many of their fixed and wireless technologies, into a single packet-based transportation system under SDN- and NFV-based (network functions virtualization) common control [21].

Over ubiquitous 3D coverage areas, 6G mobile technology is expected to enable data throughput of 100 Gbps or greater, as well as ultra-low latency. The transport network, which primarily links the cell sites to the network core, has not kept up with the cell sites. Consequently, it has been identified as a potential barrier to the implementation of high-performance and cost-effective networks. As a result, to meet the new and varied requirements of 6G and even beyond mobile networks, which are being driven by the rollout of large numbers of mobile cell sites, extremely diversified architectures, and complex services and applications [21].

The vehicular infrastructure has been built as an effective instrument to link human transportation worldwide for many years to come. However, as the number of vehicles increases, the vehicular network becomes more homogenous, vibrant, and massive. This will make it much more difficult to address the rigorous standards of the next-generation (6G) network, including ultra-low latency, higher performance, heavy stability, and enormous interconnection.

Lately, ML has evolved into a powerful artificial intelligence methodology for increasing the efficiency and adaptability of either machine or wireless technology. Inevitably, incorporating ML into spectrum sharing and network is a contentious issue that is being extensively researched through industry and academia, clearing the way for the new intelligent action throughout 6G sensor communications.

Every new generation of communication technology introduces new and intriguing features. The 5G communication technology, which will be formally implemented around the world in 2020, has many unique characteristics. However, in 2030, 5G will not fully serve the expanding demand for wireless communication. As a result, 6G must be implemented. 6G research is still in its early stages. With both the readiness and impending commercial exploitation of 5G network technologies, an abrupt change in cellular data consumption, as well as the variable requires of vertical markets such as innovation, e-commerce, public transit, and health, can sometimes be assisted as part of a much larger global digitalization revolt. In the meantime, the installed capacity foundation of the IoT network devices, such as computers and phones, is expected to reach 75.44 billion through 2025, representing a fivefold spike in a couple of years. As wireless networks reach their physical size restrictions, this transition will pose a massive challenge to future wireless communications. Concerns are also growing about the considerable demand for energy to maintain such systems [21].

As a result, it is critical to creating new innovative platforms to improve this growing number of devices in smart, energy-efficient forms. To recognize this, modern world 5G and beyond (6G) mobile systems will be a game-changer for socioeconomic development.

5.8 Applications

The characterization of 5G revolves around latency, throughput, reliability, energy, and costs. Various configurations of 5G networks address mobile broadband and latency issues. However, the configurations of 6G aim toward meeting stringent network demands in a holistic sense. Hence, their architectures are quite different from the 5G networks because they chase increased reliability and efficiency and reduce latency. Thus, 6G offers particular uses based on the network's characteristics. Figure 5.3 shows some future use cases for 6G.

➢ Augmented reality (AR) and virtual reality (VR)
   Since the inception of 4G networks, streaming and multimedia services have increased in the spectrum and needs because of the ever-developing data-hungry applications. Consequently, 5G developed new systems such as mmWaves to sustain the higher capacities. However, the "rooms for improvement" for the 5G network is presented when applications

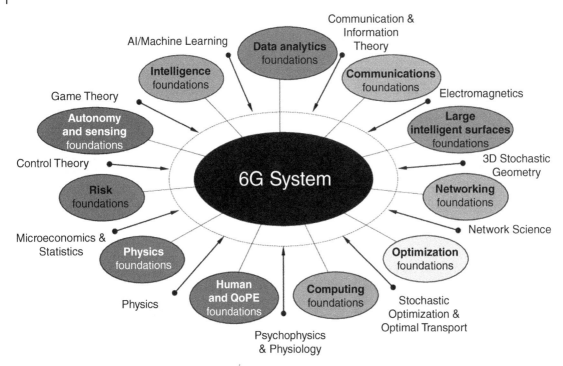

Figure 5.3 The use case for 6G technology [22].

requiring multi-Gbps are developed. Hence, 5G essentially paves the way for the early adoption of augmented reality and virtual reality. Full adoption of these two technologies will overpower the 5G spectrum because they require a capacity of approximately 1 Tbps. This capacity can only be achieved by developing 6G networks because 5G has a limit of 20 Gbps. Moreover, the latency requirements for 5G cannot effectively compress augmented reality and virtual reality.

➢ eHealth

The healthcare sector will greatly benefit from 6G because of telehealth and telemedicine technologies. These technologies are critical in a fast-evolving world because doctors can dispense healthcare services such as surgery and pharmacy remotely to their patients. Such eHealth services demand QoS. The QoS requirements includes stable and continuous connectivity, mobility support, reliability, and ultra-low latency. Thus, developing 6G networks that can achieve such requirements enhances the full actualization of eHealth services. This is because the increased spectrum and the combined intelligence of 6G networks will ensure spectral efficiency.

➢ Pervasive connectivity

The degree of growth of mobile technology is very high, meaning that mobile traffic is on the rise. Research estimates that approximately 125 billion devices will be connected by 2030. Based on such high estimates, it is probable that 6G network transfigurations will support such extremes. For instance, 6G will connect all smart and personal devices. Consequently, this makes them require high-energy frequencies to enhance scalability and better coverage.

➢ Industry 4.0 and Robotics

Technology specialists project that 6G is likely to materialize the 4.0 industry revolution initiated by 5G transfigurations. Industry 4.0 is characterized by physical (cyber) systems and the services of the internet of things. 6G will be cost-effective for industries because their architectures are built to increase the cyber computational space. Furthermore, 6G materializes reliable isochronous communication, which is critical in addressing upcoming disruptive technologies.

➢ Unmanned mobility

Connecting autonomous vehicles and machines will be one of 6G's major focuses. This means that data rates will increase, thereby demanding high levels of reliability and low latency. For instance, it is estimated that reliability should be about 99.99999%, whereas latency rates should be below 1 ms. Besides, safety is also a key factor when dealing with unmanned mobility because of the increasing data rates. Drones currently have practical uses in the military and science, but their technology is being explored for use in several industries. These drones are often unmanned, and they are controlled and connect various systems of the world.

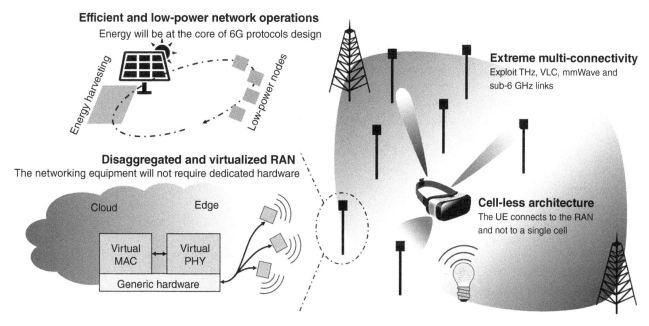

Figure 5.4 Architectural innovations of 6G networks [5].

5.9 Innovative 6G Network Architectures

6G will be characterized in many ways as substantively advanced as compared to 5G in prior versions of wireless evolution including: Terahertz-level data, extreme focus on short-range communication, AI is inherent within the network, extremely high radio density, very high network heterogeneity (extreme HetNets), and transformed radio topology (mesh, hops, and peer-to-peer). These characteristics will be manifest in the technologies, capabilities, and ultimately the applications, services, and solutions that distinguish 6G from all generations that preceded it. Figure 5.4 shows the architectural innovations of 6G networks.

The following paradigms are likely to be introduced and deployed in 6G architectures:

- Tight inclusion and integration of various frequencies and communication technologies and cell-less architectures
- 3D network architecture
- Disaggregation and virtualization of network equipment
- Advanced access-backhaul integration
- Energy-harvesting approaches

5.10 Conclusion

Disruptive communication technologies are coming. As suggested in this chapter, innovation and technology are inevitable. Thus, moving from 5G to 6G ensures that extremely higher frequencies are achieved, such as the mmWave band to support the higher spectrum technologies. It means that 6G networks will be able to work effectively to limit or prevent disruptive technologies.

New architectural paradigms are necessary for the heterogeneity of future networks, making 6G an important element in telecommunications. 6G would push us into using terahertz. This means we could potentially get a band of 275 GHz to 10 THz, which drastically compares to the 5G band of 5.15–5.85 GHz. 6G is also expected to have unlimited bandwidth, which would mean that there would be constant full speed. 6G would have microsecond latency, one terabit per second at a latency of around 100 μs – about 50 times faster than 5G is. The architecture created specifically for 6G would allow all of this to be possible and all the promises of 5G would be met with 6G. The architecture of 6G would allow AI, specifically ML, to grow and mature because of the absolute power that 6G would have. Integration of intelligence in network systems, AI, and IoT is critical in advanced communication systems.

References

1 Yazar, A., Tusha, S.D., and Arslan, H. (2020). 6G vision: an ultra-flexible perspective. *ITU Journal on Future and Evolving Technologies* 1 (1): 121–140. https://doi.org/10.52953/IKVY9186.

2 Tariq, F., Khandaker, M.R.A., Wong, K.-K. et al. (2020). A speculative study on 6G. *IEEE Wireless Communications* 27 (4): 118–125. https://ieeexplore.ieee.org/document/9170653.

3 Alsharif, M.H., Kelechi, A.H., Albreem, M.A. et al. (2020). Sixth generation (6G) wireless networks: vision, research activities, challenges and potential solutions. *Symmetry* 12: 676. https://doi.org/10.3390/sym12040676. http://www.mdpi.com/journal/symmetry.

4 Chowdhury, M.Z., Shahjalal, M., Ahmed, S., and Jang, Y.M. (2020). 6G wireless communication systems: applications, requirements, technologies, challenges, and research directions. *IEEE Open Journal of the Communications Society* 1: 957–975. https://doi.org/10.1109/OJCOMS.2020.3010270.

5 Giordani, M., Polese, M., Mezzavilla, M. et al. (2019). Towards 6G networks: Use cases and technologies. https://gmoein.github.io/files/Towards%206G.pdf.

6 Zong, B., Fan, C., Wang, X. et al. (2019). 6G technologies: key drivers, core requirements, system architectures, and enabling technologies. *IEEE Vehicular Technology Magazine* 14 (3): 18–27.

7 Chowdhury, M.Z., Hossan, M.T., Islam, A., and Min Jang, Y. (2018). A comparative survey of optical wireless technologies: architectures and applications. *IEEE Access* 6: 9819–10220.

8 Chong Han, Yongzhi Wu, Zhi Chen, and Xudong Wang (2019). Terahertz communications (TeraCom): challenges and impact on 6G wireless systems. https://arxiv.org/pdf/1912.06040.pdf.

9 Hoschek, M. (2021). Quantum security and 6G critical infrastructure. *Serbian Journal of Engineering Management* 6 (1): 1–8.

10 Gritsenko, V., Babak, O., and Surovtsev, I. (2021). Peculiarities of interconnection 5G, 6G networks with big data, internet of things and artificial intelligence. *Kibernetika I Vyčislitel'naâ Tehnika* 2 (204): 5–20. https://doi.org/10.15407/kvt204.02.005.

11 Chen, S., Sun, S., and Kang, S. (2020). System integration of terrestrial mobile communication and satellite communication—the trends, challenges, and key technologies in B5G and 6G. *China Communications* 17 (12): 156–171. https://doi.org/10.23919/jcc.2020.12.011.

12 Ilchenko, M., Uryvsky, L., and Globa, L. (2021). *Advances in Information and Communication Technology and Systems*. Cham: Springer.

13 Zhu, J., Zhao, M., Zhang, S., and Zhou, W. (2020). Exploring the road to 6G: ABC—a foundation for intelligent mobile networks. *China Communications* 17 (6): 51–67.

14 Hoole, P. (2021). *Smart Antennas and Electromagnetic Signal Processing in Advanced Wireless Technology*. Aalborg: River Publishers.

15 Sergiou, C., Lestas, M., Antoniou, P. et al. (2020). Complex systems: a communication networks perspective towards 6G. *IEEE Access* 8: 89007–89030. https://doi.org/10.1109/access.2020.2993527.

16 Slalmi, A., Chaibi, H., Chehri, A. et al. (2020). Toward 6G: Understanding network requirements and key performance indicators. *Transactions on Emerging Telecommunications Technologies* 32 ((3).

17 Wild, T., Braun, V., and Viswanathan, H. (2021). Joint design of communication and sensing for beyond 5G and 6G systems. *IEEE Access* 9.

18 Swaroop, A.S., Gori Mohamed, J., and Niaz Ahamed, V.M. Full duplex radio-wave transmission for 6G internet (6G connectivity). (2017). *International Journal of Recent Trends in Engineering and Research*, 3(7), 201–208. https://dl.icdst.org/pdfs/files3/fad5fc0743e036119efbea51fd8c5b4c.pdf.

19 Shafin, R., Liu, L., Chandrasekhar, V. et al. (2020). Artificial intelligence-enabled cellular networks: a critical path to beyond-5G and 6G. *IEEE Wireless Communications* 27 (2): 212–217. https://doi.org/10.1109/mwc.001.1900323.

20 Mathew, A. (2021). Artificial intelligence and cognitive computing for 6G communications & networks. *International Journal of Computer Science and Mobile Computing* 10 (3): 26–31. https://doi.org/10.47760/ijcsmc.2021.v10i03.003.

21 Yang, P., Xiao, Y., Xiao, M., and Li, S. (2019). 6G wireless communications: vision and potential techniques. *IEEE Network* 33 (4).

22 Saad, W., Bennis, M., and Mingzhe Chen (2020). A vision of 6G wireless systems: applications, trends, technologies, and open research problems, *IEEE Network*, 34, 3, 134–142, https://dl.icdst.org/pdfs/files3/fad5fc0743e036119efbea51fd8c5b4c.pdf.

6

6G: Architecture, Applications, and Challenges

6.1 Introduction

5G service is available in North America and East Asia today, and new launches are still being released in Europe. By 2022, we should see the final launches of 5G in those areas. Interestingly enough, China actually accounts for over half of the world's 5G connection. In 2021, nearly a thousand new 5G devices were announced, including phones, modules, modems, tablets, and many other devices. That fact makes one anxious to see the future of devices, as these will not be fully functional or released to the public until 2025. The 5G we use today is still not at its full capacity of balancing broadband operators with growth into new markets – it will still be a few years until we reach its full potential.

As noted in Chapter 1, 5G technology has gone through two releases, Rel-15 and Rel-16. The 5G technology is assumed to result in a high data rate that aims to enhance mobile broadband, support the ultra-reliable, and possess low latency services and massive numbers of connections. The essential functions of the Rel-15 are initial access (including beam management), channel structure, and multi-antennas, which partially fill the requirement of performance of International Mobile Telecommunication (IMT-2020).

In order to support the performance requirements of IMT-2020, several new technologies and scenarios such as non-orthogonal multiple access, ultrareliable, and low latency communication, vehicle to X communication, unlicensed band operations, integrated access, backhaul, terminal power-saving, and positioning have been introduced, which can also help in widening the use of 5G networks. Unlike 1G, 2G, 3G, as well as 4G, the 5G communication applications possess diverse abilities, including three main features: high Gbps speed of the mobile broadband, a million connections of massive machine type communication, and a 99.99% of the microsecond delay of ultra-reliable low latency transmission to meet society's demands in the coming decade.

The 5G network is expected to have a massive impact on the life of humans, the global economy, and culture. The 5G communication network was expected to be on its evolutionary path after 2020 for the maximum optimization of the features and expand deployment scenario as the nonterrestrial network (NTN), unmanned aerial vehicle, and was expected to increase the operating band up to ~114 GHz. This can help to enhance the participation of the vertical industries and emerging enterprises. Figure 6.1 shows that the fulfillment of 5G requirements implies a dramatic change in the design of cellular architecture and wireless technologies [1].

In this era of distributed intelligence, technology and connectivity are swiftly shifting from 5G to 6G networks. In some countries, the 5G rollout has not commenced, yet the world is already talking about 6G networks. The number of 6G networks initiatives is increasing globally, ultimately providing interesting prospects for the future. Various public and private sectors are currently focusing on research and innovation (R&I) to get ready for the massive potential demands of 6G networks. Research suggests that the commercialization of 6G will begin in 2030 [2]. Since 5G is still in the rollout phase, a lot is still to be achieved and realized. This extends architectural and functionality pressure on 6G. Figure 6.2 shows the envisioned timeline for 6G.

Most of the current 5G availability covers the enhancement of mobile networks. The future of 5G is not only improvement in this area but also expansion into new areas. Much of the new releases of 5G will focus on research and improvements that prepare for the coming of 6G networks in the future. Industrial IoT is one of the focuses of the next releases of 5G wireless. Improving reliability is key. We will also see many new 5G private networks with new security improvements and connection methods. Also included in the future releases of 5G is a focus on critical IoT systems directed to time-sensitive communication, sidelink enhancement, multimedia broadcast services, and NTNs such as those used by satellites.

From 5G to 6G: Technologies, Architecture, AI, and Security, First Edition. Abdulrahman Yarali.
© 2023 The Institute of Electrical and Electronics Engineers, Inc. Published 2023 by John Wiley & Sons, Inc.

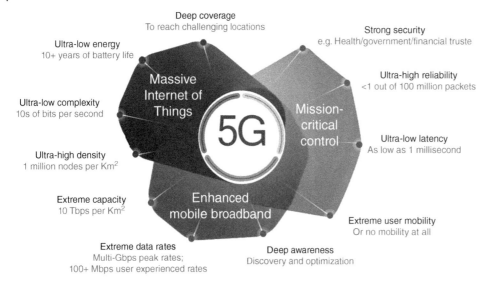

Figure 6.1 5G requirements imply a dramatic change in numerous directions [1].

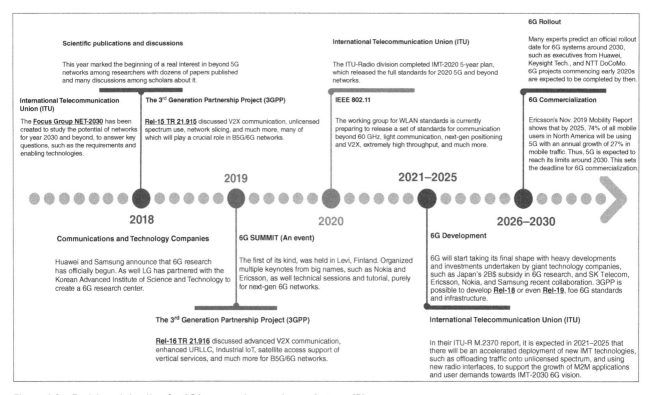

Figure 6.2 Envisioned timeline for 6G by researchers and manufacturer [3].

Future releases will also see an expansion of the technologies that can actually take advantage of 5G. 5G will continue to evolve, but eventually we will start to see the transition to 6G system. While 6G is still close to 10 years away, the research into 6G continues to grow and could speed up the time for the move to 6G. Some of the more advanced scenarios expected to be covered by 6G networks include ubiquitous services, cyberphysical systems, manufacturing and transportation services, and tactile and haptic communications. Figure 6.3 shows performance indicators from 5G to 6G evolution.

Furthermore, it is important to note that the latency constraints currently imposed on 5G networks demand quick solutions. The increasing developing technologies ranging from extended reality (XR), telesurgery, and autonomous driving to

Figure 6.3 6G and 5G comparison in multiple specifications [3].

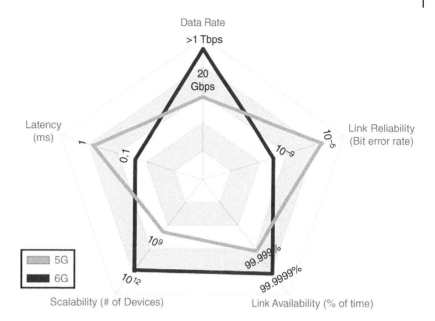

telepresence require very high data rates. Hence, maintaining efficiency might be a problem because of the increasing use of smart devices and the general establishment of the Internet of Things (IoT) [4].

If all these factors are put together, 5G networks become oversaturated. As a result, technology and communication companies such as Vodafone and AT&T are working tirelessly to develop 5G improvements. Consequently, in the fast-growing world of the internet, discussions about 6G are right behind [5]. Conclusively, the main goals of 6G networks include the satisfaction of massive data rates, latency, and reliability across multiple and various smart devices. Therefore, in this chapter, we intend to explore 6G networks from various angles such as architecture, polarization and virtualization, security and privacy, green networking, applications, intelligence, core network, D2D, mobile internet, and visions.

Figure 6.4 shows the individual efforts by different countries toward the materialization of 6G networks. For instance, in Europe, 2020, research and innovation programs focused on three areas of 6G development: Hexa-X, RISE-6G, and NEW-6G. According to the European Commission (EC), R&I of 6G requires about €900 million to cover the full 5G deployment [2]. Meanwhile, Australia, Finland, China, South Korea, Japan, and the USA are following suit by allocating huge budgets toward 6G development.

6.2 6G Network Architecture Vision

The gradual but steady evolution of wireless networks from the first generation to the fifth generation is consistent when it comes to data rate, latency, services, capacity, and coverage. The massive increase in mobile data traffic has revolutionized the new face of the internet and multimedia technology. The future is to become a data-driven society in which things and people will become universally connected in just a matter of seconds or even milliseconds. Many of the aspects of the 6G wireless network communication are still under scientific exploration. Mature technological study and development of the physical layer are still needed to make them practical and feasible in engineering.

The 6G network will support an average of over 1000 wireless nodes per person in the coming 10 years and is assumed to provide a terabit rate per second. One goal of 6G wireless communication network is also to provide instant holographic connectivity – anytime from anywhere – and create a quick link between people and things

Figure 6.5a,b show the popularity and increase in the development of smart devices have led to the emergence of dynamic applications such as real-time video streaming, telemedicine, and online gaming. The figures show the constant digital data consumption increase.

Regarding the 6G 2030 vision, Soldani said, "All intelligence will be connected following a defense in-depth-strategy-augmented by zero trust model through digital twinning, using beyond 5G (B5G)/6G wireless, and machine reasoning will meet machine

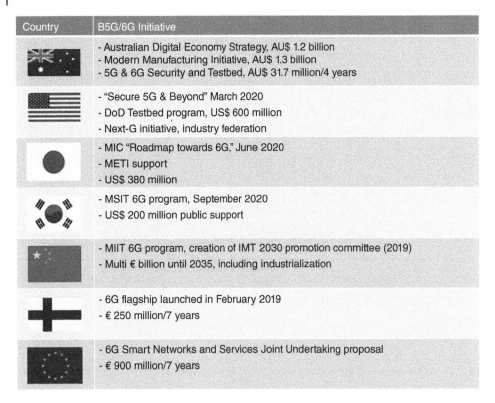

Country	B5G/6G Initiative
	- Australian Digital Economy Strategy, AU$ 1.2 billion - Modern Manufacturing Initiative, AU$ 1.3 billion - 5G & 6G Security and Testbed, AU$ 31.7 million/4 years
	- "Secure 5G & Beyond" March 2020 - DoD Testbed program, US$ 600 million - Next-G initiative, industry federation
	- MIC "Roadmap towards 6G," June 2020 - METI support - US$ 380 million
	- MSIT 6G program, September 2020 - US$ 200 million public support
	- MIIT 6G program, creation of IMT 2030 promotion committee (2019) - Multi € billion until 2035, including industrialization
	- 6G flagship launched in February 2019 - € 250 million/7 years
	- 6G Smart Networks and Services Joint Undertaking proposal - € 900 million/7 years

Figure 6.4 Global examples of Beyond 5G (B5G) and 6G initiatives [2].

learning (ML) at the edge" [7]. Hence, Soldani believes that the underlying necessities and challenges of future networking motivate this vision. These necessities and challenges include:

- The flaws in machine learning (ML), such as pattern recognition algorithm.
- The need for digital twinning. This refers to the virtual representation via a real-time process.
- The shift to connected intelligence demands further security based on the zero trust model.
- The need for deeper edge intelligence.

6.2.1 6G Use Cases, Requirements, and Metrics

The current generation of technology, 5G, is an important stepping stone in the progression and evolution of technology. It boasts upward of 25 Mbps data rates, supports VPNs, and more. It is the testbed for the viability of a large, pervasive mobile network, but does not come without its own problems. It is not cheap to implement, being one of the more expensive ways to build out a mobile network. Its security is poor but getting better, and is still being actively researched for its viability.

6G, on the other hand, may be the new big milestone in human civilization and technology as a whole. Boasting a whopping 11 Gbps data rate, as well as being cheap to implement, manufacture, and use, 6G could literally change the world. Communications from land to space would become faster and easier to maintain. Energy could be transmitted wirelessly, which would allow for better use of energy sources outside of the Earth. Security for homes and businesses could be improved several times over with new, lightning-fast response times and real-time monitoring of resources. 6G could be the biggest leap in technology yet. It, however, has its own problems. With a network as large as a 6G-enabled one would need to be, privacy and data ownership would need to be a big talking point before implementation. Security around these new links would need to be improved several times before it would become viable. Thorough planning and research will have to be done before any of these things can be implemented, but I believe that this technology could be the turning point for the technological world.

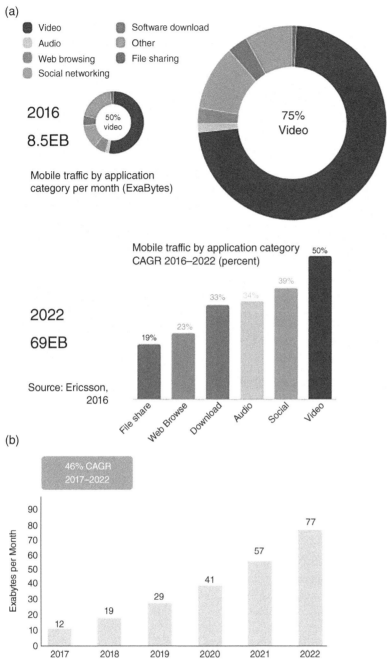

Figure 6.5 (a) Rising demand for data volume and the diversity of requirements [6]. (b) Rising demand for data volume and the diversity of requirements [1].

6.2.2 What 5G Is Currently Covering

About 130 operators have received spectrum to launch 5G services. We expect to see significant growth in 5G connections across the Asia Pacific and Europe. China will dominate 5G connections by 2025. According to Global Mobile Supply Association (GSA) [8], the number of announced 5G devices rose by 1.3% between January and February 2023 to reach a total of 1840 devices which is an increase of 53.7% in the number of commercial 5G devices since the end of February 2022 [8]. including smartphones, fixed wireless access customer premises equipment (CPE) devices, hotspots, notebooks, and tablets. The third-generation partnership (3GPP) 5G to 6G high-level roadmap is depicted in Figure 6.6.

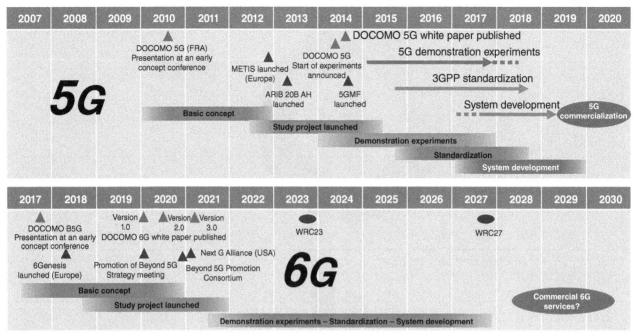

ARIB 20B AH: Association of Radio Industries and Businesses 2020 and Beyond Ad Hoc
5GMF: The Fifth Generation Mobile Communication Promotion Forum
FRA: Future Radio Access
METIS: Mobile and wireless communications Enablers for the Twenty-twenty Information Society
WRC: World Radiocommunication Conference

Figure 6.6 5G development history and 6G schedule [9].

The sixteenth and seventeenth releases of Release 16 will focus on the definition of new use cases, study items, and work items toward 6G. Release 16 of the LTE Standard offers a new radio in the 5 and 6 GHz frequency bands, and a sidelink for direct communication between devices. Release 16 adds support for IoT, WWC, FRCCS Phase 2, IAB, device power savings, mobility enhancements, and enhanced massive multi-input, multi-output (MIMO) with multiple transmission and reception points. Release 17 will include a larger ecosystem with Critical Internet of Things (CIoT) services. It will also include support for RAN slicing, network automation enhancements, device power saving, multiple universal subscriber identity modules (USIMs), and unmanned aircraft systems (UAS). GEO, MEO, LEO, and HAPS satellites are in the sky in relation to the Earth's rotation. They act as relay nodes and/or base stations and can be used for a variety of purposes. Broadband IoT, based on narrow band IoT and long-term evolution/machine-type communications (LTE/MTC), will contribute nearly 34% of IoT connections by 2025. The release of versions 16 and 17 will make it easier to deploy and operate 5G networks. Release 18 of 5G is expected to enable more use cases and enhance its capabilities. 6G is still under development and many use cases and related enabling technologies are still under discussion. 6G will likely provide a new air interface or simply enhance the existing 5G Advanced new radio. While the deployment of 6G systems still lies ahead, many ongoing programs and related investments provide a compelling prospect for the significant acceleration of 6G studies. 5G is enabling the development of applications including holoportation (real-time holographic communications), tactile/haptic communications, medical/health vertical, government/national security, imaging and sensing, public safety services, and transportation.

The commercialization of 5G is taking shape gradually in many countries as it attempts to improve the performance of the existing networks. The rise of smart devices is massively increasing, ultimately increasing the data rates [10]. Many users can attest to the satisfaction they receive from using smart devices. Furthermore, organizations and companies can boost their revenues because of the multiple business ideas that exist. There is increasing demand in use-cases because of the high-end networks of 5G and 6G networks. The increase in network capacity is the reason why companies are striving to upgrade networks' capacities. Currently, 5G networks are game-changers because they connect a massive number of devices and machines. For instance, according to Ericsson Mobility Report (2019) [11], the 5G New Radio (NR) can cover approximately 65% of the total population by 2025.

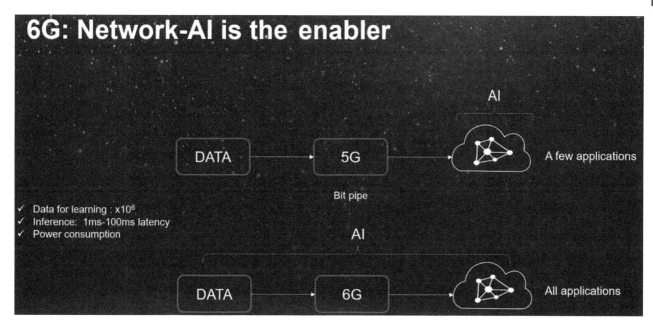

Figure 6.7 Artificial intelligence implementation in 5G and 6G communication networks [13].

Consequently, this generates up to about 45 percent of all mobile data worldwide. Hence, the 5G NR network is built to offer a flexible network that supports applications using various systems such as enhanced mobile broadband (eMBB), ultra-reliable-low latency communications (uRLLC), massive machine-type communications (mMTC), and enhanced vehicle to everything (eV2x). With this kind of trend, 5G will increase the densification, ultimately increasing the network capacity. Moreover, NR covering both uplink and downlink operations improves the performance of various applications, including industrial IoT, through the extension of bandwidth resources [12].

Artificial intelligence (AI) and Big Data have paved the way for developing applications like XR. Extended reality requires huge amounts of data for efficiency. For instance, 16K virtual reality needs a throughput of approximately 0.9 Gbps, so VR applications are likely to exhaust existing 5G wireless capacity. The seamless streaming and holographic display used by this application require extremely high data rates. Therefore, operators and wireless vendors predict that by around 2030, 5G will begin to reach its data limits. This increases the pressure to speed up the development and implementation of 6G networks. AI will be used to optimize networks from end to end (Figure 6.7), maximizing user experience as the overall goal. 6G could use AI to find the air interface and to write the underlying network software [13].

6.3 6th Generation Networks: A Step Beyond 5G

Even though 5G networks are still in the deployment phase, there are increasing discussions surrounding 6G networks. The increasing innovations of high-data-intensive applications such as brain-computer interaction (BCI), flying vehicles, e-healthcare, and extended reality shapes the need for 6G networks. A plethora of research argues that 6G networks will revolutionize the evolution of cellular networks from "connected things" to "connected intelligence" (A research program conducted by the University of Oulu focused on the adoption of the 5G ecosystem and innovation of 6G). The resulting "connected intelligence" will adapt and control key performance indicators (KPIs). Thus, 6G vision's summary can be primarily focused on achieving "ubiquitous wireless intelligence" [14] by possessing multiple and new aspects of network requirements such as services, and technologies, research problems, and operational improvement. Figure 6.8 shows a hierarchical approach for discussing these aspects of 6G' [3].

The term *ubiquitous* is used to define the omnipresent coverage and seamless accessibility to services. In retrospect, 6G aims to exploit the concept of AI, ML, and Big Data techniques, which support autonomous decision-making in smart devices and applications. For instance, while utilizing these applications, networks can modify themselves based on the experiences of the users. Moreover, ML and Big Data techniques help to support data usage in real-time, which is always a major boost in internet technology.

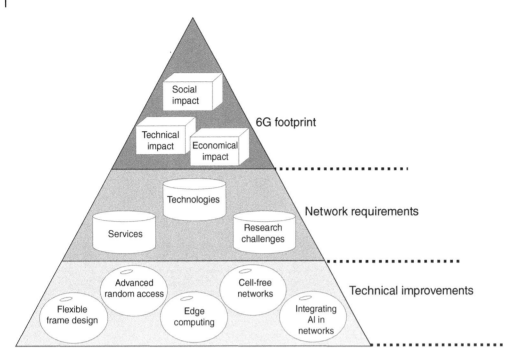

Figure 6.8 A hierarchical approach for the discussion of 6G's aspects [3].

It is paramount that these networks deliver huge data rates characterized by high reliability and extremely low latency to enhance simultaneous service delivery – both uplink and downlink. Operators stress significant rate-latency reliability because it paves the way for the exploitation of frequencies beyond the GHz range, ultimately improving the wireless network. According to network operators and affiliated companies, the emergence of 6G is set to improve data rates, user experience, latency, and three-dimensional coverage [15]. For instance, some of the network's approaches aim to include sensing, communication, positioning, navigation, and computing, among others. Furthermore, quantum computer systems are likely to enhance personal data security and privacy.

Looking at these discussed 6G applications and use cases, the networks must adopt new sets of requirements. Formalizing 6G is essential in the existing global trends. Consequently, researchers and academicians are working tirelessly to transform the face of wireless technology. For instance, some renowned institutions partaking in this course include the University of Oulu in Finland and the Terabit Bidirectional Multi-User Optical Wireless System (TOWS) for 6G LiFt by the UK government (a research group focused on increasing wireless capacity). Furthermore, worldwide telecommunication companies such as Samsung, AT&T, Ericsson, Nokia, and SK Telecom and communication experts have started to shift focus beyond 5G networks. This paradigm shift to 6G networks comprises key enablers such as large intelligent service (LIS), super-massive multi-input, multi-output (SM-MIMO), terahertz communications (THz), laser communication, visible-light communication (VLC), quantum computing, and orbital angular momentum multiplexing (OAM) [12].

6.3.1 6G and the Fundamental Features

5G has shown the need for better flexibility from its need and success in meeting diverse requirements and accommodating different technologies. To achieve flexibility in a network you need three things: awareness, availability of options, and adaptation/optimization. Despite the success of 5G in meeting flexibility needs, it's still falling short of truly optimal flexibility, and 6G will be even more difficult to accommodate due to new applications and requirements. To prepare for 6G there needs to be a better exploration of awareness regarding the various facets of communication environments by using artificial intelligence, expanding the technological options, and providing proper utilization of existing options. Most published literature on 6G doesn't touch on the flexibility perspective.

The major objective of most research studies for 6G is to identify future applications, services, and requirements. There are three main categories that these items fall into for 6G. eMBB, ultra-reliable and low latency communications (uRLLC), and massive machine-type learning (mMTC). There is also a list of potential applications for 6G like unmanned aerial vehicles (UAV), vehicle-to-everything (V2X), VR, AR, holographic conferencing, smart objects, healthcare, and more.

Flexible multi-band utilization is the first key enabler for 6G. With 5G utilizing the new millimeter wave spectrum, with all the benefits it provides, it is currently limited by IMT regulations. The THz band also plays into the flexibility of spectrums by offering more channels. To better summarize the rest of the section sources can be put in three categories: multi-band flexibility, information source flexibility, and spectrum coexistence flexibility.

Ultra-flexible physical (PHY) and medium access control (MAC) layers are possible for the multi-numerology systems pioneered by 5G. Multiple waveforms can be used in a single frame for the purposes of 6G, which can serve as radar sensing for multiple networks. The waveform domain NOMA techniques improve resource allocation possibilities for 6G. For the MAC layer, two issues require flexibility: waveform parameter assignment and numerology scheduling.

Ultra-flexible heterogeneous networks are the next key enablers. This feature covers flying access points that allow for flexible positioning and optimizable trajectories for different functions. Improvements to satellite networks are also covered. Network slicing and cell-free networks allow for user-centric systems with the flexibility of multiple networks under one umbrella. Blockchain technology and quantum communication technologies can also contribute to the future.

Integrated sensing and communications emphasize sensing or the observable data that can be used to improve communication systems. Technologies like spectrum sensing and awareness, location awareness, and context awareness fall under this category. Radio environment maps (REM) are used to obtain environmental information, but as technology progresses may eventually be able to gather a complete awareness to form a database. The drawback to such a complete awareness is the sheer volume of data generated as well as the need for immense processing power.

Intelligent communications in 6G comprise the usage of AI and ML. AI is already very prevalent in 5G, being used for better on the edge processing and sorting of the massive thoroughfare of data in 5G. As a key enabler of 6G, it has a variety of uses in sensing and communication as well as REM. The massive data and processing demand from REM would benefit from distributed intelligence and AI. AI can also be used to better control wireless networks, creating an intelligent network.

Green and secure communications are the next two key enablers. Green communication involves technologies like radio frequency energy harvesting, backscatter communications, symbiotic radio, wireless power transfer, and zero energy IoT systems. The goal of green communications is simply to make 6G more energy efficient by recycling energy and lowering the energy needed to run systems. Secure communication for 6G is dealing with cryptography-based security, which exploits PHY layer properties to secure things. One of the security challenges is securing the vast amounts of private data generated by REM and ISAC concepts. Figure 6.9 depicts a general outline of 6G key characteristics.

6.4 Emerging Applications of 6G Wireless Networks

Normally, the categorization of 5G network major services includes eMBB, URLLC, and mMTC. These applications are vital in ensuring optimal service delivery for technologies relying on 5G. Besides, with the early focus on 6G networks, some additional applications include holographic telepresence, remote surgery, XR, and autonomous driving. Thus, these applications define the required criteria and standards useful for 6G success.

5G already supports some of these applications, but it cannot keep up with the rising data demands. Therefore, the entry of 6G is vital in solving the rising data constraints likely to be experienced by 5G architecture. It is worth noting that the 6G network system will expand the communication environment by extending into space and underwater. Since 6G technology is still in the early stages, it is hard to predict its full potential.

6.4.1 Virtual, Augmented, and Mixed Reality

As its name suggests, virtual reality refers to services that create an immersive environment, allowing the users to experience the first-person point of view. From a practical perspective, 2020 saw many companies, businesses, and people worldwide embrace online systems due to the COVID-19 pandemic. Human contact became limited to stop the spread of the virus. Consequently, in an attempt to maintain life's normalcy, the world settled on using online platforms such as Zoom and Microsoft Teams to mirror face-to-face interactions. Thus, virtual reality allows geographically separated individuals to have effective communication. However, in this kind of virtual reality, the emerging 6G networks will manipulate the natural human eye by supporting the real-time movement in extremely high-resolution situations. This kind of resolution is supported by electromagnetic signals that help to display thoughts and emotions.

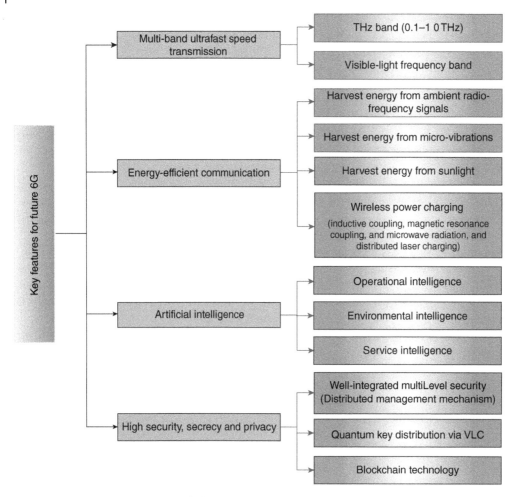

Figure 6.9 Key features for future 6G [16].

Consequently, XR with 4K or 8K resolution support high-definition imaging suitable for applications such as video gaming and 3D cameras. Besides, the application is vital for office environments because it supports virtual physical meetings, social experience, and education and training [17]. Since these new applications use over 1 Tbps, they can saturate the current 5G networks, thereby enabling a smooth transition. Conclusively, virtual, augmented, and mixed reality will support minimum latency and ultra-high reliability.

6.4.2 Holographic Telepresence

It is undebatable to state that human beings are highly dependent on technology. Be it in business or social life, technology has currently taken center stage in human lives. In recent times, one significant development that has happened in technology is holographic telepresence. Through this technology, an object or a person in a different location may appear right in front of their intended subjects. This technology will massively transform the nature of communication. For instance, media and news companies mainly use it. Interviews are done through holographic telepresence, thereby making it look as real as possible to the viewers. In addition, some of the life-changing applications of this technology include robot control, telehealth, and telemedicine. This technology transforms remote experiences, thereby limiting the need to travel. Since it supports "real-time" physical presence and stereo voice, holographic telepresence needs massive amounts of data rates of approximately 4.32 Tbps to maintain a reliable connection [18]. In addition, the synchronization of the viewing angles requires low latency. Ultimately, this requirement put a lot of pressure on the existing 5G network.

6.4.3 Automation: The Future of Factories

The advances in robotics, augmented reality, industrial IoT, AI, and ML are transforming the manufacturing scene. The concept of Industry 4.0 is slowly being embraced as AI is enhancing the manufacturing process. The vision for 2030 is to embrace the concept of Industry 5.0 through the help of the 6G network. For instance, human intervention in industries is gradually reducing because machines are taking over through automated controls. Consequently, these industries will require a high standard of functioning. This includes reliability, accuracy, security, maintenance, and effective management, among others. Also, this type of environment will be highly dependent on robots and drones to support movements and transportation. With a 5G network, the robotic communication will likely face instabilities, thereby affecting various processes. Thus, 6G is set to solve industrial reliability, delays, and security issues, among others.

6.4.4 Smart Lifestyle with the Integration of the Internet of Things

There have been talks about smart cities and how technology can change the world into a utopian/dystopian setup. Millions of applications are now connected via the internet. Some of the areas that benefit or are bound to maximally benefit after the 6G rollout include transportation, healthcare, residential surroundings, shopping, and security. Smart cities are working toward developing flying taxis, which will demand stringent connectivity. Consequently, the quality of life is set to improve due to the ubiquitous connectivity of various smart devices.

Furthermore, smart homes are taking center stage in the smart lifestyle. The Internet of Things is revamping homes through the integration of home appliances. For instance, appliances like refrigerators are now operated like smartphones. Naturally, such systems require improved data rates and high security. Thus, 6G envisions fulfilling this need by developing better infrastructures that enhance autonomous decision-making.

6.4.5 Autonomous Driving and Connected Devices

Transportation industries are set to experience major transformation due to 6G development. Concepts such as smart driving, remote driving, self-driving, and intelligent roads are bound to revolutionize the transport sector [19]. This means that upcoming 6G networks should be extremely reliable, have low latency, high precision positioning, and V2X communication massively to support this type of environment. Besides, safety on the roads is paramount. This should be one of the major focuses of 6G regarding transport. For instance, continuous updates of live traffic and hazard information should be continuous to keep road users safe. Hence, 6G should support efficient vehicular communication to improve safety. When 6G and AI combine, the traffic control systems will improve movement and communication. Thus, the goal of 6G in autonomous driving is to ensure public safety.

6.4.6 Healthcare

Ongoing colossal changes in the healthcare sector supported by 5G networks pave the way for further improvements. The concept of telehealth and telemedicine are revolutionizing the quality and service delivery in healthcare. For instance, through AI, patients can receive treatment and get medication remotely. Besides, the aging population requires intensive medical care due to underlying conditions. Consequently, this poses a huge burden on the healthcare system because of the continuous need to conduct various tests. To control the traffic in healthcare, the current 5G technologies have established e-health systems where patients can send their samples online and receive prescriptions online [20].

Moreover, the ongoing 6G development is developing a ubiquitous health system that will support telehealth and telemedicine systems. Also, it is working toward developing remote surgery systems. As a result, this will require stringent quality of service (QoS) requirements. 6G is motivated to enable remote healthcare deliveries.

6.4.7 Nonterrestrial Communication

Disaster management is a critical tool that helps handle natural calamities such as fire outbreaks, tornadoes, and floods. Therefore, networks must support nonterrestrial communication (NTC) to limit the loss of life and property damage. NTNs should be built to offload traffic and reach remote areas. Some of the technologies supporting NTC include UAVs, high-altitude platform stations (,s), drones, and satellites, which require ubiquitous connectivity and low latency systems [21].

As a result, these communication systems will decongest communication channels, support high-speed mobility, and increase throughput services. Ideally, NTCs are useful in supporting surveillance, meteorology, navigation, and remote sensing. Besides, 6G can support laser communication, which is vital for long-distance communication.

6.4.8   Underwater Communication

Developing underwater network communication is vital for security reasons because it helps monitor activities in the deep sea. Compared to land or air, communication in water needs more acoustic and laser communication to achieve high data transmission speeds. Underwater network communications are made up of submarines, sensors, and divers, among others. Since they can integrate with other terrestrial networks, underwater communication will require stringent connections with low latency [22].

6.4.9   Disaster Management

As previously discussed, the primary focus of 6G network technologies is to improve latency, deep coverage, and reliability, among other advantages. Because of the potential massive improvement of 6G networks, disaster management procedures will likely realize fast and effective results. For instance, 2019 saw the outbreak of COVID-19, which changed the general outlook of activities in the world. Globally, people were able to receive safety messages and updates about the pandemic. Hence, those who were in red zones became more informed and prepared about the virus. Information was shared via major internet sites such as Facebook, Twitter, and CDC websites to mitigate the spread of the virus.

Moreover, drone technologies have always been at the center of disaster management. Drones use can be used to monitor disasters remotely, thereby helping organizations to offer the appropriate assistance. Ultimately, 6G technologies used in disaster management, such as NTN and UAVs, aim to promote public safety.

6.4.10   Environment

Through NTN and UAV satellites, 6G is set to maximize environmental coverage and access to remote areas. Consequently, this is essential in monitoring environmental conditions such as weather and agriculture. Impending environmental disasters and hazardous places can be located using NTN and UAVs. Besides, 6G network architectures will support sensors to detect pollution and emission of toxic gases in real-time.

6.5   The Requirements and KPI Targets of 6G

The upcoming 6G technologies require the creation of innovative use-cases to improve the existing 5G networks. Therefore, the requirements of 6G networks will demand stricter and more diverse technologies. It is likely that 6G will largely carry information related also to nontraditional applications of wireless communications, such as distributed caching, computing, and AI decisions [23]. Network developers agree that 5G technology will not satisfy various requirements as users' needs keep increasing. Figure 6.10 shows a comparative study of KPIs between 5G and 6G networks.

Where 5G is already operating on high mmWaves frequency, 6G is set to increase these frequencies to very high levels. Thus, higher peak data will be achieved as well as seamless, ubiquitous connectivity.

6.5.1   Extremely Low Latency

The emerging new applications require sub-milliseconds or zero latencies for effective service deliveries. These real-time applications such as tactile internet and virtual reality need extremely low latencies to function properly. Furthermore, the inclusion of other high-tech systems such as THz communication and quantum computing will ensure the ideal latency requirements are met. Besides, other vital applications such as ML, Big Data, and AI are useful in delivering low latencies for 6G networks.

6.5.2   Low Power Consumption

Research indicates that smart devices and phones using 4G and 5G networks are constantly charging. The existing computation systems consume a lot of power, leading to the quick depletion of batteries. Therefore, one of the core goals of upcoming 6G networks is to reduce power consumption and increase battery life. The question is, how will these networks

Figure 6.10 Comparative study of KPIs of 5G and 6G [23].

Key Performance Indicator (KPI)	5G	6G
Peak data rate	20 Gb/s	1 Tb/s
Experienced data rate	0.1 Gb/s	1 Gb/s
Peak spectral efficiency	30 b/s/Hz	60 b/s/Hz
Experienced spectral efficiency	0.3 b/s/Hz	3 b/s/Hz
Maximum bandwidth	1 GHz	100 GHz
Area traffic capacity	10 Mb/s/m^2	1 Gb/s/m^2
Connection density	10^6 devices/km^2	10^7 devices/km^2
Energy efficiency	not given	1 Tb/Joule
Latency	1 milli sec	100 micro sec
Jitter	not given	1 micro sec
Reliability	$1-10^{-5}$	$1-10^{-9}$
Mobility	500 km/h	1000 km/h

achieve this? 6G is set to offload the computing tasks to smart bases with reliable power storage and supply. In addition, other energy-harvesting techniques of 6G will ensure the longevity of battery life. For instance, electromagnetic signals and micro-vibrations can be utilized to achieve this. Moreover, the development of wireless battery charge is vital in promoting its life.

6.5.3 High Data Rates

One of the main focuses of 6G is to offer extremely high data rates as compared to the existing 5G networks. It is expected to deliver peak data rates of over 1 Tbps per user on average. This means that the upcoming networks will need very high bandwidths in the THz band. Besides, it will explore OAM multiplexing, laser communication, and over 1000 antennae elements. Ongoing research is also focusing on integrating quantum communication to increase the data rates. Consequently, upcoming innovative technologies are bound to benefit from this.

6.5.4 High-Frequency Bands

Data rates keep going up, especially on cellular wireless networks. Wider bandwidths and higher frequencies are required to match the fast-increasing data rates. For 6G, network developers and researchers are focusing on sub-THz and THz bands that can offer rich spectrum resources for electromagnetic and light waves communication. The utilization of such frequencies will support multi-Tbps data rates useful in transmitting information for outdoor and indoor scenarios. Besides, THz frequencies are likely to solve inter-cell interference. Despite its ability to support high-frequency bands, it is vital to note the arising challenges for further studies.

6.5.5 Ultra-Reliability

Reliability is set to increase by 100% from the existing 5G networks. Applications such as telehealth, industrial IoT, and smart cities need very high reliability to function optimally. Hence, 6G networks should attain QoS requirements for efficiency and effectivity.

6.5.6 Security and Privacy

All networks, including 3G, 4G, 5G, and 6G, require security and privacy. Since 6G is set to support massive data rates, it demands improved security. Research shows that quantum computing is likely to secure these networks. However, blockchain technology will come in handy due to the potential challenges as they are difficult to manipulate.

6.5.7 Massive Connection Density

Great connectivity is all about increasing the number of connected users. Literature shows that millions of people and devices will be interconnected due to autonomous devices. The Internet of Everything (IoE) concept will take shape as smart devices such as smartphones, drones, and vehicles will become interconnected. This will increase the connection density of 6G networks. Thus, developers believe that applications such as THz communication and VLC will be able to meet the demands of expected connections.

6.5.8 Extreme Coverage Extension

Currently, the 5G rollout is focusing on metropolitan and urban areas. This means rural areas are neglected, thereby creating a gap that 6G is yet to fill. Many people in rural areas still lack access to better internet connectivity. Thus, 6G is set to penetrate these rural areas because its objective is to enable worldwide connectivity. Besides, it looks to support underwater communication by developing effective electromagnetic signals.

6.5.9 Mobility

One of the expected characteristics of 6G networks is their dynamicity. It is expected to come at extremely high speeds of approximately 1000 km/h. However, it is important to note that such speeds would require frequent handovers. However, AI applications will come in handy and support complex decision-making processes that might challenge mobility.

6.6 6G Applications

The dominant service and application of the 6G wireless network will be AI, and sensing and learning will be the dominant services and applications of the 6G wireless network. Virtual reality for everything will provide VR, haptic feedback, and AI services to the end-user. The primary applications of the 5G mmWave network are sensing and collecting big data for ML. The primary application of the mmWave network is collecting data to train neural networks. AI services will be provided through the neural edge, the deep neural node, and the neural center. Quantum key distribution (QKD) will be deployed for the fiber-optic link between the neural edge and the neural center. The 6G wireless system will use satellite and fiber for the front haul, backhaul, and mid-haul, as well as terrestrial nodes for direct access. 6G applications comprise the following [21].

- ➢ Holographic communication
- ➢ Tactile internet
- ➢ Industry 4.0 and beyond
- ➢ Teleoperated driving
- ➢ Internet Bio-Nano things
- ➢ Multisensory XR applications
- ➢ Blockchain and distributed ledger technologies
- ➢ Connected robotics and autonomous systems (CRAS)
- ➢ Wireless brain-computer interface (BCI)
- ➢ Digital replica

Figure 6.11 and Table 6.1 show a breakdown of applications and a brief description of their roles and how they relate to 6G networks.

6.7 Challenges in 6G: Standardization, Design, and Deployment

There are many challenges faced by the push to 6G and a "connected intelligence." Power and energy consumption costs in new machines will continue to rise, which will cause us to rethink how diverse our current systems are. The latency of current networks must be improved to allow for intelligence at the edge devices. Machine learning must converge with

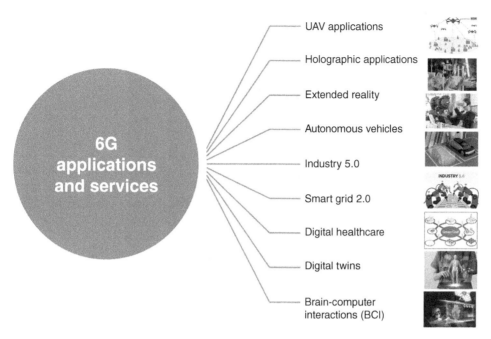

Figure 6.11 The most essential of 6G applications [24].

Table 6.1 Brief description of 6G applications [21].

Applications	Brief descriptions
Holographic communication	This enables human communication through holographs-3D images in thin air. To improve the experience of remote communication as we embrace a borderless workplace. Latency and high bandwidths are some of the challenges associated with holographic communication. 6G will solve these challenges.
Tactile internet	Enables human-to-machine interactions and machine-to-machine interactions.
Industry 4.0 and beyond	Comprises cyber-physical systems, IoT, and cloud computing. Additionally, AI and ultra-fast wireless networks will drive the fourth industrial revolution. This enables smart cities, factories that are some of the vision for 6G.
Teleoperated driving	Allows cars to be controlled remotely. These cars are also referred to as semi-autonomous vehicles. Semi-autonomous cars require a fast and ubiquitous wireless network with ultra-low latency.
Internet Bio-Nano things	An interconnection of biological nano-sized objects (nanomachines). Takes application largely in healthcare. 6G is proposed to provide the perceptual requirements and ultra-low latency required by IoBNT.
Multisensory XR applications	AR/MR/VR that incorporates perceptual experience. Supported by URLLC and eMBB and perceptual factors to be supported by 6G. An excellent candidate to provide a better gaming experience.
Blockchain and distributed ledger technologies	Blockchain is postulated to provide security for 6G networks. They also require low latency, reliable connectivity, and scalability, which 6G networks will provide.
Connected robotics and autonomous systems (CRAS)	CRAS is required to improve industrialization through the use of robots and autonomous systems for industrial operations. They require a high rate and reliability and low latency.
Wireless brain–computer interface (BCI)	BCI enables the communication between the brain and electronic devices. This requires ultra-low latency, high reliability, and high data rate.
Digital replica	These are also called digital twins, and they create a digital copy to replace people, places, systems, objects. This requires a very high data rate, which 6G will enable.

machine reasoning. The impact of this network on our ecosystem is also important, lowering the impact of greenhouse gasses and CO_2 is key. We are moving from IoT to a massive connected intelligence, which will necessitate the need for much stronger security

Security is the next big component that will need to be improved on as 6G networks begin to be rolled out. Mobile networks have relied on SIM cards for identifying devices as part of their security. This method causes issues in IoT from the size of the cards to the symmetric key encryption that they use. As the bandwidth increases in 6G networks along with the moving of much more data to the cloud, the need for end-to-end security will become even more important. Software-defined networking, AI, and NFV will work together in 6G networks to provide that end-to-end security. Some of their suggested approaches to 6G security challenges include authentication of remote endpoints, certification of the platforms, support of remote attestation, and secure properties. Physical security is another important consideration for 6G networks.

Privacy protection in 6G networks is another important consideration and one that may concern people the most. Protection of personal information when so much data is available on the network will be very important. This is another global issue but one that will be viewed differently depending on where in the world you are. Standardizing regulations pertaining to privacy will be difficult due to the difference in governmental views on what should be private.

Physical and digital worlds will become more and more interconnected at the network technologies continue to improve. With this come concerns about people's health and safety when they spend such a large amount of time connected. Many countries are enacting laws and regulations that dictate how much time they can be expected to work and what level of network connection is considered working. Liability issues are just one more potential issue related to the growth of 6G networks that will have to be solved as the technology moves forward.

The benefits of 6G are massive. However, it is critical not to overlook the impending challenges. The enabling technologies of 6G are therefore tasked to develop solutions for these challenges. Some of the issues likely to be faced by 6G networks include:

- Global coverage
- Limited system capacity
- Challenge in deploying computing and AI capabilities
- Hardware constraints on Edge AI
- The trade-off between high intelligence and privacy
- Security
- Noncompatibility of hardware and software nongenerality

In terms of challenges and hurdles for 6G, multiband and utilization requires advanced front-end hardware, spectrum coexistence creates interference problems and, waveform coexistence also creates new interference problems. Heterogeneous networks face issues with overburdening. ISAC systems also suffer from an excess of burden.

Ultra-flexible framework must have three key points. First, new tech should be combined into communications through a flexible key enabler platform without having to wait for a decade. Second, a key enabler tech needs to work together to meet various requirements, so a flexible cognitive engine could optimize different aspects. Lastly, flexibility should be quantified while optimizing with some sort of flexibility performance indicators.

In conclusion, 5G was defined by its variety of requirements and applications, and 6G is expected to continue in its footsteps. Flexibility should be the focus of 6G to better build off what 5G has developed by using key enablers. These enablers include ultra-flexible PHY and MAC, ultra-flexible heterogeneous networks, integrated sensing and communications, green communications, and secure communications.

References

1 El Hassani, S., Haidine, A., and Jebbar, H. (2019). Road to 5G: key enabling technologies. *Journal of Communications* 14 (11).

2 Castro, C. (2021). 6G Gains momentum with initiatives launched across the world. *6G World Exclusive* https://www.6gworld. com/exclusives/6g-gains-momentum-with-initiatives-launched-across-the-world.

3 Salameh, A.I. and El Tarhuni, M. (2022). From 5G to 6G – Challenges, technologies, and applications. *Future Internet* 14 (117).

4 Zhao, Y., Yu, G., and Xu, H. (2019). 6G mobile communication networks: vision, challenges, and key technologies. *SCIENTIA SINICA Informationis* 49 (8): 963–987.

5 Rastogi, R., Saxena, M., Chaturvedi, D. et al. (2021). Kirlian experimental analysis and IOT. *International Journal of Reliable and Quality E-Healthcare* 10 (2): 29–43. https://doi.org/10.4018/ijrqeh.2021040104.

6 Dixon, C. (2016). Social drives video to dominate mobile data traffic by 2022. *nScreenMedia* (Nov. 16): https://nscreenmedia.com/social-drives-massive-growth-mobile-video.

7 Soldani, D. (2020). On Australia's cyber and critical technology international engagement strategy towards 6G – How Australia may become a leader in cyberspace. *Telesoc* 8 (4): https://jtde.telsoc.org/index.php/jtde/article/view/340.

8 GSA (2023). 5G-Ecosystem March 2023 member report, https://gsacom.com/technology/5g/.

9 Kishiyama, Y., Suyama, S., and Nagata, S. (2021). Trends and target implementations for 5G evolution & 6G. *NTT Technical Review* 19 (11): 18–25.

10 Al-Ansi, A., Al-Ansi, A., Muthanna, A. et al. (2021). Survey on intelligence edge computing in 6G: characteristics, challenges, potential use cases, and market drivers. *Future Internet* 13 (5): 118. https://doi.org/10.3390/fi13050118.

11 Ericsson (2019). Ericsson Mobility Report: 5G uptake even faster than expected. https://www.ericsson.com/en/press-releases/2019/6/ericsson-mobility-report-5g-uptake-even-faster-than-expected.

12 Wu, Y., Singh, S., Taleb, T. et al. (2021). *6G Mobile Wireless Networks*. Springer Link.

13 Rowe, M. (2021, May 12). 5G, B5G, 6G: A tale of two similar conferences. 5G Technology World. https://www.5gtechnologyworld.com/5g-b5g-6g-a-tale-of-two-similar-conferences/.

14 Letaief, K., Chen, W., Shi, Y. et al. (2019). The roadmap to 6G: AI-empowered wireless networks. *IEEE Communications Magazine* 57 (8): 84–90. https://doi.org/10.1109/mcom.2019.1900271.

15 Lopez, O., Mahmood, N., Alves, H. et al. (2020). Ultra-low latency, low energy, and massiveness in the 6G era via efficient CSIT-limited scheme. *IEEE Communications Magazine* 58 (11): 56–61. https://doi.org/10.1109/mcom.001.2000425.

16 Alsharif, M.H., Kelechi, A.H., Albreem, M.A. et al. (2020). Sixth generation (6G) wireless networks: vision, research activities, challenges, and potential solutions. *Symmetry* 12: 676. https://doi.org/10.3390/sym12040676.

17 Bastug, E., Bennis, M., Medard, M., and Debbah, M. (2017). Toward interconnected virtual reality: opportunities, challenges, and enablers. *IEEE Communications Magazine* 55 (6): 110–117. https://doi.org/10.1109/mcom.2017.1601089.

18 Xu, X., Pan, Y., Lwin, P.P., and Liang, X. (2011). 3D holographic display and its data transmission requirement. In: *2011 International Conference on Information Photonics and Optical Communications*, 1–4. Piscataway: IEEE.

19 Ray, P., Kumar, N., and Guizani, M. (2021). A vision on 6G-enabled NIB: requirements, technologies, deployments, and prospects. *IEEE Wireless Communications* 28 (4): 120–127.

20 Giordani, M., Polese, M., Mezzavilla, M. et al. (2020). Toward 6G networks: use cases and technologies. *IEEE Communications Magazine* 58 (3): 55–61. https://doi.org/10.1109/mcom.001.1900411.

21 Imoize, A., Adedeji, O., Tandiya, N., and Shetty, S. (2021). 6G enabled smart infrastructure for a sustainable society: opportunities, challenges, and research roadmap. *Sensors* 21 (5): 1709. https://doi.org/10.3390/s21051709.

22 Zeng, Z., Fu, S., Zhang, H. et al. (2017). A survey of underwater optical wireless communications. *IEEE Communications Surveys & Tutorials* 19 (1): 204–238. https://doi.org/10.1109/comst.2016.2618841.

23 TechPlayOn (2020). 6G wireless access – key performance indicators (KPIs) https://www.techplayon.com/6g-wireless-access-key-performance-indicators-kpis.

24 Shimaa A. Abdel Hakeem, Hanan H. Hussein, and HyungWon Kim. (2022). Security requirements and challenges of 6G technologies and applications. *Sensors* (Basel, Switzerland), 22(5), 1969. https://doi.org/10.3390/s22051969.

7

Cybersecurity in Digital Transformation Era: Security Risks and Solutions

7.1 Introduction

The online world has brought with it its own share of both good and bad. No one can deny the overhauling advantages of the internet while looking at the world today. But there are many issues underlying all that has been achieved, and day-after-day specialists across the world are seeking solutions for them. Some problems are abstract and emanate from human nature itself. But online communication has embedded itself into the human consciousness to such an extent that any day without it would cause overwhelming and world-changing events, so it is imperative to consider some perspectives: the ubiquity of the internet, its base and defining characteristics, the essential nature of change or evolution, as well as its innate relationship with the digital world at large.

The very last perspective accounts for the jeopardizing of essential data with the presence of intervention brought about by cyberattacks. These activities, of which there are many, seek to undermine the entire concept of online technology and threaten the well-being and interests of people henceforth. Moreover, one of the primordial elements of computing technology, data, is affected, influenced, altered, or damaged beyond any point of use. The players who undertake a deep understanding and research of the online world relate these actions to real-world crime. The actions taken to investigate cyberattacks and catch those responsible resemble traditional forensic evaluation of the crime scene and discovered evidence quite closely, and in some cases, under a single banner or concept [1]. However, the third perspective and, by extension, all the other factors create rampant challenges and restrictions that could result in some terrible consequences by the way of cyberattacks, as well as the larger institution of cybercrime, suggesting that the traditional approach does not go far enough in preventing or stopping cybercrime.

The *digital transformation* is a perspective of immense consequence in terms of the resulting state of cybercrime, and what options are available or need to be developed for security and protection. Such a "transformation" is occurring all across industries and governments as other issues and internet issues have seemingly arisen out of the widespread use of the internet [2]. Moreover, the motivations of digital transformation are affecting the outcome of innovative technologies, as well as the nature of the digital world brought about by new technologies. Therefore, the first and foremost need for addressing the overlying topic at play would be properly identifying and analyzing the winds of change [1]. But, one thing is certainly clear: While innovated digital and online technologies are making the entire scenario vulnerable to attacks, the application of the theoretical understanding holds the most potential for cybersecurity and protection measures.

7.2 Digital Transformation and Mesh Networks of Networks

Tackling the state of the digital world, which includes the online capabilities and functions, certainly is a challenge. It is mainly because digital transformation is a continuous predicament, and by definition, it also relates to "the novel use of digital technology" to resolve problems that are already present in the larger human experience [2]. However, this form of "use" may not include the increase of efficiency by the way of automation by digital technologies. Instead, the consequential occurrences of innovation and creativity are supposed to be the indicators of digital transformation, and it eventually becomes the standard with the replacement of the traditional model [3]. There can be a cavalcade of different reasons and motivations behind digital transformation, and varying degrees of adoption and implementation by the organizations and institutions promulgating it.

However, "digital transformation" must not be confused with "digitization," which refers to the transformation of political, business, industry, trade, and media into a digital format. The term is mainly referring to the conversion of analog infrastructure and methods to an increasingly digital format. Such a social transformation could still be occurring, but digital transformation is attributed mainly refer to such bodies and individuals who are already "digital," meaning the dominant presence and influence of digital technologies [4]. These bodies are undertaking and, in the process, undergoing innovation and creativity due to a variety of different reasons – it might be to present some service or product in a better way, to solve a longstanding issue that has remained since the past, or to improve their own capabilities according to plans and objectives [5]. Taken on the basis of every individual change, the term that is most applicable is the process of "digitalization."

Digitization has promulgated and ensured that the process of digitalization, but digital transformation is something of a "societal change brought about by digitalization." As a result, digital transformation brings about significant changes in the very business model, nature of consumption, socioeconomic dynamics, legal and policy undertaking, organization perceptions, cultural transformations, and many more. Now, digital transformation has many consequential technologies with different terminologies, technologies, and even the theoretical reasoning behind it [2]. Technologies like the Internet of Things (IoT), Industry 4.0, electronic payment, office processes automation, and cryptocurrencies seek to effectively shift from the past "monolithic" architectural designs to essential micro-services that facilitate agility, flexibility, and speed [5]. The applications are becoming self-contained, highly capable, and unique in their own right, which directly requires the characteristics of networking to transform accordingly.

The high applicability and customizations raised in the overall cases across software identity and its capabilities also reflect a highly compartmentalized approach. The internet has always been this significantly enlarged and obfuscating continuously proliferating set of connections [4]. However, this has also resulted in some notable limitations, which most definitely need to be overcome. The constant change or transformation over the different parts or locations of the internet refers to its apparent slow reaction against the requisite mutability of the network itself. It could even take months on end to effectively consolidate a particular change across a segment of the network, something that businesses in the digital domain just cannot afford [6]. The vector requirements of change also refer to the networks themselves becoming "cloud-like," with its variety of functions getting triggered instead of the costly and burdensome of loading the entire defined stack one at a time; see Figure 7.1.

The digital transformation, as a result, is happening on the basis of the shift of previously monolithic applications to something that is distributed and facilitated on an ad-hoc basis. This is known as the *mesh network of networks* or the *service mesh* [8]. This particular state of the network is a theoretical understanding that has been present for quite some time but only after cumulative applications has it come to such prominence. Such a layout can essentially be referred to as a

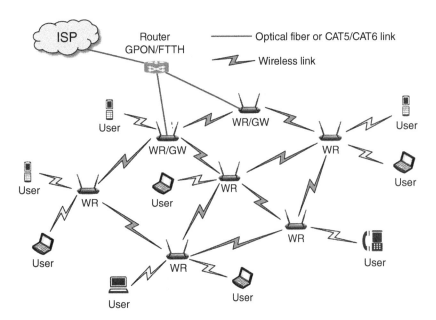

Figure 7.1 A general depiction of what a mesh network of networks looks like [7].

self-contained network, operating on the fringes of the internet at large. Under such a networking concept type, every particular node or device relays the information required through the effective functioning of the entire network, and it does not need to be connected to the internet to work properly [1]. However, prominent and successful implementations of this "mesh" plan require at least a single point of entry and exit, respectively, to the internet.

But, by every means, one can see the obvious advantages of such a measure. After all, the network does not have to be this significant and all-encompassing platform, and it may operate well and without far fewer problems coming along the way. The requisite nature of delineations from the internet, first of all, is the easy and effective possibilities of undertaking something like that and making them highly customized [4]. This results in not just the environment around some software getting changed but instead allows for the change and adaptability with the prevalent conditions to become apparent and quite actual.

Moreover, it also ensures that one of the longstanding internet issues is effectively corrected and resolved: data privacy. Over the last few years, there has been significant controversy around this issue in particular. Before increasing the digital footprint, people want to make sure their private data cannot be made public [3]. The loss of data and its apparent effect on the social world that exists online is quite obfuscating all such issues [8]. Moreover, such measures are increasing the reach of online facilities to places where it was simply unimaginable before. However, the most advantageous and consequential feature that the mesh networks operating within the internet is that they are, by definition, highly adaptable. Such a measure shall ensure that complete and effective network functioning should proceed even when there is a fault or error within it.

Thus, digital network functions through the topography defined as a mesh network of the network is almost a certainty at the moment. Despite still being in a transitional state, their apparent disadvantages and invited issues speak about an effective and consequential conjoining with the state of cybercrime. By definition, such a system exposes not just the effective network but all the mesh networks associated with any possible threat that may arise [2]. These attacks can be devastating, and may also seriously affect the entire state of the network from a multitude of perspectives. The entire state of such an established theoretical network topology does have a number of considerations as to why security pulls to the opposite end when digital transformation is at hand.

7.3 Security as the Enemy of Digital Transformation

The advent of technologies like IoT, Industry 4.0, and cryptocurrency do not just indicate the apparent facilitation and improvement over the previous technologies. They also indicate a scenario where networking itself is going through change. This was the takeaway from the last section, and it is certainly essential to discuss the changes in networking capabilities. Since this software requires a lot of data, it is evident that storing them on the cloud is preferable to the traditional mode of storing them in the secondary memory. Basically, the existence of these technologies indicates the rise of an extremely decentralized, cloudlike environment, which shall define the networking environment itself in the future [5]. Although they have not become overwhelmingly prevalent as of yet, many would surely feel the advantages of doing so given that vast amounts of data footprints are left behind by each individual even today.

With the rise of the above-mentioned technologies, the footprints are only going to increase in size, and that demands even greater decentralization and application of a variety of cloud technologies. The apparent effect on the internet, as well as the digital and online world as a whole, is going to be momentous. But, to be exact, many fundamentals of online connectivity are going through a transformation due to such a state of circumstances [2]. Therefore, since security and protection issues are so inherently connected with the underlying concepts that have governed online technologies up until this particular moment in time, one can say that the prevalent and contemporary security and protection measures would soon become completely outdated. More than anything, these security measures will be rendered quicker than anyone might imagine [6]. In the aftermath, the amount of private and essential data that is going to be left vulnerable is simply astounding to properly comprehend and understand.

The first point of open vulnerability to cyberattacks rests squarely with the fact that the "surface" for the attack. All the applications, data, and processes occurring across a highly de-centralized domain open up a lot of opportunities for attack. Cybersecurity measures have always rested on the diagnosis of such an attack, and its quickness and efficiency open up the possibilities to launch some effective recovery techniques. With the hackers and their malicious programs finding all new and vulnerable points of entry, keeping a track of the entire system to present and promulgate the entire scenario becomes more improbable. The attackers or the virus program's lateral threat across the software system network is also an important point of consideration. This is almost a theory-like viewpoint, which essentially states that any "network is only as strong as its weakest link" [9]. Security services and products, available widely at the moment, relate to monolithic silos, in which case proper visibility and assessment of the issue would be negatively limited.

By all means, digital transformation also significantly increases the magnitude of damage to occur. It is important to remember that the cloud environment of the mesh network of networks, which is how networking is going to look, represents a number of critical operations taking place all across the board [8]. Perhaps, more importantly, the value of data for businesses has greatly increased over time, and it would seem that beyond anything the vulnerability increases as the size and the complex interconnectedness among them should be the main point of focus. The increasing size and complexity also make the standard security approaches and concepts of the past become ineffective, or partially effective. The latter is not an improvement upon the former as it has already been established that any network is only as strong as its "weakest link" [10]. The potential for damage being incurred upon businesses, families, and even individuals becomes greater than any time beforehand. Of course, without any effective measure in place, the overall "damage" indirect to the main case of the cyberattack would be easily devastating for everyone involved [2]. For instance, the healthcare sector must run round the clock; without the presence of essential facilities at hand, the situation could result in a large number of fatalities. In a system that is wholly dependent on delivering all possible facilities and interventions to the sick, the situation seems to speak a significant amount of damage that does not hold against a lengthy recuperation from illness.

For businesses specifically, the case for digital transformation is certainly more pronounced when compared to anything else. In the past, the networks were self-contained if businesses happened to be large enough to be spread across a geographical area that could be considered, in every way, significant [6]. Under such circumstances, all equipment was decided to be on-site, resulting in software facilitations to be present across any unit instead of the cloudlike support and derivation of services of today and the presented future [1]. These networks, as a result, could not be considered as being *contained* – instead, the terms *dispersed* and *interconnected* would be more appropriate. As a result, the chances of properly diagnosing an attack upon such a distributed network become increasingly harder for the appropriate personnel [3]. Even with the self-functioning security measures at the place, the attack may commence and successfully infiltrate the network well before it is properly identified, diagnosed, and recovered.

Another major point of concern among many lies with the fact with the increasing rate at which the digital transformation itself is taking place. Many experts working in the field of information technology (IT) claim that while the transformation rate can be attributed to many essential technological innovations, the same cannot be said for security [5]. It is quite evident that an almost overwhelming majority of security measures are exactly what many might comprehend them to be: software. For a majority of time in the past, the entire point of developing new security strategies and installations has been more or less on an anticipatory basis [2]. In some cases, the acceptable format of security innovation and implementation took place when a specific kind of attack first took place. Therefore, the challenges in proper diagnosis at a digital transformation pace, which is faster than it has ever been before, certainly make such an approach completely outdated in every sense. In a situation wherein cybercriminals are creating or profiling revolutionary approaches for attacks, the security change measures implemented by any particular entity [6], whether it is an individual or a business, must be drastic and game-changing.

In enterprises, measuring security coverage and utility is not enough to maintain secure systems. Common security measurements are becoming inadequate since cybersecurity problems are often complex and hard to manage. Due to these complex issues, information ecosystems, government regulations, and critical infrastructure require a more in-depth analysis beyond the basic coverage provided from a simple security standpoint. The business strategy, planning, and performance must all be incorporated within the cybersecurity ecosystem. Aside from the base security protections in place for more enterprise organizations, the cybersecurity programs need to incorporate other business analyses within the program. This may include efficiency of security, resiliency, current process of IT, and gap analysis.

All forms of risk and all aspects that may affect internal systems must be incorporated within the cybersecurity program. The overall culture of the organization needs to change its mindset toward a security approach in that all assets of the organization are on board. Cybersecurity practitioners and enterprises face many challenges related to cybersecurity challenges. These challenges include risk management, due diligence and negligence, operation efficacy and efficiency, system prioritization, security operations, skill sets and training, and budgeting:

- Risk management for companies can be summarized in three simple statements. They are not doing it, they do it poorly, or they think they do it but don't. Even if there are regulatory requirements, companies typically don't do a good job managing risk. Risk management is one of the problematic areas in security. The larger an organization grows, the harder it is to manage it. Enterprises need to consider all scenarios and reduce as many outliers of the organization as much as possible.

- Due diligence and negligence suffer from not having a standard of care, or norms. Enterprises are generally secretive about operations and security controls or goals. They typically curtail their security programs based on their operational or customer base. Regulatory guidelines provide guidance but only provide a bare minimum for security coverage and generally do not implement every aspect of an organization. Due diligence must work alongside risk management to keep track of potential sources of vulnerabilities.
- Operational efficacy and efficiency can be classified in multiple ways. Efficacy relates to "if" the security controls are sufficient and "if" they are working as intended. For organizations to analyze efficacy, they must evaluate the controls in place. The analysis must take into consideration if the control is sufficient, works to reduce risk, and is performing as expected. Alongside efficacy, efficiency must be assessed. However, in different enterprise environments, goals may differ depending on cybersecurity programs or the issues related to implementing security controls. Resources must be determined and implemented so that the controls in place do not affect overall organization efficiency.
- Prioritization aids in determining regulatory compliance and controls that must be prioritized and implemented for the organization. This must be determined by the enterprise's desired security controls and/or goals.
- Security operations is a difficult area for many organizations since many require necessary funding and lack qualified personnel. With these issues, security operating procedure (SecOp) departments struggle with maintaining a stable department due to the lack of funding and staffing shortage. These two issues put pressure on SecOp departments to find the required expertise and maintain enough funds for the department. However, these departments face other issues such as strategic planning and business planning to maintain secure systems while directly or indirectly increasing revenue. Overall, SecOp departments face many internal issues as well as external issues from outside attackers. Since external threats only need to find a single opening, the SecOp departments must defend and reduce every possible point of entry.

The increasing sophistication of cyberattacks known as polymorphic attacks, can be seen as a feedback loop that is itself promulgated by digital transformation and is affecting such theoretical presentation of technologies that made such an attack possible in the first place [10]. In this case, actual artificial intelligence (AI) technology is applied for the attack to take place, and when it does happen, the computer virus program can mask itself according to the security measures at the place. Thus, the entire case of digital transformation, while bringing innumerable improvements and advantages to the table, is also giving rise to many threats and dangers. These threats do not just undermine the specific sets of technology for which the transformation is actually taking place, but the entire logic and concept behind it. If businesses are to be left crippled in every sense, then they might very well question bringing the state of digital transformation in the first place [6]. Moreover, such socially dependent systems as healthcare, transit, and education cannot be left unprotected in any sense because the damage happening cannot be compared to the positive points brought forth by digital transformation.

7.4 The Current State of Cybercrime

Cybercrime comes in many shapes and forms with significant and inordinate deviations occurring at any level of observation and analysis. Perhaps the most consequential aspect of the current state of cybercrime can be stated as an increased rate of mobile use for performing these attacks. The attackers who are utilizing mobile phones to promulgate their advantageous goals, as well as to hide their tracks as much as possible [11]. Despite this, the observable data shows a record increase of almost 680% with regard to attacks taking place [12]. More specifically, the entire volume of cyberattacks in the entire year of 2018 amounted to about 70% of all that have actually taken place [13]. The most obvious use of attacks taking place at the moment was by rogue mobile applications to effectively defraud consumers with even old and primordial phishing and malware attacks upon devices, respectively.

It would seem apparent that a majority of cybercrimes are taking place with the continuing rise of mobile phones in performing everyday tasks. As people are increasing their mobile phone usage, so do the majority of attackers and fraudsters. In addition to the exposure and, by extension, the vulnerability of people at large, this has become a reality. The main cause of such a great increase could be attributed to the unprecedented development of mobile phone technology over the last decade or so [12]. As a result, a higher section of the population is utilizing mobile phones, often with not much in the way of proper security and protection measures. Apart from all of these rather straightforward, and yet incidental attacks taking place, special consideration for cross-channel vulnerabilities is also becoming rather obfuscating and prominent.

Cross-channel attacks have also become far too common as consumers tend to demand services and products with only a bare minimum of actions taking place. The growth of digital channels is not just a happenstance, but a very obvious reality as companies or brands facilitate easy and fully accessible cross-connections happening all across the board [6]. The new application program interface (API) economy specifically indicates this obvious growth of interconnected channels, and it relates to users sharing personal and essential information with multiple platforms or service providers [2]. As referred to before, it widens the probable surface for attack greatly, leading to nothing but overwhelming vulnerabilities to proliferate like never before. The challenge is obviously a cross-platform protective strategy that works across every platform. Such a measure can only be possible through widespread and influential policy making, as well as their effective implementation.

Another obfuscating major threat in the current scenario presenting cybercrime also includes the attackers utilizing legitimate platforms to promulgate and reinforce their activities. The most obvious one among all possible platforms is a highly popular social media platform with undeniable global reach. Platforms such as Facebook, Twitter, WhatsApp, etc. observe hundreds of millions, and even billions of users getting online and interacting with each other [14]. With such an overwhelming individual presence, and an even greater data footprint left behind, it becomes quite obvious why cyberattack specialists and fraudsters have found many advantages on these platforms [12]. In particular, the messaging service often acts as the discreet pathway that leads them to converse with one another. It also helps them in selling private credit card data, exchanging stolen identities that they siphon off from various databases, and many other illicit takeaways taken from their cyberattack escapades.

If deriving everything that is observable is really relevant, then it would seem that the entire situation is only going to get worse with time. By 2018, the fraudulent activities happening through social media surged by about 43%, and other ways to commit fraud through social media continue to emerge [12]. Self-functioning bots are also fitted into this equation, making the traceability of the person behind the attacks close to zero. Social media is such a haven for cyberattackers due to ease of use, absence of fees, and many other benefits [13]. AI makes it even easier for attackers to use mobile apps to hide their tracks in real-time.

This can be directly seen as a detriment of the cloud environment of networking that is slowly but surely becoming prevalent all across the world. Central to this discourse, however, lies the question of how exactly security measures can tackle such a scenario. As already pointed out, the situation seems to have given a greater power of anonymity and reach to the attackers. Additionally, the threat is only going to spread and become even more damaging. The fraudulent and malicious activities mentioned above indicate that these could easily be carried unnoticed through the most legitimate of ways, and hosted for free across well-known platforms [8]. It certainly brings the entire orientation of these platforms into question. However, the necessity to understand the core of the issue is paramount since the solution might need to be of a similar magnitude as well.

Moreover, the presence of blockchain systems has seemingly compounded the problems. Alluded to before, these are one of the most consequential technological innovations in recent times, but beyond everything else, it seems to have given a much-necessary edge to attackers [13]. Without the presence of any proper regulation, cyberattacks can easily distort the fields in any single block, and the whole chain – no matter how numerous or complex it is – will get affected [2]. Moreover, the latest discovery has revealed a number of domain name systems that seemingly give the attackers enough refuge to drive their internal conversations and other illegal activities without any kind of fear. This is mainly because there is no effective oversight for these systems, unlike the legitimate systems, which are openly operated over the internet on a daily basis. But, the above scenario seems all too plausible for the future.

At present, most people underestimate the threat of cybercrime. It is not just an incidental case anymore, since down the line it also indicates a very dark future for the online world – something that people actively avoid entering, due to the high potential of it affecting everyone's lives and well-being [15]. The rise of the cloud environment, as well as the complete irrelevancy of security and protection measure silos, present a host of problems well out of their grasp [8]. Everything points to starting from its most abstract and creating policies and laws that may reverberate across this postmodern online world. That is why there is also an essential need to address the issue with regard to the relation between security and technologies [1]. Not only does it need to happen on a general basis for complete and thorough understanding, but it also needs to take place from the perspective of digital transformation.

7.5   Security and Technologies of the Digital Transformation Economy

First, it is important to note that essential security considerations in this present age of digital transformation are something that goes completely unidentified. That is because the digital transformation scenario contains many essential technologies that may very well result in the complete overhaul of how businesses operate, as well as how exactly they present their services and products to clients and consumers [15]. However, it is also certainly important to note that the entire question has resulted in a vision that is a direct extension of digital transformation, and how businesses and other entities may approach security [4]. Referred to as *security transformation,* the aim of the entire abstract planning is also a complete overhaul of understanding security and protection tactics and strategy to present some groundbreaking and reverberating innovations. Over half the companies all over the world, by an estimate, define security challenges to be their most obvious and challenging issue for properly enacting security and protection measures [8]. Additionally, the same percentage of organizations also agree there is another overwhelming issue regarding hiring IT security professionals at the level they may need now.

Many are regarding security transformation as more than just an issue, which could be effectively by technology. Businesses are looking for their strengths beyond anything else like learning properly how to make different teams work. And, these just do not include those concerned with security, but also application technical personnel and also networking experts [11]. It can also refer to a complete shift in the perception of how security measures get operated, architected, and implemented all across the board [12]. Moreover, another major idea involves the complete elimination of the idea behind isolationist development security measures for every individual technology out there. The need for uniformity in access and identification in a proper fashion is one of the highest priorities in enacting proper security measures all across the board.

Another popular thrown about the idea all across the board is the consequence of organizations and businesses coming together in order to exchange information everything in relation to security over the threats presented by a digital presentation [4]. Ideally, the suggestion is the solution that should be aimed if policy making and legislation are to be properly and consequentially involved with the entire scenario. The exchange of information the technology would present a free yet equitable market environment for everyone – even the not-so-financially secure ones – can become aware of what is actually happening [14], and prepare themselves against such a possibility. It would also promulgate the commercially available IT security firms to be at a pace that shall be universally available to anyone purchasing their products and services [8]. The rise of needs with the oncoming of a new breed of attacks and strategies with malicious intent would be driving the pace at the same rate as a technological commodity.

However, to say that this particular solution has no issues that have already been presented would be completely nonfactual in every way. After all, there are far too many exigencies involved when it involves companies, especially the large ones, in working toward a common goal to achieve something definitive [1]. It cannot also be denied that if a free market is to be operated then the ones who get hold of a piece of information would be needing some returns to justify and reinforce their research capabilities. Moreover, the need for an oversight body would require something obvious; however, the effectiveness and the influence of the said body veer toward evidence that requires practical measures and points of reference [12]. Additionally, nothing could be more pointless than the infiltration of attackers into the entire body under some kind of trick or strategy behind. These individuals or entities have already proved that they can thrive within an extremely legitimate environment. So, there should be a constant doubt if the access and identification of the activities and their reactions are available to both sides.

However, above everything else, the most talked-about implementation of technology across the entire sphere of security transformation is the question of automation. Automation is, of course, one of the primordial reasons why computing technology exists in the first place [4]. The fascination with the inner workings of the human brain and the question of how to exactly work faster, more efficiently, and with greater capacity can be described as the main motivator behind almost all major innovations in the technological arena for the past 70 years or so. But, the realization of effective security in this day and age with the help of very sophisticated automation that rivals and even, in some cases, surpasses the things put into use by attackers [16]. Many experts fully believe that automation should be the driving force behind security programs of the future – driving and promulgating all processes and operations effectively onward.

Front and center essentially talk about such game-changing technological innovations made in the fields of AI and machine learning (ML). These new-age technologies constituted just a source of fantasy or fiction in the past, but the implementation of these "intelligent" programs even in their present state would be capable of many things [12]. The primordial one is, of course, adapting itself with information available online regarding new approaches and strategies for attacking. Moreover, these technologies can also work together across multiple and distinct but nevertheless interconnected platforms and software support, respectively [17]. The effective potential for AI in the future is only going to improve, and it could very well upend the entire perception of humans' perspectives toward technology influencing their lives.

The entire scenario, accompanied by effective security provisions, certainly reflects a state of multiple opportunities that could be taken singularly, in combination, or all at once. Nonetheless, it cannot be denied that there is, in fact, a lot of confusion regarding the positives and negatives of all such considerations [8]. But, it would seem that all other discussions presented under the banner of security transformation taking place in very short spaces of time and efficiently at a constant rate could only be made possible through automation. Many also tend to believe that humans need to play a secondary role in these technologies operating on their own accord with respect to diagnosing and launching recovery measures in case of a cyberattack anywhere [16]. The case could be made strongly on this one, since it fits all the criteria boxes of what pieces of evidence are actually pointing toward.

7.6   Tackling the Cybersecurity Maturity Challenges to Succeed with Digital Transformation

The overall case with the cybersecurity challenges brought forth by digital transformation is sometimes stated to be a two-headed beast and is certainly quite interesting. As already pointed out in the past sections, increasing demands for ease of use and the full support of a digital environment is the main criteria involved in the effective propagation of the entirety of digital transformation [2]. But, on the other hand, there is also the question of full and effective facilitation that can safeguard and protect private data is also driving the business providers back into the corner. The gap in the effective security state resulting from this is perhaps the most essential discussing point about the increase in the number and sophistication of cyberattacks.

However, the proper way of tackling the issue lies with exactly how the network is going to be laid out, as well as its final infrastructure. Under such circumstances, security standardization should be one of the highest priorities out there. This could be preferably achieved by way of effective legislation, all focused on the timely standardization of new and innovative services and products [5]. One of the ways in which this can be achieved is referred to as "privacy by design, which essentially takes upon the entire matter of designing and building the architecture as the point of allocation and integration of necessary protection and security interventions" [1]. Instead of just addressing the issue by "bolting" security products and services afterward, the need for a certain extent of such programs and software is expected directly of the providers themselves. More importantly, these measures and approaches shall be taken under consideration to give the best privacy in terms of personal data in the most robust and secure fashion [12]. Bodies for standardizing technological innovations and blueprints are already present, but there is a need for addressing the submissions and plans in quicker and more efficient ways.

But, in spite of such widespread measures addressing the general prevalence of maturing cyberattacks, it is also important to address the drivers of digital transformation individually to a certain extent. It is because these technological phenomena often tend to be fairly complex in both their operations and what they can possibly achieve [6]. IoT is a prime example, and the potential for its own improvement and innovation down the road is quite high. But, it is one of the main drivers that also holds the greatest potential to give rise to the cloud environment networking topography [4]. The effective issues of cross-communication between other devices in the mesh could be under the credentials and directives of a single individual or entity, but that does not make them less susceptible to cyberattacks in any way. The present state of these attacks is constantly getting improved and enhanced, as pointed out before [18]. Thus, what these embedded security measures need to achieve is a spontaneous way to develop and implement new strategies and interventions to the effective probability of an attack.

Standardization of these kinds of security practices and allotment can lead to widespread adoption by organizations and businesses. Their products and services need to show the acceptability of their measures in order to put such items on the market, or act as a benchmark for consumer purchase appropriately [5]. In the case of healthcare technologies, however, the magnitude of threat or danger at hand would need to be completely evocative of some innovative and creative pursuits being organized and implemented. It is because, unlike the ideal IoT environment where a set of devices perform what is expected of them with respect to the user and other connected devices, health technologies need to be hosting them from the cloud. In terms of provisioning the security and protection interventions, the "instance" is only the way because the people seeking out the services, as well as their historical data in detail, are only supposed to increase with time [1]. As a result, the measures for securing these software allocations, alongside their activities, become an issue of multilevel provisions. These must occur in different ways, and they must not obfuscate one another's processes in any way. Basically, under such considerations, constructive mode of planning and sharing of essential information should become the foremost concerns.

The rise of electronic identification and trust service providers is seemingly what exactly the future has in store. In terms of consequential undertakings being spearheaded, it is certainly one of the most notable ones with the square purpose of lateral and vertical propagation of electronic time stamps, key certificates, as well as essential policies [16]. Such technologies need to get more automated, more authenticated, and dependent on the uniqueness of identity variables that users can only provide by themselves. The case for automating them all across the board, however, still remains an unresolved challenge.

Cryptography is currently the most feasible approach, and its consolidation of potentially rich solutions for security does hold a lot of weight across the most critical of processes [19]. Cryptographic algorithms are being developed to operate uniformly and effectively across the oncoming stage of networking infrastructure [14]. They are mostly involved in provisioning what can be referred to as the benchmark embedding of security and protection allocations available at large.

The quantum computing and consolidation of cryptographic algorithms all point to an effective and long-term solution. In this case, security interventions and other measures become more agile through high and complex self-functioning capabilities [14]. This would allow for effective solutions to be promulgated with respect to essential considerations and issues across the board [2]. Combined with cryptography, it can offer a chance for tackling and resolving essential issues across the board, which will effectively give rise to the situation one might be looking to actually tackle the issues in connection with spontaneous changes and transformations [6]. The issues to be tackled must be addressed and analyzed through the lens of a multipoint perspective, and the eventual result would effectively provide a chance to promulgate and capitalize on the contemporary scenario of digital transformation.

7.7 Security Maturity and Optimization: Perception versus Reality

The case for the information related to and explained above speaks of widespread implementation of only the most essential perspectives that can posit a change. These overhauling changes are supposed to represent the probable and supposed technical knowledge that needs to be adopted and effectively implemented against the maturity of cyberattacks. It certainly gives rise to the effective realization of "security maturity" or "security transformation," which has already been explained in great detail everywhere else [12]. However, the case at hand for business enterprises and organizations represents a far more insidious issue that needs addressing and correction along the way. This is all about the organization's perception – with the presence of a very wide gap between any particular company's perceived security strength, and what it actually represents.

By all considerations, the root of the problem seems to be focused on the fact that organizations deem their knowledge about security issues and protracted attacks under a qualitative experiential setting. This results in the maturity of security infrastructure and specific technologies being perceived and evaluated as a collection of layers of security measures in place that provisions essential security allocations along the way [4]. Such a viewpoint obfuscates quite enormously the differences between the perceived and actual states of security reality across the board. Quantitative metrics should, of course, be involved in the question with regards to properly keeping every defensive mechanism against similarly matured cyberattacks [19]. It becomes quite apparent since the cyberattacks have at present a wider vulnerable area for launching attacks for infiltration than any time before. If unanticipated and unprepared, there could very well be a seriously a scenario wherein the magnitude and area of coverage in relation to the attack could be left unaddressed even if proper diagnostic and recovery missions are launched.

That is why these organizations need to focus on multiple points to promulgate their security measures and infrastructure. To embed properly, in this case, means that security and protection should be integrated across every other concern in the organization existing in the form of data and implemented software, including people, the culture pursued and developed, tools, controls, operational technologies, as well as cloud applications [2]. Only through the permeability of security and protection interventions across all these domains can there be any hope of proper and consequential security strategies getting implemented [20]. However, such a requirement is hardly a commonplace aspect for organizations that claim to be the experts and leaders in the field of launching high-end and new-age security solutions.

Most seem to attribute this situation to the overall case to the lack of proactive attitude of organizations when it comes to properly dealing with security threats. Only about 24% actually engage in taking an approach that actively seeks to deal with security issues by the way of anticipating attacks that are most likely to occur [4]. The effective security and protection strategies implemented in connection with this would most definitely require inculcating and integrating a proactive approach toward everything in relation to security measures in an effective fashion [14]. If not, the entire scenario would seemingly continue to perpetuate, resulting in constantly changing share costs, inability to fulfilling the terms in a contract, the trust that exists with customers, as well as with fines in relation to proper regulation.

The first possible step toward ensuring that solution would require that security strategies and approaches mature over time. However, without the full context with respect to information provisioning in developing and updating security infrastructure and technologies, it can be accurately said that no matter the amount of investment, all efforts and overtures shall fail [14]. The basic purpose is to embed such measures and activities by first assessing and properly identifying the assets, functions, and processes being transformed appropriately and with reason to the overall issue of security optimization [21]. Starting from this, the security activities and technologies in connection shall proceed with a top-down approach to optimize by way of increasing visibility and controls. These increased capabilities shall inordinately result in nothing but the absolute realization that high-risk areas must be monitored constantly. In the case of automating such measures, the development of security and protection measures would then be automatically launched with the shortest window of time and will be available before damages start taking place [2]. However, the most essential part remains the fact with the full support and connections of every other aspect of danger that could be a consequence of the attack could be easily identified, and the type of attack would only become a partial matter of consequence for the IT security team.

7.7.1 Why Cybersecurity Maturity Is Not What It Should Be in the Digital Business and Transformation Reality

The difference between the apparent perception and reality for cybersecurity maturity has been discussed. What remains is why the maturation is not exactly what it should be, and why it specifically rest with looking into those threats after they have already been made. It was stated that the majority of the organizations and enterprises tend to focus on the qualitative values of experiential risk management instead of quantitative ones [19]. The latter has a major role to play because it presents the full picture of the threat and risks involved with the cyberattacks, as well as other similar criminal activities. So why do businesses tend to focus on the qualitative experiences, instead of the statistical nature of quantitative data?

Widespread and key security conventions have become challenged, such as the principle of least privilege and the perception of the fact that prevention is better than the cure [20]. However, with the digital transformation taking place faster and more efficiently than ever before, the entire situation at hand exposes security allocation requiring satisfying a few primordial principles regarding building trust and resilience in a holistic fashion [14]. Both these factors are the end consequence that an enterprise can bring about, especially with regards to explaining exactly why security maturity is not so effective and easy to achieve.

The very first of its kind is shown by the fact that most companies base their security development investments in accordance with a checkbox compliance approach. However, with the increased complexities and size of digital systems environments, the effective implementation of security infrastructure and controls in the traditional format is neither possible nor feasible [22]. Instead, the focus should lie on risk-based decision-making, whereby the entire case of making security investments and controls are spread across the entire organization with the effective classification of high-risk areas against the low-risk ones [19]. This would result in the business capitalizing on such areas with an equitable approach. Moreover, this also translates to the second key factor of an obfuscated perception, which claims that the focus should shift from the security architecture to business outcomes. Applying such a change would inimitably result in a system that constantly looks for improvement and enhancement over its capabilities.

The absence of an outcome-based approach also speaks a lot about the IT security professionals in relation to their roles within the organization. The traditional viewpoint holds them to be the "defenders" when they should be facilitators who have their point of concern all over the holistic interests of the organization [23], safeguarding the thin veil of balance that exists between the needs for protection and that of helping the business achieve its outcomes [6]. This also relates to the constant flow of data as the aforementioned digital businesses have increased the volume and complexities with regard to their daily availability, as well as who gets to see them and act on them [12]. But, businesses must give up this perception that they need to control the flow of information because they might very well be not feasible as time passes on. Conversely, these organizations should direct their focus on how best they can protect all their essential processes and controls accordingly. By doing this, they can achieve the best probable outcome to be hoped for if they want to have security concerns that mature and resolve attacks and threats in an iterative and continuous fashion [14]. Data could be the key difference-maker, and developing an effective team that can analyze and derive meaning from inordinate amounts of information is perhaps the most important factor at hand.

Moreover, the need for balance should also be discussed as to how people are treated in the context of security and protection. Under the theoretical basics of proper organizational management, it is well-known that motivating and driving people to perform in their jobs without any forceful or compelling qualities must be taken into account [24]. By every means, the entire point of properly tackling cybersecurity considerations brings about the question with regards to balance. This directly means that legacy compilation of security and protection contingencies must be pursued to ensure the full-range realization of identifying, processing, developing, and consolidating brand new strategies along the way [25]. It also means that proper allocation of contingencies must relate to the full retention and select application of the detection-and-response mechanism.

Understanding the true nature of security maturity or transformation is complex, and it is easy to see why business enterprises and organizations do not perceive them accurately. The entire point of the problem seemingly emanates from the question of what route must be taken [26]. By the presence of the qualitative analysis of the current state of cybercrime, it is quite evident that the old contingencies and measures are completely outdated [20]. But, in more cases than one, these contingencies could also be applied to counteract issues that may be presented in a rather innovative fashion but in its function is actually something that is traditional [21]. What true security maturity means is that the old and the new contingencies should work in balance in order to make business organizations resilient and to project a sense of trustworthiness all around.

7.7.2 Why Cybersecurity Maturity and Strategy Are Lagging

For the last few sections, it has become quite clear as to what exactly is wrong with cybersecurity maturity and its resultant strategies. However, despite opening up enough facts to facilitate inferences and logical reasoning, it is still true that an overwhelming majority of these organizations remain completely unaware, implement the measures incorrectly, or have a skewed perception of what security maturity must be [19]. Such a state of circumstances mostly points to the need for understanding what are the exact factors resulting in such security measures not reaching up to the desired levels of pervasiveness or ubiquity.

One obvious reason why that is the case might be due to the state of organizational culture. In a number of organizations, culture is defined as highly stratified with a lot of bureaucratic measures being implemented that is preventing such security measures to get implemented across the board [27]. Under the unsolicited state of culture, there might be a skewed and unhealthy transaction of information. As a result, the major consequence derived from this particular state can be defined as a limitation or fault with the powers that be in the organization itself. It could very well be the case of the issue not reaching the right ears, and even when they do the people who are responsible might have their own reasons for not giving much credence to it [28]. As a result, the proper embedding of security measures and approaches in a mature state never gets realized.

Other reasons may also include the question of the resources available at a particular organization. This is quite easily applicable to a majority of organizations all across the world, and it can also be attributed to the statistical information showing the occurrence of the main topic being discussed in this section. Small- to medium-level business companies do not have much in the way of capital, first and foremost, and they feel that it is certainly a risk to have an outdated security system in place [16]. In many cases, even when the monetary resources for purchase, as well as the capital for investing in a security maturity model, is available, the business drivers may feel that the funds are better off being spent elsewhere [2]. This risk-taking attitude can become a detriment itself when these organizations grow and increase market share. This increases their vulnerability, but implementing more mature security measures may be costly due to the ineffective core functionalities and processes, which need to co-depend on the security measures already taking place.

However, the most obvious reason for this might be due to the "perception vs. reality" consideration when threats first emerged. The effects of a skewed basis of properly taking into stock all the features and essentialities in place can result in nothing but the most damaging of results or outcomes. Even when the individuals of interest might feel that all their concerns about security and protection are completely ensured, it could easily be not a case like it at all [19]. However, the worst-case scenario is most definitely the mixed takeaways that the businesses might take in the case when an attack gets through and results in some significant damage to the entire organization's most essential assets or resources [29]. It might very well be the case they might lookout for some new technology, which will surely not produce any real results along the way.

But, the most interesting reasoning accomplished throughout the entire case of lagging security maturity lies with the lack of self-awareness. It might seem like a continuation or extension of the previous one, but its effects are truly consequential [27]. Even when an organization has become aware of a skewed cybersecurity maturity plan being implemented, they might not know what is actually happening, or what exactly might be wrong with this particular perception [30]. Under such a scenario, the exact need for security maturity models must be raised and questioned in great detail. These hold the potential for depicting exactly what and where the problems remain with respect to any integrated security plan by any organization, especially when the strategy becomes one with the technological environment, too. These are the exact reasons why cybersecurity maturity, as well as their associated strategies are increasingly left behind.

7.8 Changing Security Parameters and Cyber Risks Demand a Holistic Security Approach for Digital Business

The eventual and consequential transformation happening across the field of digital technology is certainly evocative of a fairly few influences. One of the main reasons for this is that tackling cybersecurity maturity, as well as transformation, is not panning out as they are supposed to be [3]. There are many underlying reasons as to why, but the main point of consideration and reasoning remains with the limitation of security measures and strategies being put into place [19]. The current prevalent attitude is perpetuating the skewed, qualitative perception strictly, instead of effective quantitative data. Moreover, the state of cybersecurity maturity, in association with its strategies, is consequently lagging behind the increasing sophistication and innate quality of attacks pursued [31]. Therefore, the consequential parameters of cybersecurity are also getting shifted toward being considered as mostly transformed fundamentals of widespread perception.

One of the primordial factors of consequence in the entire question of cybersecurity depends on the structure of the organization. The key decision-makers across these bodies recognize the importance of both organizational structure and business infrastructure at the place [2]. Unfortunately, the key decision-makers for security maturation often view the presentation of the data as too-detailed, often inconsistent, and devoid of efficacy in terms of what can generally be considered professional and expert insights [26]. Therefore, it is vital to convince the key players as to the priority of putting in place the parameter of structure that intervene across each and every level, process, and data point inside an organization.

Another factor of consequential security maturity assurance lies strictly with the overall determinants at play with respect to clarity. As the technological or digital environment is getting too complex, as well as disseminated all across the cloud, this particular factor certainly has a lot of roles in terms of "correcting" the current scenario at play [27]. The overloading of technological shorthand, as well as acronyms, dissuades decision-makers from learning what they need to know. If they are not completely aware – clearly and fully – of everything in relation to cyberattack risks and threats, then this will have a denigrating effect on the overall security in any organization [19]. Information flow must be at a maximum distance of technicalities for the decisions to be consolidated quickly, and actions implemented accordingly.

Another significant security parameter at play is the availability of real-time data. The main issue with the current form of data security is that it is not at all effective or consistent. This is because the information currently being considered as the base reason for launching cybersecurity implementations might signal a need to intervene too late to protect the data at risk [3]. Under such circumstances, the contemporary scenario expects security allocations to identify, diagnose, and come up with a solution to a problem in the shortest time possible. If automation is the case, one can see how effective and consequential real-time data could actually be [28]. This is especially relevant if an attack happens at some critical point, and the security interventions respond to the critical point but the malicious software has already done the damage and moved elsewhere.

However, a possible solution would indicate the implementation of a holistic approach with regards to tackling the problem as it occurs and providing solutions spontaneously [29]. The holistic approach encapsulates a hierarchical model of various modes of players, activities, and strategies to give an effective solution, as shown in Figure 7.2. The entire thing is regulated by the topmost management to accurately overview the entire risk landscape [33]. The governing principles are ultimately what determines what, how, or when security strategies are going to be applied. But, the end objective of the overlying parameters at play rests with the complete realization of striking a balance between resilience and operability.

Moreover, a tight and rigid form of control exists in the entire scenario of only the most critical assets. And, the entire risk-mitigation pathway in four distinct phases. Identifying risks and risk appetite is the first point of concern, wherein a complete list of complete assets, intimated risks, and the potential ones are required [20]. This would lead to the creation of a risk appetite for every individual risk with a complete analysis of effects and influences upon the said assets [34].

Holistic cyber risk–management approach

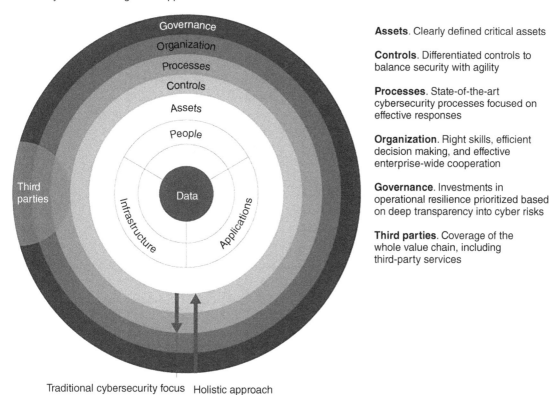

Traditional cybersecurity focus   Holistic approach

Assets. Clearly defined critical assets

Controls. Differentiated controls to balance security with agility

Processes. State-of-the-art cybersecurity processes focused on effective responses

Organization. Right skills, efficient decision making, and effective enterprise-wide cooperation

Governance. Investments in operational resilience prioritized based on deep transparency into cyber risks

Third parties. Coverage of the whole value chain, including third-party services

Figure 7.2   A depiction of hierarchy with the operability of the holistic solution model [32].

The analysis or evaluation stage marks the stage wherein the risks, in accordance with their respective appetites, are measured alongside their likelihood of occurrence, as well as the holistic nature of their overall impact.

After the complete evaluation and analysis are finished with regards to any risk, the probable interventions to act on those risks should they present themselves must be consolidated. Additionally, a great point of concern is all about the evaluation of the mitigation or corrective measures put forth, which shall reflect exactly what needs to be done [33]. The final step of the holistic process of risk mitigation involves proper and effective monitoring. This shall not just work on its own with associated technological enhancements, but will also work in accordance with the full information getting related to everyone within the organization, including the topmost decision-makers [11]. The entire effectiveness and design of a strategy like this would need to happen only when the organization or enterprise is fully ready to undergo change. Other influences over cybersecurity must be considered one by one, and its influences require to be correctly predicted even if they are simply potential at best.

7.9   Cybersecurity Challenges and Digital Risks for the Future

The situation at hand is mostly of a critical nature, but it would be left off in a greater peril if there is no effective predictions as to what challenges may become apparent and of critical nature. Cybersecurity challenges are somehow going to rise seemingly out of nowhere, and anticipatory discourse is the best hope the overall industries and individuals may have in properly countering it [29]. However, in this case, no one can also deny the obvious influence of digital environment risks involved with respect to the entire predicament at last.

Cybersecurity challenges in the future are going to be interesting, as most of the policymaking in recent times and across this particular field of concern often need to be reactionary in nature. After all, the digital transformation is progressing at an unprecedented pace, and there could be no denying that this will only magnify and increase down the line [35]. However, the challenge mainly rests with increasing concern over government and business interventions over people's privacy through their online identities. Another essential challenge with respect to cybersecurity is that of the *security divide* [13].

In this particular case, the increasing sophistication of security and protection can lead to the disenchantment of a large portion of the society, leaving some being actually able to protect their data, while others cannot. Moreover, the mesh network of networks with different security and protection details could also dissuade users from overall internet use.

Similar is the case for digital risks, which might not be directly related to cybersecurity but can give rise to issues that affect it nonetheless. The first is the case with the agile technologies getting developed and released in short orders of time. Due to such a small window, such a technology might be implemented or utilized somewhere that might not have been anywhere under the consideration of the creator [2]. Under such circumstances, especially if the implementation is illegal or illicit in nature, it can cause a lot of problems for the developer, which will not easily go away. The risk of not having enough manpower resources in the face of quick transitioning of digital technologies also present a significant problem or risk that needs addressing under the entire digital domain [34]. A combination of these factors can easily increase the presence of vulnerable systems all across the public and open online world. As a result, the incidence of cybercrime always indicates something increasing in the future.

7.10 Conclusion

Security issues, if not looked at carefully with the development of the right solutions, can become the main barriers to users and service providers who are participants in the sharing economy. The whole problem of securities has been brought about by the coming of large data accounts, which are shared among a significant number of customers who, in all cases, have varying motives. It has been considered that the new technologies should be decentralized so that their uses are controlled entirely by individuals, groups, or small enterprises. Emerging technologies show that there are changes in the relationship between science and social, cultural, and economic factors. Different policymakers, however, are trying to outshine each other in the market so that they can win the trust of their customers. This is only done through the success of the technology, and that provokes a healthy completion among the developers.

Digital security is essential and is raising issues more than ever, as discussed in this chapter. From this, it is clear that there are many different types of security threats to individuals, businesses, or the government. This is associated with all the things we do with the current technologies daily, which range from simple home devices to company and government technologies. For example, if a person shops online with a hacked computer, then the attacker can access the user's sensitive information, which is usually used for fraud or other crimes that may land an individual in danger. The use of credit cards while shopping is also putting people at risk, especially when pin codes are stolen and personal information is lost. Security threats are now infringing on human rights because they allow unknown individuals to access people's private communications.

Privacy and security are all intertwined – a threat to one is a threat to both. Breaches to security disturb the trust and are a significant violation of ethical principles. The need to protect data is urgent and complex. It is a necessity to ensure that the security functions are protected so that cooperative management is not compromised. The securities, however, are overstretching given that hackers and other security threats sources are also growing along with social improvements.

The now-commonplace digital and online world coalesce over several factors that give rise to the entire scenario of unsolicited or unlooked-for cyberattacks. This has been a constant story for the majority of time in which digital technologies have existed. The entire scenario of digital transformation results in an overwhelming shift that is proliferating the presence of vulnerable systems and activities to an almost exponential extent.

However, the challenges that exist for security maturation are also a very real possibility. The situation involves a factor of issues that permeate every single aspect of the business enterprise or organization. Without a holistic perspective of ensuring a better and more efficient response, the situation of additional cybersecurity challenges and digital risks will be representative of everything and more challenging roadblocks in this field across every single facet of everyday life.

References

1 Ciardhuáin, S.Ó. (2004). An extended model of cybercrime investigations. *International Journal of Digital Evidence* 3 (1): 1–22.

2 Assante, M.J. and Tobey, D.H. (2011). Enhancing the cybersecurity workforce. *IT Professional* 13 (1): 12–15.

3 Craigen, D., Diakun-Thibault, N., and Purse, R. (2014). Defining cybersecurity. *Technology Innovation Management Review* 4 (10): 13–21.

4 Barclay, C. (2014, June). Sustainable security advantage in a changing environment: The Cybersecurity Capability Maturity Model (CM 2). In: *Proceedings of the 2014 ITU Kaleidoscope Academic Conference: Living in a Converged World – Impossible Without Standards?* 275–282. IEEE.

5 Gavas, E., Memon, N., and Britton, D. (2012). Winning cybersecurity one challenge at a time. *IEEE Security & Privacy* 10 (4): 75–79.

6 Kane, G.C., Palmer, D., Phillips, A.N. et al. (2015). Strategy, not technology, drives digital transformation. *MIT Sloan Management Review and Deloitte University Press* 14: 1–25.

7 Binh, L.H. and Truong, T.K. (2022). An efficient method for solving router placement problem in wireless mesh networks using multi-verse optimizer algorithm. Sensors 22 (15): 5494. https://doi.org/10.3390/s22155494.

8 Berman, S.J. (2012). Digital transformation: opportunities to create new business models. *Strategy & Leadership* 40 (2): 16–24.

9 Ali, S., Bosche, A., and Ford, F. (2018). *Cybersecurity Is the Key to Unlocking Demand in the Internet of Things*. Bain and Company.

10 Norris, D., Joshi, A., and Finin, T. (2015). Cybersecurity challenges to American state and local governments. In: *15th European Conference on eGovernment*, 196–202. Academic Conferences and Publishing Int. Ltd.

11 Massacci, F., Ruprai, R., Collinson, M., and Williams, J. (2016). Economic impacts of rules-versus risk-based cybersecurity regulations for critical infrastructure providers. *IEEE Security & Privacy* 14 (3): 52–60.

12 Anderson, R., Barton, C., Böhme, R. et al. (2013). Measuring the cost of cybercrime. In: *The Economics of Information Security and Privacy* (ed. R. Böhme), 265–300. Berlin, Heidelberg: Springer.

13 Kshetri, N. (2010). *The Global Cybercrime Industry: Economic, Institutional and Strategic Perspectives*. Springer Science & Business Media.

14 Kaplan, J.M., Bailey, T., O'Halloran, D. et al. (2015). *Beyond Cybersecurity: Protecting Your Digital Business*. Wiley.

15 Miron, W. and Muita, K. (2014). Cybersecurity capability maturity models for providers of critical infrastructure. *Technology Innovation Management Review* 4 (10): 33–39.

16 Dunlavy, D.M., Hendrickson, B., and Kolda, T.G. (2009). Mathematical challenges in cybersecurity. *Sandia Report,* February.

17 Yar, M. and Steinmetz, K.F. (2019). *Cybercrime and Society*. SAGE Publications Limited.

18 Prislan, K., Lobnikar, B., and Bernik, I. (2017). Information security management practices: Expectations and reality. In: *Advances in Cybersecurity* (ed. I. Bernick, B. Markelj, and S. Vrhovec), 5–22.

19 Aradau, C., Lobo-Guerrero, L., and Van Munster, R. (2008). Security, technologies of risk, and the political: guest editors' introduction. *Security Dialogue* 39 (2–3): 147–154. https://doi.org/10.1177/0967010608089159.

20 Matt, C., Hess, T., and Benlian, A. (2015). Digital transformation strategies. *Business & Information Systems Engineering* 57 (5): 339–343.

21 Payette, J., Anegbe, E., Caceres, E., and Muegge, S. (2015). Secure by design: cybersecurity extensions to project management maturity models for critical infrastructure projects. *Technology Innovation Management Review* 5 (6): 26–34.

22 Pathak, P.H. and Dutta, R. (2010). A survey of network design problems and joint design approaches in wireless mesh networks. *IEEE Communications Surveys & Tutorials* 13 (3): 396–428.

23 Muttik, I. and Barton, C. (2009). Cloud security technologies. *Information Security Technical Report* 14 (1): 1–6.

24 Wirkuttis, N. and Klein, H. (2017). Artificial intelligence in cybersecurity. *Cyber Intelligence, and Security Journal* 1: 21–23.

25 Moss, M. and Endicott-Popovsky, B. (ed.) (2015). *Is Digital Different?: How Information Creation, Capture, Preservation and Discovery Are Being Transformed*. Facet Publishing.

26 Maughan, D., Balenson, D., Lindqvist, U., and Tudor, Z. (2013). Crossing the "Valley of Death": transitioning cybersecurity research into practice. *IEEE Security & Privacy* 11 (2): 14–23.

27 De Bruin, R. and Von Solms, S.H. (2015, November). Modelling cyber security governance maturity. In: *2015 IEEE International Symposium on Technology and Society (ISTAS)*, 1–8. IEEE.

28 Loader, B.D. and Thomas, D. (ed.) (2013). *Cybercrime: Security and Surveillance in the Information Age*. Routledge.

29 Mehravari, N. (2013, November). Resilience management through use of CERT-RMM & associated success stories. In: *2013 IEEE International Conference on Technologies for Homeland Security (HST)*, 119–125. IEEE.

30 Tisdale, S.M. (2015). Cybersecurity: challenges from a systems, complexity, knowledge management and business intelligence perspective. *Issues in Information Systems* 16 (3): 191–198.

31 Pavone, V. and Esposti, S.D. (2012). Public assessment of new surveillance-oriented security technologies: beyond the trade-off between privacy and security. *Public Understanding of Science* 21 (5): 556–572.

32 Broehm, J., Merrath, P., Poppensieker, T. et al. (2018). Cyber risk measurement and the holistic cybersecurity approach: Comprehensive dashboards can accurately identify, size, and prioritize cyberthreats. Here is how to build them. *Risk Practice*, November. © 2018 McKinsey & Company. https://www.mckinsey.com/~/media/McKinsey/Business%20Functions/ Risk/Our%20Insights/Cyber%20risk%20measurement%20and%20the%20holistic%20cybersecurity%20approach/Cyber-risk-measurement-and-the-holistic-cybersecurity-approach-vf.ashx.

33 De Bruin, R. and Von Solms, S.H. (2016, May). Cybersecurity Governance: How can we measure it? In: *2016 IST-Africa Week Conference*, 1–9. IEEE.

34 Mohammed, D., Mariani, R., and Mohammed, S. (2015). Cybersecurity challenges and compliance issues within the US healthcare sector. *International Journal of Business and Social Research* 5 (2): 55–66.

35 Oswald, G. and Kleinemeier, M. (2017). *Shaping the Digital Enterprise*. Cham: Springer International Publishing.

8

Next Generations Networks: Integration, Trustworthiness, Privacy, and Security

8.1 Introduction

Wireless communication has been developing since the 1980s with significant changes and advancements with each generation. We have recently entered the fifth generation of wireless communication (5G). The adoption and deployment of this newest generation are underway but will likely take three to five more years to see fully realized deployment. 5G has brought about the move to the cloud for software-based networking bringing about on-demand, automated learning management of networking functions. However, even with these advancements, researchers have already begun to look at the next generation (6G) and how the networking landscape will change with its adoption.

Through a variety of technologies, 6G is speculated to bring about a shift to completely intelligent network orchestration and management. The current evolution of 5G networking is helping to visualize the architectural framework of 6G. Heterogeneous cloud infrastructure should be an expected part of 6G as the existing cloud infrastructure and that of future generations will not simply disappear with the introduction of a new infrastructural component. Where it will begin to diverge from the present is the increased variety and flexibility of specialized networks for personal subnetworks and things such as flexible workload offloading.

As with every next step, 6G communication networks will have their own security and privacy concerns that need to be addressed. Mobile security continues to be a very significant issue with each generation of wireless communications and seems to only get more difficult with each technological advancement. Past generations have dealt with everything from authentication issues to cloning. With the availability of computing resources and the sophistication of attackers now, it is expected that 6G networks see some of the same attacks such as zero-day attacks and physical layer attacks, but also an increase in advanced attacks such as quantum-based attacks and artificial intelligence/machine learning-based (AI/ML-based) intelligent attacks. The addition of AI into the networks of 6G will work for and against its security, it seems.

We must build on fifth-generation security research to gain insight into what will be coming with the later generations of communication. Currently, there is not much insight and research that looks at the holistic picture of 6G security.

Sixth-generation applications will have much higher requirements and required functionality for their networks. Networks may be required to reach Terabits per second (Tbps) connection speeds while also facilitating connectivity for extremely high counts of devices due to IoT and the plethora of mobile communication devices that use these networks. With applications that are extremely sensitive to latency, end-to-end latency of a network will require latency reduction down to microseconds. These extreme jumps in speed, density, and latency requirements will still require the 6G network to be 10–100 times more energy efficient than current generations. This may be accomplished through extremely low power communication in some cases. These all breed causes for concern.

Latency-sensitive applications will have to consider the impact of implementation causing increases in the latency of the system, while the extremely large bandwidth of networks will offer up many challenges for ingesting and processing the network's data as it flows. Some applications will likely require processing locally to avoid these issues. As these networks evolve and become less homogenous, they will also introduce another security issue with the diversity of devices on a given network and their ability to change their edge networks frequently and quickly. The hypothesized key performance indicators (KPIs) to help address these requirements include protection level, time to resolution (TTR), coverage, autonomic level, AI robustness, security AI model convergence time, security function chain round-trip time (RTT), and cost to deploy security functions.

From 5G to 6G: Technologies, Architecture, AI, and Security, First Edition. Abdulrahman Yarali.
© 2023 The Institute of Electrical and Electronics Engineers, Inc. Published 2023 by John Wiley & Sons, Inc.

Additionally, the benefits of intelligent radio and radio access network core (RAN-Core) convergence offer up further considerations. As these technologies enable the more rapid and automated management and deployment of radio networks via AI/ML, they will add an attack vector through AI/ML training manipulation. Detection of these types of exploits will be key to securing things. RAN and core functions being virtualized and able to be implemented close to the edge as well as at the core for higher-layer RAN functions will improve low-latency communications, but also add complexity to securing these new networks.

Edge computing married with AI will be essential, but will also offer up its own challenges. Edge intelligence relies on the hierarchal training of systems to allow analysis and prediction to happen closer to the edge, making way for real-time analysis and prediction. However, its algorithms and training are highly data-dependent, meaning that it is very important that the data is aggregated from reliable sources that can authenticate and guarantee data integrity. This may be a great space for the implementation of blockchain to keep consistency and authenticity. Another very interesting topic mentioned here is homomorphic encryption.

The specialization of networks is also a potential area of concern. 6G network capabilities will allow for use of subnetworks in many parts of society. The implementation of smaller-scale specialized networks will require the use of lightweight but well-tested authentication and encryption. The creation and utilization of trusted execution environments offer another potentially more feasible solution for some closed networks. As with 5G, the continued use and expansion of the use of open application program interfaces (APIs) in 6G should also be recognized. These have been found to be vulnerable to multiple types of attack. The flexibility and computational capabilities of 6G networks could also allow for entirely closed-loop networks which are zero-touch and autonomous but will likely still have to consider things like distributed denial of service (DDOS) attacks and man-in-the-middle attacks.

And as with all technology, the endpoints connecting to the network must be considered a threat to the security of these new and advanced networks and functions. The move to enhanced Subscriber Identity Mobile (eSIMs) and Integrated SIMs (iSIMs) for some devices may continue to help secure communication, but longer-term moves away from traditional symmetric key encryption appear to be the only way to move forward securely. It must also be considered that this communication encryption may need to be able to deal with quantum computing and its advantages in encryption breaking.

Some use cases of 6G technology bring up concerns for potential security concerns as well. Unmanned aerial vehicles (UAVs) may begin to be used for all kinds of new purposes. In addition, as such will be a target of malicious attacks because they are unmanned, and they are much more susceptible to being jammed and captured to harvest data from their systems. Holographic telepresence could lead to not only enormous amounts of data to deal with but also a need to secure the holographic images being presented and keep them private. Especially when used in presenting to a remote location. Another potential vector of consideration is the world of extended reality. It will lead to an increase in storage and collection of personal data as well as require trust in the data relayed to not discredit the system's perceived quality. However, this is a field that will likely involve a wide variety of devices, and this will need consideration when designing security mechanisms.

A field that will become a very large staple in the world of wireless communications networks is connected autonomous vehicles. The interconnectivity already being seen between cars and their manufacturers for updates and patching of software and monitoring will only increase with autonomous vehicles. This is a huge area for a potential attack. The author also mentions that these vehicles will also be susceptible to the same kind of physical attacks mentioned for UAVs. The networks utilized by these systems will be very heterogeneous and rely on multiple third-party providers like communication service providers (CSPs). National Institute of Standards and Technology (NIST) has already begun filling out the recommended areas on which the connected and automated vehicle (CAV) security framework should focus. These same considerations expand to the introduction of the next generation of smart grid technologies as well. Protecting supervisory control and data acquisition (SCADA) systems are connected through hundreds and thousands of IoT monitoring points that encompass a huge variety of devices and scenarios. Industrial applications are also an area of great potential as well as intelligent healthcare and the use of digital twins.

Blockchain has been in the spotlight for its potential and all of its applications. The advantages of blockchain such as disintermediation, immutability, nonrepudiation, proof of provenance, integrity, and pseudonymity are particularly important to enable different services in a trusted and secure manner in the 6G networks. This leads to a focus on distributed ledger technology (DLT) to offer some layer of trust to the developing and potential uses of 6G. This could be in the form of protecting the integrity of AI data for example. Blockchain is not without its potential problems, though. While not easily attainable, a 51% attack is something that has been of concern for even current usage scenarios, along with many other forms of attack we are already seeing on current iterations of blockchain.

Another developing technology that will greatly affect future security and implementation is quantum computing. While there is active development in quantum-resistant hardware and encryption, this vein of technology will likely have a very substantial impact on the level of security and trust a network is considered to have.

As with all technology, users' privacy must be heavily weighed and considered in 6G applications. As the amount of data and interaction continues to increase, so will the amount of data collected and stored in these systems. With connectivity up to 1000 times that of 5G, these new systems will have even more stringent requirements as well as performance minimums. Securely transmitting and receiving that private data in such large pieces will have its own challenges as again, authentication, ingesting, analyzing, and properly protecting that information will grow exponentially in most cases.

There are many promising and developing avenues to solve the 6G issues already being discussed by researchers – solutions that address the concerns and help to develop the security and adoption of 6G communication into systems and applications. As we are already moving into the implementation of 5G technologies and applications, 6G will be here very quickly and the need to identify and begin to consider how the security framework of these future networks will change, as it is inevitable that the advancements will bring new attacks.

8.2 The State of 5G Networks

5G networks have been in the works for years and the development stage is gradually coming toward its end. Once 5G networks are perfected, it will bring a very huge performance increase from the fourth generation. The emergence of these 5G networks is also going to require new network technology combined with other technologies for new applications and also realize the great vision of the Internet of Everything (IoE) [1]. The state of 5G networks in 2022 is actually complex. It has been growing as more countries gain access to it worldwide, but it is actually slowing down a little. So as connectivity increases for 5G, its performance worldwide seems to be diminishing; not many subscribers experiencing "true" 5G at all. There has even been a study on how much it is slowing down [2]. Over the past year from Q3 2020 to Q3 2021, the median global 5G download speed fell to 166.13 Mbps, down from 206.22 Mbps in Q3 2020. Even though it is growing and spreading, there are still many countries that are unable to use 5G networks. Out of all the countries, though, the US is the country with the highest 5G availability. South Korea and Norway have been in the lead of 5G speed with 530.83 and 513.08 Mbps; see Figure 8.1. In the US, the three major providers have some form of a low-band and high-band spectrum. Using these bands, providers are able to get much better throughput and data rates.

Looking at these facts and the problems that come with 5G technologies, such as low latency and high return bandwidths and slowing down over time, it is important to look at the solutions to these problems and the key technologies that will be needed for them. To optimally deploy fifth-generation communication networks where they need to be deployed, it is important that computers are used. Cloud networks and applications are provided by synchronizing network-side applications and deployment abilities to the RAN and UE (user equipment). With more computing power, deployment of 5G networks will be more flexible, and better meet the requirements of applications like low latency needed and the high return bandwidth. Once this is achieved, there will be easily accessible HD video, UHD video, and more. There are a lot of services that are high-bandwidth, and computing and caching, can save bandwidth and lessen delays, improving the quality of services (QoS) of the whole network. Services that are low-latency will allow network administrators to configure the network to allow third parties to flexibly and quickly deploy applications and services for mobile users, enterprises, and verticals Industries [3].

5G has been under work plenty of time, as it has been improving upon the 4G network architecture, and so the key technologies have been decided for it already. One major key technology that would be good for 5G would be the use of the Internet of Things. The internet of things is a collective network of many devices that gather data and use them for many different reasons. Another key technology for 5G networks is the radio millimeter waves. Using these the 5G networks are going to have a lot higher bandwidth and transmission speed compared to older LTE transmissions. There will also need to be a big change in mobile networks such that they can accommodate super-high traffic while also being extremely mobile. There are a few basic models for 5G and its development, and there are many areas with 5G networks already deployed but they are far from perfect. Figure 8.1 is a picture of one of the basic models.

In 2021, there was a decent amount of progress toward the deployment of 5G networks, and one big part of going to be global is standardization. The standardization of 5G networks, begun in 2016, cleared a lot of hurdles in 2021. Many groups worked on the standardization of the structure, elements, gateways, capabilities, configuration, etc. Overall 2021 was a great year for the deployment of 5G and the development of key technologies for it.

Figure 8.1 Speediest intelligence 5G data rate from Q3 2021 [1].

8.2.1 Applications and Services of 5G Technologies

5G technologies are no doubt more powerful than the fourth generation, and with this new technology and speeds will come new applications and services that couldn't be used in the previous-generation networks. 5G technology will power a wide range of future industries from retail to education, transportation to entertainment, and smart homes to healthcare [4].

These new applications and services will affect the quality of service for many people, and it will also impact the economy. There is a lot of expected revenue from these technologies and services. For example, the COVID-19 pandemic exposed the value of fast, reliable wireless communications. With offices closed, healthcare professionals overwhelmed, and schools on lockdown, virtual options were fast-tracked. A tiny company called Zoom saw its stock price soar, and other companies were quick to jump on board. Rather than drive to an office or fly to a conference hours away, people logged into virtual meetings. Doctors met patients online. Students received virtual instruction – and many districts showed huge gaps in quality learning environments, technological access, and family support. Now that the pandemic is over, there is no turning back. Employees have mostly returned to the office at least part of the week, but we have seen that things we thought had to be done in person can be done more effectively virtually – and also, that some things still require in-person connection.

One of the biggest and most relevant reveals was that if we are going virtual with multiple aspect of life, we will need ubiquitous high-speed mobile networks. 5G networks are already showing that they are very innovative and very fast. They can support 10–20 Gbps internet speeds, which is very similar to fiber optic internet connections which are amazing. Ultra-low latency comes with high-speed internet, which can be used in a lot of applications like self-driving cars and the like. With 5G networks, there can be latencies as low as less than a millisecond. 5G networks are supposed to be really reliable, and that means with 5G downloads will be faster, and uninterrupted. Another service that will be provided by 5G networks is secure access to cloud applications and powerful processing power. This will allow end users to do many more resource-heavy tasks at higher speeds, improving the means of production for many different tasks. Speaking of means of production, 5G networks will provide new opportunities for device manufacturers and software developers, which will in turn provide more job opportunities for people. 5G services are supposed to also include "better cell coverage, maximum data transfer, low power consumption, cloud access network, etc." [4].

Entertainment and multimedia are big factors for 5G networks. Most of the internet traffic is used to stream videos and entertainment. Since the fifth-generation networks are a good deal faster than the fourth generation they will allow very high-definition videos. It will be able to stream HD videos in 4k at high speed, which will improve the QoS. This will allow for a multitude of different operations from end users using their phones to watch HD videos to large corporations being able to stream high definition live events. High fps, resolution, and dynamic ranges on video streaming will be available with little to no interruption.

One huge aspect of entertainment when dealing with networks is augmented reality and virtual reality. They usually require resource-heavy HD videos and low latency, which is something 5G networks are able to provide with exceptional performance. Video games will also be more complex, as there will be networks that are able to hold all of their data and have high-speed internet for online play.

The internet of things (IoT) is another amazing service that 5G networks are able to enable, at least partly. The IoT is a huge network that would need a good powerful network like 5G to operate. It connects almost everything – every object, appliance, sensor, device, and application to the internet. All these sensors and devices being connected will collect crazy amounts of data and use it to analyze and process information in many different ways. Using the IoT, 5G networks can enable smart homes to be like never seen before with applications and devices connected and personalized to you. It also allows monitoring of all the devices you have connected. These smart devices in homes are very cool and remotely accessible, which can be good for entertainment, security, and just overall improving quality of life.

Another use of the IoT that 5G will enable would-be improving healthcare. It could support doctors and nurses in looking at medical information using a reliable wireless network. Things like remote surgeries would be possible from far away by the most knowledgeable doctors using devices. It would also be able to provide remote classrooms for people learning healthcare, they could watch procedures in HD and real-time. There is also real-time monitoring for patients with serious conditions like chronic diseases or heart problems. With this connection of everything, you could have a doctor whenever you need one. The healthcare industry as a whole will need to integrate smart devices and sensors, IoT, analytics, and HD video for this to work, which is all possible with 5G due to its cost efficiency and powerful internet speeds. There will be real-time alerts for patients or people who are injured.

There are still so many other uses – such as smart farming, smart cities, and security. Farming is a huge part of human society as a whole because everybody needs to eat. Improving it in any way possible is always a good thing for everyone. 5G technology can do just that by making it smart using sensors and GPS technologies. This makes it easier to track livestock. It also improves the ability to control irrigation, access to the farm, and energy consumption. Smart cities are cool, as there would be a whole city under a network instead of something small like a smart home. This can improve the quality of life for an entire city, which seems amazing. It has services like traffic management, instant weather update, local area broadcasting, energy management, smart power grid, smart streetlights, water resource management, crowd management, and emergency response. Security is important in every aspect of human life. IoT services provided by 5G networks will provide much greater security and surveillance around the globe due to its higher bandwidth.

With IoT being a hot topic in the tech world, there is an aspect of it that could become surprisingly useful if handled correctly and effectively. IoT forensics is the use of IoT to gather legal evidence in the event of a crime or investigation, and it is paramount that we gather this data legally and ethically. How the evidence is procured matters if it is to be considered legitimate in the eyes of the courts.

There are three levels of IoT forensics:

1) *Device level.* An investigator may need to find and use data from local memory.
2) *Network level.* An investigator may look at network logs to determine the guilt or lack of guilt of a defendant in an investigation. The network-level consists of many different forms of networks, but vital evidence can be pulled from any of them.
3) *Cloud level.* That which could be the most helpful due to its accessibility on demand and its vast capacity.

With all these benefits, challenges still arise to impede IoT forensics. Issues that affect IoT forensics include:

- Inability to pinpoint the origin of the data and where it is stored
- Use of traditional techniques in areas that are not applicable
- Ensuring secure chain of custody

Soon IoT will be present in the lives of all people but its massive presence brings issues related the diversity and complexity of IoT devices. Different companies using different standards and different offers to their customers on vastly different devices that are connecting to a plethora of different networks make the IoT overwhelmingly complex. These compatibility issues could leave devices vulnerable to manipulation and exploitation. In the effort to combat this complexity, files within the forensics perspective are proprietarily becoming more standardized with their file formats and often require reverse engineering efforts to access. There are also legal limitations during the investigation as to what data investigators can access and what they cannot. Another issue arises when looking at the storage capacity of IoT devices. Most of these devices carry a very low threshold for data storage and it becomes very apparent that evidence could be lost and have no way of recovery. Switching this storage to the cloud seems like an easy fix for this solution but with this switch, it becomes difficult to prove that the evidence was not altered or modified. This leads to the overall issue of chain of custody. This is a vital part of evidence integrity as the chain of custody provides the chronological history of the evidence during every stage of the investigation process. Providing an accurate chain of custody ensures the validity of the data and makes it less likely to be tossed out.

Another great application of 5G would be using it to make autonomous vehicles. This is something that everybody dreams about while driving. This is hard to do and complex because of all the variables that could happen during a drive. This is why AI and machine learning (ML) are being used to have cars learn different possibilities and how to react. This requires a crazy amount of new technology, network speed, data throughput and bandwidth, ML, and most of all reliability. If a car were to suddenly lose connection and it caused it to crash, it would be bad. This is why 5G networks would be good, as they would have good reliability and low latency. Fourth generation latency is much higher than 5G latency, which is why it would be better for self-driving cars as they would need to react fast. Overall, there are many uses for 5G networks, especially when you look at the IoT applications and services.

8.3 6G: Key Technologies

Even though 5G is good and all, the human race and its need for the internet keeps evolving and needs to keep adapting with that. 6G networks are already in the works and are being researched and will allow for even more powerful networks. Sixth-generation networks are fully expected to support broadband services, global connectivity, and great infrastructure, and promote the development of networking as a whole. It will need to improve mobility, networking, and intelligence in almost every aspect of human life and transform societies everywhere. Naturally, with all of these new crazy new network possibilities, a bunch of crazy new technologies will emerge. This includes virtual reality, augmented reality, mixed reality, smart cities, autonomous driving, ultra-smart healthcare, holographic communication, IoT, Internet of Everything, and even wireless brain to computer communication. All of these innovative technologies will require crazy network performance. This includes improved rate, delay, coverage, access, energy consumption, interaction, reliability, and degree of intelligence. At the same time, some new performance indicators will be demanded, such as context awareness, delay jitter, and the convergence of communication, sensing, control, and computing functionalities. 5G may not be able to meet all of these requirements so we need to go even further beyond 6G networks. 5G mainly uses sub-6 GHz bands so there might not

Table 8.1 Key performance indicators comparison between 6G and 5G communication systems [8].

	Key performance indicators	5G parameter values	6G parameter values	Ability to ascend
1	Maximum transmission rate	20/(DL)/ 10/(UL)	>1Tbit/s	100 times
2	Experience rate	0.1–1	10–100/	100 times
3	Flow density	10 Tb/s/km^2	10^2–10^4 Tb/s/km^2	1000 times
4	Positioning accuracy	1 m (indoor) 10 m (outdoor)	0.1 m (indoor) I m (outdoor)	10 times
5	End-to-end delay	1 ms	< 0.1 ms	10 times
6	Error rate	500 km/h	1000 km/h	10,000 times
7	Mobility	450MHz–6GHz(FR1)	95 GHz–3 THz	—
8	Spectrum bandwidth	FR1:100MHz FR2:400MHz	20 GHz	50–100 times
9	Spectrum efficiency	30–100bps/Hz	200–500 b/s/Hz	2–5 times
10	Connection density	10^6/Km^2	10^8/km^2	100 times
11	Network energy consumption	100 bits/J	200 b/J	2 times
12	Base station computing capacity	100–200T ops (operation per second)	1000T ops	5–10 times

be enough bandwidth to use all of the applications wanted. In the future, it is predicted that Tbps level rates will be required for new technologies such as holographic communication. In addition, 5G networks have nowhere near the coverage needed, as 6G networks are planning to have nigh-global coverage for everyone on the planet. 6G networks also want to incorporate AI /ML in almost every aspect of the network instead of just a few parts like 5G networks. Table 8.1 shows the differences between the key performance indicators of 5G and 6G and why 6G networks would be better.

With all of these crazy high expectations for 6G networks, there are obviously going to be some new technologies needed to meet them. 6G networks are going to need ultra-high rates, ultra-wide coverage, ultra-large connections, ultra-low latency, ultra-high reliability, ultra-precise positioning, and ultra-low energy consumption.

One of the major key innovations that will help on the road from 5G to 6G networks is the technologies of the sub-THz band. With all the new technologies being introduced with the thoughts of 6G networks including holographic communication, virtual reality, intelligent networks, and many more, they are going to need ultra-high-speed communication technologies to keep up. The predicted data rates for these services are supposed to be greater than 100 Gbps, which is huge. Current fourth-generation and fifth-generation technologies cannot keep up with this at all. This means that sixth-generation networks are going to have data rates that top out at least 100 Gbps all the way to 1 Tbps. This is not an easy goal to realize. Today's technology is nowhere near this; the data-rate processing capabilities of today's advanced digital baseband systems are limited to a few tens of gigabits per second and are power-consumption bonded [6]. There is going to need to be an increase in the frequency of the carrier, and that is where the sub-THz band comes into play. Specifically, the 200–300 GHz (0.2–3 THz) frequency band is what has drawn the interest of many researchers due to its ability to be amplified and low attenuation coefficient. It looks promising as in the 200–300 GHz frequency band; several transmission results with a data rate of ~100 Gbps have already been shown using diverse technologies.

With data rates in the Tbps using the sub-THz band frequencies, the sixth-generation technologies seem actually feasible. Generating the signal for the sub-THz band, however, is quite difficult but it can be generated using high-tech electronics and photonics-based approaches. Moreover, once it is achieved, it will have a myriad of uses. One problem with this is that it is still in the very early stages of research and development, and it seems to be short-range. There is a problem with free space path loss in the sub-THz band signals. There is still plenty of research to be done on this key technology and its uses.

Another key enabling technology for 6G wireless communication networks is VLC (visible light communication). VLC is another option like sub-terahertz frequency bands for fiberoptic level speeds of data rates. VLC relies on the visible light spectrum for wireless communication. If VLC were used in short-range communications instead of the widely used RF bands, it would allow for ultra-high bandwidth in the THz levels. It would also be completely free of any electronic interference because it uses visible light. Its spectrum is also abundant and mostly unused and unlicensed and would have a very high-frequency reuse rate, which is very nice. The current research and development on this technology are lackluster as it is limited to around 100 Mbps over a range as small as 5 m. There is new innovative micro-led technology that will drastically improve this though. Using these micro-LED diodes, there have already been data rates in the 10 Gbps range and very high bandwidths have been realized in a laboratory setting. Its potential is vast and promising for the future of 6G networks. By the time that 6G networks are realized it is assumed that VLC will break through the terabytes per second range of data rates.

Another important key enabling technology for 6G networks is satellite communication. This is a broad technology and it has to deal with the overall infrastructure of sixth-generation networks. Without satellite communication 6G networks may not be able to be realized. Not just satellite communication but also the combination of satellite and terrestrial communication and their ability to cooperate well. Throughout the history of wireless communication, satellite and terrestrial communication have not really been working together. They were in more of a competition for the first four generations. Now it is time for them to come together to provide coverage for the world. Satellite communication is complex and is mostly used with broadband communication. Over the last 20 years, it has become more affordable to create and research and development have steadily been increasing. Many different countries have networks of LEO satellites, which are low Earth orbit (LEO) satellites. These orbits are closer to earth and faster than other satellites. These can be used to provide coverage as they are not too far away. If we are able to integrate satellite and terrestrial communication it can help in a lot of ways. Terrestrial communication can be used to provide coverage to people near their hubs and satellites will be able to reach even further beyond. If one of them goes out there can be another to replace it to provide reliability in the connection that many different innovative technologies will require. However, LEOs alone will not be good enough to do this. There are going to need to be high and middle orbit satellites along with the LEO and they must all work in harmony with each other. Once there are high throughput satellites complementing the coverage of the already good terrestrial hubs the coverage will be like never seen before. The integration of these satellites must be compatible with 5G networks and be able to integrate into the sixth generation. This is one key step to going above and beyond 5G networks, as most LEOs used in the fifth generation are used for television broadcasting. Currently, the huge networks of satellites like Elon Musk's Starlink use DVB-S2X. Starlink seems to be successful, as it is able to provide internet coverage to many places across the globe, and it has even been providing internet to places like Ukraine, which is at war, and across the globe from America.

One problem with 5G and satellite communication is the fact that it is just not very compatible. 5G is superior to DVB-S2X in many ways such as uplink synchronization, hybrid automatic repeat request (HARQ), control channel, discontinuous reception (DRX), radio resource management (RRM) measurement, mobility management, and core network, also the spectrum efficiency is lower [7]. It is just hard to implement all of the standards of 5G onto satellites, which is where 6G comes in, and a bunch of research and development is necessary. One problem with the integration of 5G with satellite communication is channel propagation. There is a problem with satellite communication as aspects like weather and other atmospheric conditions easily interfere with them. LEOs can also travel at considerably fast speeds, which can interfere with signals. It comes with a multitude of problems including time synchronization and high amounts of Doppler shift. The Doppler shift is when something is moving at fast enough speeds to significantly affect the frequency of the signal that it is emitting. Since satellites are so high in the sky, their signals also have a large distance to travel, resulting in huge amounts of path loss for the signals. However, with directional beams from satellites, a solution may be provided for the high path loss. This is called beamforming that we will talk about later as a key technology. The cell radius of satellites is also many times larger than that of terrestrial communication systems, which can be bad for end-users. All of these challenges have their own key technologies for the bigger key technology, which is Terrestrial-Satellite integration to make 6G networks possible.

One enabling technology for this is wave forming. Wave forming is used by leveraging the natural multipath propagation of electromagnetic waves. It is proposed as a promising paradigm that treats each multipath component in a wireless channel as a virtual antenna to exploit the spatial diversity. With this technology, it will be an improvement on current beamforming techniques. There is also going to need to be an improvement upon current coding and modulation if terrestrial and satellite are going to merge. One of the major problems with the satellites is their speed and Doppler shift. There is going to need to be compensation for this, especially with LEOs as for altitude 600 km, Doppler shift arrives ±720 KHz at carrier 30 GHz, and the maximum Doppler shift variation arrives at 8.16 KHz/s [7]. However, this key technology is achievable by predicting the effect of the Doppler shift before it happens and adjusting accordingly.

Going beyond 5G with satellite and terrestrial integration is where it starts to get interesting. Using this will provide coverage to all of the Earth – sky, ocean, and even parts of space. With LEOs, GEOs, and MEOs all working in perfect harmony this is achievable. The overall idea behind 6G is to provide seamless communication with anything anytime anywhere. 6G is supposed to be a universal mobile communication system that will provide the compatible standards needed for the interaction of satellites with ease. This includes terminal, frequency, and network architecture, platform, and resource management integrations as well. However, similar to the deal with 5G networks trying to implement this, there are still some technical problems when dealing with integrating the satellites with terrestrial systems for 6G. Since the network is going to be three-dimensional when you integrate terrestrial and satellite communications, it is going to need to be managed well. Network coordination and management are very important in this aspect, and it is going to have to be multilayered. This means it needs to cover management for each layer, from the highest satellites to the lowest terrestrial and end users. Satellites also have a limited load capability, which will be difficult to get around. This means that all network nodes cannot possibly fit on a single satellite; there is going to have to be a balance of the capability of the satellite and the quality of its service, as the more it holds, the worse it gets.

The communication between satellites is also an issue, especially with the routing aspect of it. If they end up being an end-user who is unable to access a certain terrestrial station, then their message might have to hop between many different satellites, which could be a problem. This means there needs to be research into superior routing for links between satellites.

Just the satellites alone are not enough. There are also going to be high altitude platform systems (HAPS), aircraft, and drones that act as stations and terminals for the network. See Figure 8.2 for an amazing depiction of 6G network architecture using all of the satellites, HAPS, and other technologies.

Another key enabling technology for 6G networks would be good distribution security mechanisms. The vision for 6G networks is to have everything connected and with that comes many personal data on the same network. This poses major concerns for the security, privacy, and overall trustworthiness of the network. There are going to have to be creative and innovative cryptographic techniques to accommodate the vast amount of vulnerable data that will be on the network. It would be completely irresponsible to allow the network to be put up without any preventative measures in place. There is also the use of blockchain technology to help with security for 6G networks. Blockchain is a DLT where cryptography and hash functions are used to form a chain of data blocks, created when an event occurs [9] and verified in a decentralized way using consensus algorithms. Blockchain technologies were originally made for cryptocurrencies to provide safe and secure transactions. Using it in 6G networks will no doubt prove useful and allow for the safe transmission of data. Blockchain

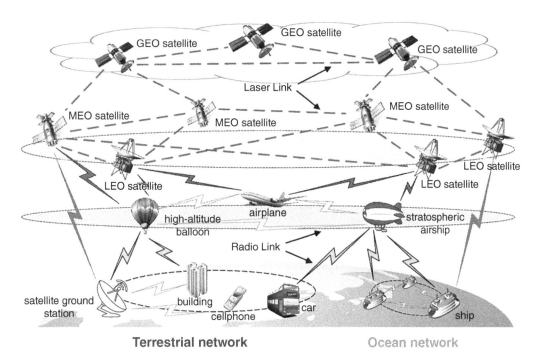

Figure 8.2 The architecture of the aero-space-ground-ocean integrated information network [8].

technology would be able to provide integrity, auditability, and security. Blockchain allows asymmetric PKI-based cryptography and the inclusion of privacy preservation frameworks for greater data privacy & confidentiality [9].

Another key enabling technology for 6G would be orbital angular momentum (OAM) technology, which has been researched recently and looks promising in 6G networks. It would improve spectrum efficiency by quite a bit compared to previous technologies and generations. The main advantage of OAM is the ability for multiple channels to be able to significantly increase their transmission capacity and spectrum efficiency due to the characteristics of its waves. Due to this innovative method of increasing these aspects, it seems like a good choice for 6G networks. Recently, the usage of higher frequency wavebands like the sub-THz band has been thought of as being used with OAM. Recent experiments on this topic have proved the possibility of using OAM for wireless communication.

The next key enabling technology mentioned is without a doubt one of the most important, if not the most important technology for 6G networks; Pervasive artificial intelligence throughout the entire network. It is so important for the road to 6G. Fourth-generation networks had no AI in them at all, while fifth-generation networks had partial inclusion. Going beyond the fifth generation to the sixth is going to require full integration of AI. Artificial intelligence in 6G will be used to automate almost every part of it. This automation will simplify many aspects including the transmission of data and will speed up processes significantly. With the analytics of the vast amounts of data coming through the network, it will be able to automate many network tasks like configurations. It would save enormous time by automating a bunch of tedious tasks. There is going to be AI in almost everything from devices, configuration, and details to communication.

8.4 6G: Application and Services

Now that we have discussed the key enabling technologies for 6G technologies, we can finally get into what we are going to be able to do with them! There are tons of futuristic applications and services that have been thought of those 6G networks may be able to offer.

Figure 8.3 shows some of the applications and services that can be provided with 6G technologies. It also shows a model of the different aspects of the network itself. One of the most researched and desired autonomous devices that 6G will be able to provide is self-driving cars. 6G networks will bring autonomous vehicles near perfection. Drone delivery UAV systems are also feasible with 6G technologies, and they will save a lot of time transporting packages and would boost the economy. Both of these technologies use sensors and GPS technologies within the current networks, but they are just not good enough. 6G networks will change that. Smart UAVs also have a variety of different uses including military, commerce, science, agriculture, recreation, law and order, product delivery, surveillance, aerial photography, disaster management, and drone racing [11].

The next application/service is interesting. 6G networks should be able to provide wireless computer-brain communication and interaction. This sounds straight out of a sci-fi film. There is an interface called the brain-computer interface that is supposed to allow control of smart appliances. How it works is having a direct communication path between the brain and the smart devices by actually getting signals coming from the brain. These signals are then transmitted to devices where they are analyzed and interpreted into commands. With all the features and crazy improvements from 5G to 6G, this should be possible.

Another possible service coming from 6G networks would be haptic communication, allowing communication using the sense of touch. This will allow end users to feel different textures and temperatures, such as feeling the texture of a certain surface through the screen if they wanted to. This is all possible due to the high amounts of data being able to go through 6G networks and the new technology that will advance with it.

A huge application of 6G services would be smart healthcare. Healthcare being smart would have a huge impact on the world and 6G networks would be able to provide it. It would allow remote surgeries from anywhere in the world. This would decrease the cost of healthcare as it becomes more accessible and would allow more people to get the help they need, especially in remote areas. There would be a doctor at your fingertips anytime you need one due to the advanced network. Network stability would be essential because if the signal were to cut out during something like surgery, it could be fatal. This is why 6G networks are going to have to be ultrareliable and have low latency.

Another exciting service that could be provided by 6G networks would be five-sense information transferring. Humans experience the world using only five senses: touch, taste, sight, smell, and hearing. 6G networks would allow for complete immersion of the five senses. The BCI technology mentioned earlier will definitely help this technology thrive.

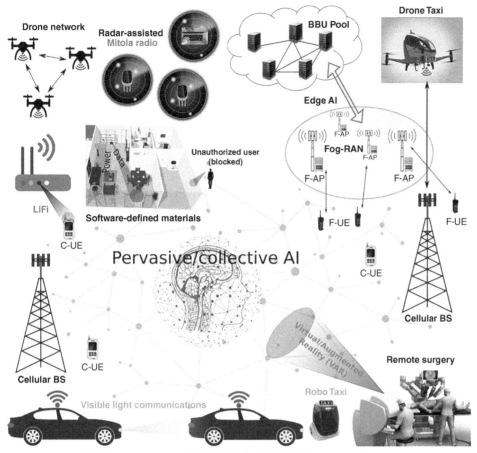

Figure 8.3 Possible 6G communication architecture scenario [11].

Another interesting application coming from the 6G networks would be the Internet of Everything (IoE). It is an upgraded version of the IoT. IoE is the seamless integration and autonomous coordination among a very large number of computing elements and sensors, objects or devices, people, processes, and data using the internet infrastructure. Under 6G networks and the IoE, everything will be bound under one network and be connected.

8.5 Benefits of 6G over 5G: A Comparison

6G networks blow the fifth generation out of the water. 5G networks stand no chance in almost every single way. 6G will be 50 times faster than 5G, 100 times more reliable, offer wider coverage, and support 10 times more devices per square kilometer [12]. Once 6G networks are fully realized, then 5G will most likely become obsolete. 6G will provide better opportunities for services and applications, too.

8.5.1 Artificial Intelligence in 5G and 6G: Benefits and Challenges

6G AI is supposed to be fully integrated, and this will offer multiple benefits:

- It will save time by handling tedious tasks like configuration.
- It will save money because it will automate many tasks that cost money to complete.
- It will provide faster network speeds as artificial intelligence learns to work at inhuman speeds.
- It will provide security to the network as it will be able to detect threats and handle them with great speed.

AI will play a crucial role in the new network. To address the security demands of 6G networks, SDN (software-defined network) and NFV (network function virtualization) should be complemented with AI techniques for proactive threats discovery, intelligent mitigation techniques, and self-sustaining networks.

6G envisions IoE, but its security is based on SIM cards, and this makes resource-constrained IoT devices vulnerable to attacks. Local 6G networks are smaller than 5G networks and interconnect with wide-area connectivity when needed. 6G cell sizes will shrink from small cells to "tiny cells," a greater number of vulnerable devices will be deployed, a new security strategy is required, and security provisions for the massive number of devices within each sub-network by the wide-area network are far from practical. 6G networks will rely on AI to enable fully autonomous networks; however, attacks on AI systems, especially ML systems, will affect 6G. Privacy issues are related to attacks on the data stored on blockchains and to the use of VLC. AI's role in 6G systems involves providing security for systems, architectures, technologies, and privacy.

Multilayered intrusion detection and prevention are effective in SDN/NFV-enabled networks, but rule-based detection systems are ineffective against attacks that evolve using AI techniques. Anomaly-based intrusion detection systems in Industrial IoT (IIoT) can identify malicious packets based on their behavior. These systems are more effective than conventional intrusion detection systems for security in large-scale IoT. 6G sub-networks can be considered an extension of local 5G networks and can use ML-based security techniques to prevent malicious traffic. To maintain network security, 6G will rely on edge intelligence compared to present centralized cloud-based AI systems. AI at the tiny cell level can prevent denial of service (DoS) attacks on cloud servers and at the wide-area network, the level can use behavior-based approaches to detect intrusions. Using AI in network security, we can predict attacks such as 51% attacks on blockchain before they occur. We can also use quantum computers to perform tasks much faster. In the 6G network, multiple base stations allow simultaneous communication for devices. Edge-based machine learning models could be used to detect privacy-preserving routes, rank them, and allow devices to transfer data via privacy-preserving routes. AI has privacy, security, and ethical issues, and can be used to launch intelligent attacks. Issues arising from the use of AI in 6G are poisoning attacks, evasion attacks, and API-based attacks. Potential countermeasures include adversarial machine learning, moving target defense, input validation, differential privacy, homomorphic encryption, and adversarial training. Solutions for privacy preservation include edge-based federated learning, homomorphic encryption, and differential privacy techniques. AI-based 6G systems need less human intervention, are less ethical, and learn differently than humans. AI can uncover network vulnerabilities, convert vulnerable IoT devices into bots, and launch DDoS attacks against critical nodes. Using AI to empower defense systems, distributed intelligence can be used to weaken the learning process of the attackers.

There are challenges that come with integrating the network with AI. This means that there needs to be an AI smart enough to control the network at high speeds without failing. It also means that a lot of time and money is going into creating such an AI. It won't just be configuring the network, either – there are going to be many AI controlling smart devices, smart cities, smart societies, etc. This is going to be very costly and some of it may not be compatible, so there is going to have to be standardization with it.

8.5.2 Artificial Intelligence and Cybersecurity

Artificial intelligence in the new generations of networks is crucial for cybersecurity. If AI is able to learn and keep up to speed with malware that is being produced and all the different kinds of attack vectors it will protect better than we have ever had it before. Using AI with blockchain technology there will be insane security and cryptography for data. In addition, AI will be able to detect threats as they are happening and immediately start taking action to solve them. In addition, the AI runs 24/7, unlike humans so there will be protection around the clock. This goes both ways though. There is no doubt that malicious people will use AI to their advantage and attempt to use it to hack systems and steal data. This means there is going to be a completely new kind of attack to look out for.

8.5.3 Benefits and Challenges of AI and 6G for Cybersecurity as Defense and Offense

There are certainly benefits to AI being used in cybersecurity. This includes crazy amounts of protection that we have not seen before. Around-the-clock protection from AI is certainly going to be nice as it will respond to threats and detect malicious files. It has to be reliable as there is certainly going to be a lot of sensitive data on the network. Another problem is that AI can also be used for offense as stated earlier. Learning the offensive techniques with AI in cybersecurity is definitely going to be important so they can be neutralized.

Overall, the road from 5G to 6G is not going to be an easy one. It is, however, going to be a fulfilling one that is worth all of the trouble. We have gone over the optimal deployment and key technologies of current 5G networks; including the applications and services they provide. We have also discussed the key technologies of 6G wireless networks and their services and how high the standards are going to have to be for 6G networks. The comparison of 6G networks to 5G is like night and day because of how powerful 6G networks are. AI will play an important part in making 6G networks secure and reliable.

8.6 6G: Integration and Roadmap

Although 6G is still yet to become a new item on the shelf, the speculation as to what this new and improved generation will look like is certainly riveting. 5G is barely up and running, but the idea of 6G could not be more fascinating. 5G spans vastly across the world with plans of being launched across most of the Asian Pacific and European countries. With all the locations 5G will be available in, we can only imagine what the future will appear to be with 6G.

The architecture and design of 5G systems will continue to evolve before 6G is set to be released. 5G has technically released different versions, such as 15th, 16th, and 17th, which helped the vision that is what 5G was meant to be, and what 6G is expected to exceed. The 16th and 17th versions of 5G were meant to be fully what their vision was meant to become, but the 18th version, which exceeded all expectations and set the expectation for 6G.

Security and privacy are major concerns for the foundation of 6G. Security ensures trust for users while maintaining confidentiality, integrity, and availability of data at the same time. Creating an environment where users feel safe to use devices that work properly is the most important thing. There are four primary technology domains that will encompass the core of security features which include: Physical layer security, blockchain/DLT for security, quantum security, and distributed AI/ML security. Each of these domains has its own benefits, but they also possess challenges to their creation and use. For example, quantum security would provide "unbreakable quantum-safe security," but there is a lack of processing power in current networking devices. These challenges must be addressed and solved before the rest of the system can develop. Most of these security domains possess the common challenge of standardization and its need to be addressed before the true introduction of 6G. With revolutionary technology like 6G changing the way we communicate and interact, there will be threat actors wanting to exploit and hack into these new systems. Eavesdropping, poisoning attacks, privacy attacks, etc. are all common attacks today that will have to be planned for and addressed in the future in order to limit their success as much as possible.

8.7 Key Words in Safeguarding 6G

When considering roles of trust, security, and even privacy, they appear to be all intertwined but have different determining factors that depend on the generations. There are three key areas when discussing 6G trustworthiness, those being:

8.7.1 Trust

Any sixth-generation network must be able to support embedded trust for an increased level of information security. Trust modeling, trust policies, and trust mechanisms are needed for 6G to function properly. 6G is able to interlink the physical and digital worlds in order to make it safer to depend on information security.

8.7.2 Security

For 6G, IT and other networks that are considered to be in national security keep rising – this means that it is a continuation of what had occurred with the fifth generation. For the development between cloud and edge, the physical infrastructure must also grow. The physical layer has techniques that allow effective solutions to become present when securing fewer network segments in the first line of defense.

8.7.3 Privacy

Privacy becomes an issue when the consumer is no longer sure what is safe and what is not. There is no way to know when any linked or de-identifiable data is able to cross a threshold so that it can become personally identifiable information. Unfortunately, this problem becomes more of an unaddressed problem for many consumers of digital technologies. There is an alternative solution, which is creating a blockchain – a distributed ledger of technologies and different policy approaches.

8.8 Trustworthiness in 6G

When thinking about trust in networks, some may think that this just means the security aspect, which is not entirely wrong. Trustworthiness of a network is said to be more about expecting the unexpected outcomes of any communication with remote parties in any session. It addresses all possible outcomes that could affect the value of the communication. Trust spans all the protocol layers from the IP layer to the applications layer.

Any system of trust in a network must consider these main questions [13]:

- Can this interaction lead to a loss of data?
- Should flows from remote networks be served under heavy loads?
- Would it be best to drop flows and devote any resources to other flows?
- Is it possible that a source address in a packet is spoofed?
- When communicating with a remote party, how does the host minimize exposure to any possible future attacks or any long-term privacy loss?
- How do the parties protect their communicated data from any leakage?

8.8.1 Is Trust Networking Needed?

In retrospect, 6G may be able to be used for building a wide physical/digital world by understanding and programming it. Without 6G trustworthiness, data loss, loss of control of devices, breach of information, loss of money, and many more things could occur. Kantola [13] states, "At worst, during international conflicts, foreign cyberwar troops could cause havoc in a country on a level that using traditional warfare will not be needed to pressure the victim to accept the terms and conditions issued by the attacker."

National security concerns will always be an issue in mobile technology, regardless of whether it is 5G or 6G. In order to address these issues head-on, 6G networks have to be able to aid in supporting any embedded trust so the resulting level of information security in 6G networks can allow for connectivity.

8.8.2 Benefits of Trust Networking for 6G

Trust networking may be applied to specialized 6G networks in either local or nationwide usage. With using trust networks, they can be applied to packet data networks, or PDN for short, which allows them to provide remote access to specialized networks or any other critical infrastructure [14]. A simple, single administration use case can handle the whole network with all the devices – meaning all the devices are owned by independent parties and different segments. Thus, the trusted network, in theory, must be able to handle multi-admin relations. When adding a trusted network, they provide multiple trust domains where sharing the trust-related information will be set up within the domain.

8.8.3 Constraints of Trust Networking in 6G

To apply trust networks to an IP network or a mobile network, change is required. For these changes, it must be possible for the network to deploy those changes one network at a time. These changes, if needed, are to be really limited to a single administration. If a multi-stakeholder trust network is needed, no changes are allowed to be made in the host for the ease of migration.

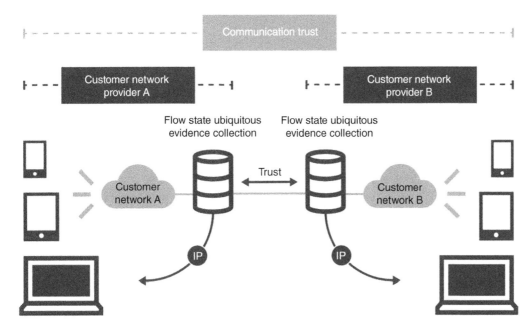

Figure 8.4 Model of ID/locator split for trust networking [14].

8.8.4 Principles for Trust Networking

When implementing trust networks for 6G, nodes act as a supporting item that supports trust networking to collect evidence of behavior for all of the seen remote network entities. This collected evidence can be used to produce a reputation for all the remote entities. The "reputation is used to make certain trust decisions, that being, asking for more information, admitting a flow, refusing to communicate, allocating resources for an incoming flow that depends on the specified load situation" [13].

In Figure 8.4, communication trust is depicted. The customer network "A" or provider "A's" networking is showing how the devices use private addresses; meanwhile, any edge node may be able to translate between the private addresses and the global unique routing locators. Policy management can be used to tailor any generic trust engine for either a use case or end user and admin needs.

Figure 8.5 discusses the first aspect, which is to control trust-related information sharing. Reputations are able to restore any state of aging by using wide-area trust management to implement a centralized system.

8.8.5 Challenges in Trust Networking for 6G

When implementing new trust networks, there will always be challenges:

- Verification of a trusted network in multiple use cases for various needs
- Trust management for wide areas across multiple different trust domains

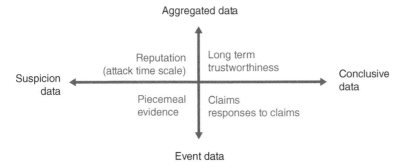

Figure 8.5 Trust-related data for reputation generation [14].

- A specific deep learning (DL) for 6G mobile networks
- Extending trust networking to span the whole end-to-end sessions

These are a few challenges that tend to arise when implementing the idea of trust networking in 6G.

8.9 Network Security Architecture for 6G

5G already spans vastly across the world, and with 6G, the launch of 6G is expected to exceed that. The first release of 5G (release 15) predominantly addressed the immediate need for enhancing the mobile broadband experience, whereas the release of the 16th and 17th versions pulled 5G toward the full vision. It balanced the needs of the mobile broadband operators with market expansion. Even better, the 18th version release went above and beyond to focus on the definition of new uses.

Beyond these releases, the 16th version supports a full 5G system resilience with new security features for service-based interfaces (SBI), transport later security (TLS), and token-based authority. For support to extend to the network automation phase two requires wireless backhauling using a radio interface with large bandwidths. For the release of versions 16 and 17, they can produce an expansion of the ecosystem to take advantage of the fifth generation. By this, it means that it will add many features to provide a full range of many functionalities that are required by the new industry segments.

Machine-type communication (MTC) is becoming one of the main players in supporting multiple technological industries. With 5G, MTC is being designed to support any digital landscape. This happens by transforming the digital society and by improving the efficiency of consumer and industrial demands. Newer models of 6G can become more researched, which will provide a more efficient method network.

8.9.1 Privacy and Security in IoT for 6G

The security phase has included many advancements that are meant to really shift the dynamic. "Big Data" is being marketed to have the potential of finding a solution that can accustom the "alphabet soup of the IT industry, IoT."

People, as of today, still operate on very old versions of the operating systems, which could lead to many vulnerabilities in their system where hackers can access their information. The big idea is that associations should be able to fuse security with their contractions toward the start instead of at the end. Fusing the security at the start could provide fewer vulnerabilities.

The Federal Trade Commission staff urges Congress to consider more protection and security enactment so it can guarantee that IoT organizations keep considering security and protection as the main issue and understand the importance to keep growing the technology that comes with IoT. The need for classified information is still very sought after and must be encoded on IoT applications and cloud administrations to avoid information breaks for the safety and security of others' information. This decision should be based on whether the expenses are essential for applications planned.

With IoT spreading its use and influence to almost every human being, the importance is to ensure that all of the devices in the network are safely guided with key elements to increase the quality of the devices. Those key elements are security and privacy that is meant to determine the effectiveness of the IoT devices.

Security is a key element because consumers have explained their concerns about how big packets of data are being handled. Most devices are sold to customers with unpatched software and operating systems, which could lead to an overlooked mistake where customers forget to change the original password after purchasing their device. Privacy is also a key element because data transmission from more than one endpoint when collected and analyzed can cause a rise in sensitive information.

In July of 2018, the International Telecommunications Union (ITU) Focus Group Technologies Network 2030 was established to explore the possibilities for system technologies for the year 2030 and beyond. This group was set in place to discuss the potential for further advancements in the world of technology and 6G just happened to be one of their concepts. As a part of their program, the 6G-Enabled Wireless Smart Society and Ecosystem were founded in 2018. The focus of this program was so they could be able to study the focus of wireless technology and the standard development of 6G communication between devices. The research areas span from reliable, near instant, unlimited wireless connectivity; distributed computing and AI; and lastly, materials and antennas for future circuits and devices. With these organizations, the furtherance of 6G and more generations to come will be closely observed for the betterment of the technological advances, as well as for the consumers.

The architect behind 6G was to have a vision of the main goal to meet the demands of what is now known as the society 10 years from now, which would significantly go beyond 5G capabilities. The main categories for the 6G vision are intelligent connectivity, deep connectivity, holographic connectivity, and ubiquitous connectivity. When saying "intelligence connectivity," this refers to the intelligence of communication systems. The connectivity is said to have to meet the following two requirements simultaneously: each of the related connected devices in a network is considered "intelligent" as well as their services, and the complex network will always need more intelligent management.

8.10   6G Wireless Systems

There are three very important characteristics that are linked to the next decade of lifestyle changes. They can be defined as a high-fidelity holographic society, connectivity for all things, and time-sensitive/time-engineered applications. High-fidelity holographic society is increasingly becoming the main communication choice and is vastly moving to augmented reality, or AR for short. It has been predicted that holograms and multisensory communications are meant to be the next thing for the virtual mode of communication. Stephen Hawking gave a lecture in 2017, via hologram in Hong Kong, which showcased the ever-growing potential for such a thing.

The system requirements for holographic communication include data rates, latency, synchronization, security, resilience, and computation. All of these requirements work together in order to ensure that the images and sound flow smoothly, without gaps and without compromising privacy.

Connectivity for all essentially means that it will include infrastructure essential for the smooth functioning of society. It will bring the necessary things to operate multiple network types, going beyond the standard terrestrial networks of today. Communication for moving platforms is required for many new applications that are emerging, such as unmanned aerial vehicle-based systems.

Lastly, time-sensitive/time-engineered applications must consider the timeliness of information delivery. This means it is considered to be a vastly interconnected society for the future.

8.10.1   Advances

The expected track of 6G is to connect humans with machines within 10–20 years. Current XR and VR/AR devices perform well but do not satiate the customers' expectations of a fully immersive experience. Integrating the technology with human bodies in a way that meets expectations would require 10- to 100-fold improvement to make sure the "infrastructure" can handle the environment.

Multiple research firms and companies have suggested that 6G will make it possible to merge implants, smart devices, embedded devices, and other forms of interconnected devices with human bodies. The aggregation of devices connected with 6G will create a more intuitive and interconnected human-to-machine mindset. Again, this is a purely hypothetical but a direction that the technology may be on track to fulfill.

The most important driving factors for sixth-generation networks that have been discussed by the ITU-T are high-fidelity holographic society, connectivity for everything, and time-sensitive or time-engineered functions. High-fidelity holographic society is implementing or integrating video communications and augmented reality by increasing the speed required for these devices to connect.

Connectivity for everything means integrating infrastructure within the network to provide connectivity to all sorts of devices. For this to work, these devices will need to be integrated within the sixth-generation network and must be within range of the wireless access to effectively operate on the network. The sixth-generation network will work to provide a more efficient and uninterrupted connection for these types of devices to continue being connected.

Time-sensitive and engineered functions operate under the premise that sixth-generation networks will provide a much faster infrastructure for these types of applications to provide a smooth and seamless process in getting users connected to their requested applications. As the phasing of fourth-generation/LTE networks begins to be replaced or upgraded to fifth-generation, sixth-generation networks provide multiple updates to their predecessor. In the case of the sixth generation, the present core networks and legacy network architecture will need to be reduced to comply with the sixth-generation case. Specifically, focusing on the TCP/IP framework, under the sixth-generation network may not provide the necessary QoS needed to provide applications with the best service.

The 6G network will seek to provide native open-source support and AI native design for its core network and its radio access network. The responses to the COVID-19 pandemic led many to come to the realization that corporate and private networks lack the high-speed infrastructure and security needed to make virtual interaction effective. Sixth-generation networks seek to provide a better solution due to the sheer number of devices that will be integrated within its network.

Sixth-generation types of networks will seek to integrate AI/ML within their applications that go beyond the current state of the 5G network. Sixth-generation networks will implement global connectivity by integrating satellite technologies with the network to provide its device base with better coverage.

8.10.2 Physical Layer Security as a Means of Confidentiality

When moving on from 5G to 6G technology, the sixth generation is meant to enhance the key performance that the fifth generation once provided. It will enable more demanding applications that range from augmented reality and holographic projection that was mentioned previously, to even ultra-sensitive applications.

The main focus of integrating ML in 6G networks is on wireless communications. Sixth-generation wireless communication networks are considered to be the backbone of digital transformation because they will provide a sense of ubiquitous, reliable, and near-instant wireless connectivity for humans and machines (Figure 8.6).

ML techniques rely on a central server, sort of like the central nervous system in the body. This tends to leave them vulnerable to critical security challenges, such as a single point of failure. Sometimes, traditional centralized machine learning techniques might not be suitable for the sixth generation, which is where federated learning can assist. Federated learning acts as a private, distributed structure.

There is a specific procedure that is needed to be followed to implement federated learning in a 6G world. The procedure of federated learning-based architecture is divided into three phases:

1) *Initialization happens in phase one.* The devices evaluate service requests and will be able to decide whether to register with the nearest cloud for the training that is required. This cloud-based training will be able to send pretrained or initialized global models to all of the selected devices.
2) *Next is the training phase.* This phase tends to deal with the selecting of each device and trains a global model using a local dataset. The meaning of this is to obtain the updated global model for each iteration.
3) *In the aggregation phase, the cloud receives updates on the model of devices for aggregation.* Afterward, a new model is obtained for the next iteration.

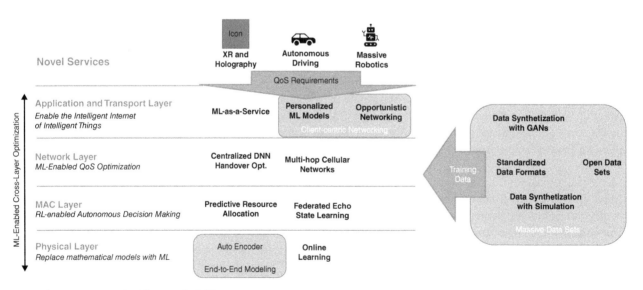

Figure 8.6 Role of ML in 6G networks [15].

8.10.3 Challenges of Implementing Federated Learning

Advancements always bring challenges. A few of these challenges that come with federated learning are the following:

- *Expensive communication:* Federated learning, unfortunately, involved hundreds of thousands of devices that participate in the model training. This, in turn, causes communication to become a critical bottleneck.
- *Security and privacy:* Devices have a number of capabilities in the network that vary with hardware, network connectivity, and energy. These issues can lead to flaws in the federated learning model and in the 6G network.
- *Federated learning protects the privacy of devices.* This even includes some procedures such as model updates, leading to raw data being disclosed.

8.10.4 Physical Layer Security for Six-Generation Connectivity

The sixth generation will take a holistic approach to implementing applications, so it will have to cope with a plethora of different systems and platforms. You can relate this to the way that the body works, with the large amount of data being collected by the network of sensors, that being environmental, and the mobility that takes into account the device capabilities, network environment, and the dynamic topology (Figure 8.7).

The physical layer features, combined with the advancements in artificial intelligence algorithms, can be exploited to:

> Enhance the classical cryptographic techniques or to meet the security requirements when dealing with simple but sensitive devices which are unable to implement cryptographic methods, e.g. devices and Nano-devices of the internet of things and Bio-Nano-things where the human inner bodies become nodes of the future internet [14].

The physical layer security addresses the most important application of the sixth generation: that being considered a "human-centric mobile communication." When increasing interest in scientific research in the wireless body area networks, this tends to include biochemical communications. According to Kantola, the "human body will be part of the network architectures, it will be seen as a node of the network or a set of nodes (wearable devices, implantable sensors, Nano-devices, etc.) that collect sensitive information to be exchanged for multiple purposes" [13].

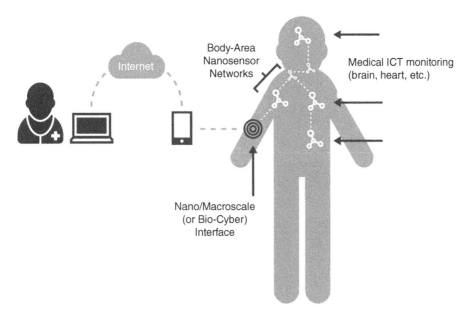

Figure 8.7 Human body as part of the global network [14].

8.10.5 Physical Layer Security Using Light Communications

The physical (PHY) layer security can play a key role in reducing latency and the complexity of security standards. The dramatic increase in the high data rate services can continue to meet the demands of the six-generation networks though.

When looking at Figure 8.8, the physical layer security (PLS) is approached in a rather conventional encryption technique that can provide the first line of defense against almost all attacks. The main idea behind this method is to utilize intrinsic properties from a VLC channel to access and enhance security.

In considering the evolution of 6G wireless communications, you must address unresolved challenges for PLS in the VLC research. This is known to include coding, jamming, mapping, subset selection, and much more (Figure 8.9).

8.10.6 Challenges for Physical Layer Security

When researching the vast topic of what is the physical layer security, there are certain questions that appear to come up. These challenges try and hinder the ability of the physical layer security to work and function properly. When asking these questions, it's best to try to think not only logically, but functionally.

1) How can artificial intelligence be exploited for tuning the physical layer security algorithms?
2) Which are the most suitable physical layer features that can be exploited?

These questions, just to name a few, can be asked to determine whether or not the PLS is working properly and effectively. These challenges or questions need to be answered in order to ensure everything is up to speed and working properly.

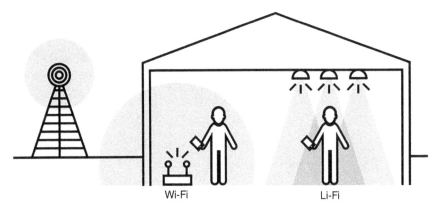

Figure 8.8 Visible light communications vs radio frequency communications [14].

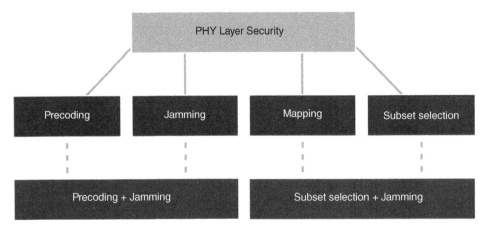

Figure 8.9 Layout of the PHY layer security for VLC systems [14].

8.10.7 Privacy Requirements for 6G

Theories of the sixth generation are expected to move technology in a direction that previous generations have yet to go to. These thoughts may have been able to create the basis for all future smart devices. This can turn into sharing large volumes of data of personal data between both individuals and organizations, or even individuals to governments.

Within a de-identified data set, certain insights about your identity, relationships, habits, and preferences can become a target. This data may be used by smart lighting providers so they can deliver more of an efficient lighting service by using a smart grid to match the energy demand.

There are certain benefits, such as "the ability to create locally optimized or individually personalized services based on personal preference, as well as an understanding of the wider network of users and providers" [16]. The wide scope of anything considered "smart" can cover a variety of instruments used in our personal lives. Some examples of smart instruments used in daily lives include:

- Smart TV
- Smart scales
- Smart phone
- Virtual assistant
- Smart grid
- Smart factory
- Smart government

These examples can benefit in terms of an improved amount of efficiency, effectiveness, and an increased amount of personalization to individual needs. It's said that in many economies, service providers and operations may be forced to think about the difference in governance paradigms so they can be supported.

With the flow of data, there tends to be a challenge in quantifying what the de-identified data is set to mean. When considering this, the developmental measures for any level of personal information are crucial to developing a threshold test for individuals who are said to be "reasonably identifiable." This opens a conversation about any trade-off between managing privacy and building the required trust at the same time.

8.10.8 Is Personal Information Really Personal?

Personal information, or PI, and personally identifiable information (PII), are often used interchangeably when discussing information frameworks. Personal information covers a very in-depth field of information, such as someone's date of birth being considered personal information, but not personally identifiable information. Something that is considered personally identifiable information has to be unique to one said person. This can be a social security number, as only one person can have this number in their lifetime, whereas multiple people can share the same name or date of birth.

In turn, there comes the question of how many features must be linked before PI can become PII for any individual. The legal tests for PI tend to relate to a situation where a single individual's identity can be "reasonably be ascertained." However, the definition of PI can be very broad and is said to cover any information that relates to an identifiable individual, at least any information is said to be covered and protected.

8.11 Fifth Generation vs. Sixth Generation

In a world where new things are always emerging, looking at newer aspects are always seen as just ideas. In 2020, fifth-generation (5G) radio networks were just implemented globally. The fifth generation featured new things such as mass connection, guaranteed low latency, and extreme dependability (Figure 8.10). As we have noted in this book, however, 5G capabilities were more than adequate in 2020, but they will fall short of future needs by 2030.

6G is predicted to have higher coverage, less energy consumption, and cost-effectiveness alongside improved security, and comprehensive spectral. For this to be accomplished, the 6G network will need to ensure these features:

- Multiple access
- Waveform design
- Channel coding schemes

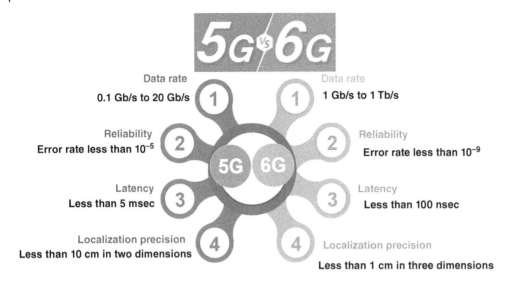

Figure 8.10 Comparison of 5G and 6G [17].

- Network slicing
- Cloud edge computing
- Antenna technologies

Proponents of 6G believe that these features will exceed the benefits of previous generations and will set the pace for generations to come.

8.12 Conclusion

It is easy to be caught up in the ideas of what 6G can do and become, how it might be proven better than the generations past, and what the bar is set at for future generations. When thinking of how great it can be, however, it is important to keep in mind some challenges and flaws that may occur, too. There is nothing wrong with being caught up in the bigger picture but thinking about the smaller things is important all on its own.

The need for wireless communications grows stronger every day, as was exposed during the COVID-19 pandemic. Now that the pandemic is over, there is no turning back. Many jobs are now virtual. People are recognizing that things they thought had to be done in person can be done more effectively virtually. Holograms and multisensory communications are the next steps. This will allow for even better virtual communication as people could perform remote activities. It will make people more connected virtually. Doctors could perform remote surgeries or technicians could perform remote repairs. Holograms would require a large amount of data to be transmitted. The transmission rate is expected to be 4.32 Tbps. This is something that 5G is not capable of providing. Virtual communications using all five senses would change the world significantly.

The development of 6G will be very important to help implement these new technologies. Holograms are one of the technologies 6G could make possible, but there are several requirements for this. Holograms will require massive bandwidths and will need ultralow latency to deliver a good experience. The user will feel simulator sickness if the latency is too high. Synchronization is important too for this technology. This is important to deliver the timely arrival of packets. It is important for the network to have knowledge of flows. Security is an important part of all technology but will become even more important as technologies like holograms are implemented. For something like remote surgery, ensuring CIA becomes essential. If someone is able to compromise the data, it could be deadly for the patient. It would also lead to potential lawsuits for the hospital, so before anything like this can be performed, security and the network will have to be perfect. Another challenge is coordinating the security of multiple flows because an attack on one flow could affect the others. This will make cybersecurity and network professionals essential, as the need and demand will only grow. Resilience, availability, and reliability are all very important parts of holographic communications. Holographic technologies will take up a large number of computer resources, so computation is a big part. Balancing computation, latency, compression, and bandwidth will be the big challenge to get holographic communications to work properly.

Tactile and haptic communications are another development that 6G will bring. It will be important for robotic automation, autonomous driving, and healthcare. The latency will need to be 1 ms as it takes 1 ms for a human to receive tactile signals. The data rates will need to be 1 Gbps. Synchronization is also important due to the reaction times of the human brain. Security will be a forefront issue for these applications as it will be for many other things in the future. Networks should be able to prioritize traffic based on how critical they are. Reliability will also be very important; depending on the application, it could need to be 99.999% reliable.

References

1 McKetta, I. (2021). Growing and slowing: The state of 5G worldwide in 2021. *Ookla Insights* 1 (1): 3.

2 Thales (2022). A 5G Progress report: Launches, subscribers, devices & more. https://www.thalesgroup.com/en/worldwide/digital-identity-and-security/magazine/5g-progress-report-launches-subscribers-devices.

3 Yuan, G. (2018). Key technologies and analysis of computer-based 5g mobile communication network. *Journal of Physics: Conference Series* 1 (1): 7. https://iopscience.iop.org/article/10.1088/1742-6596/1992/4/042001/pdf.

4 Rajiv (2022). Applications of 5G technology. https://www.rfpage.com/applications-5g-technology.

5 Zhu, Z., Hou, G., Chu, Z. et al. Research and analysis of URLLC technology based on artificial intelligence. *IEEE Communications Standards Magazine* 5 (2): 37–43.

6 Strinati, E., Barbarossa, S., Gonzalez-Jimenez, J.L. et al. 6G: The next frontier: From holographic messaging to artificial intelligence using subterahertz and visible light communication. *IEEE Vehicular Technology Magazine* 14 (3): 42–50. https://ieeexplore.ieee.org/stamp/stamp.jsp?tp=&arnumber=8792135.

7 Xu, Q., Jiang, C., Han, Y. et al. (2018). Waveforming: an overview with beamforming. *IEEE Communications Surveys & Tutorials* 20 (1): 132–149. https://doi.org/10.1109/COMST.2017.2750201.

8 Chen, Y., Liu, W., Niu, Z. et al. (2020). Pervasive intelligent endogenous 6G wireless systems: Prospects, theories and key technologies. *Digital Communications and Networks* 6 (3): 312–320. https://www.sciencedirect.com/science/article/pii/S235286482030242X.

9 Khan, A.H., Hassan, N.U.L., Yuen, C. et al. (2021). Blockchain and 6G: the future of secure and ubiquitous communication. *IEEE Wireless Communications* 20 (10): https://arxiv.org/pdf/2106.05673.pdf.

10 Tariq, F., Khandaker, M., Wong, K.-K. et al. (2020). A Speculative Study on 6G. *IEEE Wireless Communications* 27 (4): 118–125. https://doi.org/10.1109/MWC.001.1900488.

11 Chawdury, M. (2021). 6G wireless communication systems: applications, requirements, technologies, challenges, and research directions. *IEEE Open Journal of the Communications Society* 1 (1): 16. https://arxiv.org/pdf/1909.11315.pdf.

12 Mohamad, N. (2021). From 5G to 6G: What could it look like? https://www.lightreading.com/6g/from-5g-to-6g-what-could-it-look-like/a/d-id/767711.

13 Kantola, R. (2000). Trust Networking for Beyond 5G and 6G. In: *2020 2nd 6G Wireless Summit (6G SUMMIT)*, 1–6. Levi, Finland: IEEE https://ieeexplore.ieee.org/document/9083917.

14 Ylianttila, M., Kantola, R., Gurtov, A., and Mucchi, L. (2020). *6G White paper: Research challenges for Trust, Security and Privacy. 6G Flagship.* University of Oulu https://arxiv.org/abs/2004.11665.

15 Ali, S., Saad, W., Rajatheva, N. et al. (2020). *6G White Paper on Machine Learning in Wireless Communication Networks.* University of Oulu http://jultika.oulu.fi/files/isbn9789526226736.pdf.

16 Kumar, T., Liyanage, M., Braeken, A. et al. (2017). From gadget to gadget-free hyperconnected world: Conceptual analysis of user privacy challenges. In: *2017 European Conference on Networks and Communications (EuCNC)*, 1–6. Oulu.

17 Abdel Hakeem, S.A., Hussein, H.H., and Kim, H.W. (2022). Security requirements and challenges of 6G Technologies and Applications. *Sensors* (Basel, Switzerland). 22 (5): https://www.ncbi.nlm.nih.gov/pmc/articles/PMC8914636.

9

Artificial Intelligence: Cybersecurity and Security Threats

9.1 Introduction

Almost everything electronic that we see or use in our daily life will need some form of cybersecurity behind it. The potential vulnerabilities of today's digital world should not cause any paranoia, but the basics of digital security are worth learning for everyone who uses the benefits of modern civilization. Software updates and patches have become more and more important as technology has progressed and become more sophisticated over the years. They have numerous benefits like obvious security patches but can also add new features to user devices and add other new software. These updates and patches to the Internet of Things (IoT) devices and artificial intelligence (AI) software are now possible to be done at revolutionary speeds, thanks to fifth-generation network technology.

The world of AI, fifth-generation data networks (5G), the rapidly growing IoT devices, and 6G can be very helpful to us but also present numerous flaws in terms of privacy, security, and cybercrimes, as they are all new and rapidly developing technologies. It is important to note that the 5G network has better security compared to the 3G and 4G networks, but it has been said that some of the classic vulnerabilities and security flaws from 3G and 4G networks were directly carried over to the developing 5G network, thus presenting additional security flaws right out of the gate. Both AI and IoT will benefit from the development of 5G networks where businesses can use such devices, which will be tied to the growing 5G network, and can serve several purposes throughout the business market, among other areas.

When it comes to consumer-based electronic devices, they are certainly on the rise. As technology improves, these devices will consume more bandwidth and data than ever before. The increase in data and bandwidth consumption from newly emerging technologies could include things such as the ability to interpret human speech on the fly, all the way to identifying various patterns throughout data or documents from a mobile device. This data can even adjust business efficiency, increasing overall profits with endless possibilities for automation in manufacturing. This is where the development of higher-generation data networks comes into play. 5G and 6G networks will have the bandwidth needs for current and future data-hungry devices in a growing technological world.

AI technology is growing from simple tools used by ordinary scientists to as far as to use within the professional development community for higher intelligence use. These various organizations can use AI technology to fill the current gap in the data science area, which could be a big game-changer for data science! This also always gives end-consumers the ability to take their business and personal data with them wherever they go. This is a big deal when it comes to making end consumers happy while retaining their privacy with their online data.

When it comes to data processing with the help of the most recent wireless generations of 5G and 6G, AI, IoT, and modern technology, and its data often demand that data management and processing capabilities have the data possessors brought to the data itself rather than sending the collected data off for processing elsewhere. This may also include the data processing and distributed data stores all to be included within the data management process to guarantee support is offered where the received critical data is being stored. Fifth-generation network technology will be revolutionary to the end consumers, businesses, as well as data-processing centers around the globe.

AI, 5G, 6G, and IoT advances are now enhancing each other, making the fifth wave of computing.

There have been some significant developments in cybersecurity. One of the significant developments is in the techniques and tools that redeveloped with the support of AI and machine learning (ML). With this partnership between AI

From 5G to 6G: Technologies, Architecture, AI, and Security, First Edition. Abdulrahman Yarali.
© 2023 The Institute of Electrical and Electronics Engineers, Inc. Published 2023 by John Wiley & Sons, Inc.

and cybersecurity, the current and future application possibilities for cybersecurity are endless. The detection of threats is one of the significant concerns in cybersecurity and has been emphasized substantially. Officials, authorities, and organizations have always been keen on developing newer and more advanced ways to detect a potential threat or to be prepared for any advanced threat and attack that can take place. ML has played a vital role in this aspect of cybersecurity and strengthened the overall relationship as it has provided effective approaches and outcomes in terms of threat detection in cybersecurity. It has been instrumental in detecting threats by analyzing data and identifying threats in the initial stages.

Each year, cyberattacks are on the rise and are getting worse as compared to their impacts in previous years. The sharp increase of cyberattacks has resulted in more and more security threats and made it complex due to the evolvement of AI in cyberattacks or the use of AI by attackers. As a result, more sophisticated cyberattacks that are more complex than ever are taking place, which is one of the most significant challenges in AI and its implementation and usage in cybersecurity despite all the promises it has fulfilled and is going to deliver.

9.2   5G and 6G

5G is a great improvement over 4G LTE and will occur an increase in bandwidth, capacity, and reliability, but with this, cybersecurity has been a big issue. The 5G supply chain is vulnerable to malicious software, counterfeit components, and poor design and manufacturing. This increases the risk and could negatively impact confidentiality, integrity, and the availability of data. There is also the issue of traveling, and it is possible that even though US networks are secure, data breaches can happen when traveling to other countries.

5G has an increased attack surface compared to the other generations of wireless networks. This makes it especially vulnerable. There also are more vulnerabilities being found every day. If equipment is poorly maintained, there is an increased risk of security problems. Some 5G networks will be integrated with existing 4G LTE networks that have legacy vulnerabilities.

Another huge problem for 5G networks is IoT connectivity. As everything begins to become connected to the internet, huge security issues can play out. Someone could hack into a smart device and listen to conversations. This is a breach of privacy and a concern for many governments and people. IoT also brings a scary concern for safety. Someone could hack into an autonomous vehicle and cause wrecks. Therefore, it is so important for the 5G network to be as secure as possible. Although the technology may be developed to make cool IoT devices like fully automated power plants and self-driving cars, security and privacy concerns have to be addressed before this can occur. Security is not only an issue for cybersecurity specialists. It is an issue for everyone. Everyone needs to be hyperaware of online security and take measures to protect themselves. This includes creating strong passwords, watching what is being shared online, and watching what kinds of websites they are interacting with. People must be aware of common frauds that attackers are using to gain access to their personal information. Even though basic information has been shared for years about cybersecurity, people are still making these careless mistakes that will slow the development of interesting IoT devices.

The fifth generation of communication is spreading very fast worldwide and has been launched in more than 38 markets. It has given significant growth around Asia, America, Europe, and China. The global mobile supply associations have utilized the 5G feature in different electronic devices by the start of 2021. Its benefits and functions are still being discovered and explored until the development of 6G. Almost 16 versions of 5G services are utilized. Further extensions and improvements of the 18th version would lead to the launch of 6G in 2030. All of these versions of 5G are very reliable in saving data and privacy.

The 6G is a new generation with better communication and services. Billions of investments on the internet of intelligence would lead to this wireless system. This system was invented to enhance the ability to present knowledge, its execution, and implementation without the involvement of human beings. This new invention is proved very useful for the security and privacy of data.

The 6G gains in communication still need many years and a lot of investment to accelerate. Most of the advanced techniques need the emergence of a 6G system. Different public sectors and industries have started to invest in 6G. The European Commission has announced three joint projects for the early launch of the 6G system and has built pressure for other countries to join them. Australia has invested a lot to increase the value of its digital economy. The digital economy system requires cybersecurity, trust, and safety to secure the future of inventions, which need a $1.3 billion investment and the availability of 5G and 6G systems.

The author has explained that all the intelligence agencies must collaborate to work on a defense strategy. The digital growth and switching of 5G services to 6G would then be more manageable. They will collaborate formulation of the following visions:

- They will work to reduce the cost of operations.
- All cybersecurity functions must be placed at the edge of the networks.
- Learning with machines requires proper reasoning methods and architectural references to increase positive outcomes.
- Digital systems are very effective and beneficial in reducing carbon dioxide and boosting productivity.
- Technology must be safe, and privacy must be protected.

The communication and information systems will be improved, and the processing of knowledge will be easier. The global system of digitalization will adopt the 6G services as their nervous systems. It would be having more secure and private systems with various designs. This design can be implemented by increasing the function ability of 5G systems 10 times. Different initiatives are taken for the launch of 6G systems. It is fascinating to imagine the level of reliability and security in the future, which can be achieved with many investments.

Cybersecurity is still a very juvenile subject that is a proper shame considering how necessary it is. It will only continue to be more necessary and complicated even in the next 10 or so years. All you must do to appreciate cybersecurity is turn on the news and witness all the talk about the up-in-coming IoT, which is the striving of the entire world toward a greater connected network of life and technology. Look at smart homes, an interconnected network of devices designed to provide a more convenient and efficient home experience. With the prospect of artificial intelligence and subsequently ML on the horizon, cybersecurity is debatably the discipline that benefits the most from AI integration. Traditional cybersecurity measures have the reputation of being rather sluggish and insufficient in many ways.

However, looking to the future, AI is a promising improvement for such an important aspect of life. One that provides defense systems with strong, flexible, and efficient workability. Especially when you consider that, we will face an equally as dynamically growing and sophisticated barrage of cyber threats as information and communication technologies continue to advance. It is important to note that as cybersecurity continues to mature, so too do the adversarial entities in pursuit of undermining precious data. We see this in its purest sense, an arms race between two opposing sides that has no visible end. Maybe it is a little too far to label some of these actions – political activism, crime, espionage, and subversion – as war-like, as we know it. However, take everything that cybersecurity is in as a whole and it is easy to see that it is an environment of constant gray-zone conflict. A battle between those intending to compromise internet-connected systems and those tasked with preventing, coming up with better strategies to defend, and lessen the impact of attacks on internet-connected systems.

As it goes in this environment, there is much need for better ways to combat cyberattacks. Consequently, cyber-attackers are doing the same thing but for the opposite effect. That is where we find ourselves with the emergence of AI in cyberspace because it presents itself as a solid avenue to achieve these goals set forth by cyberspace actors. This research chapter serves the purpose of highlighting the shortcomings of traditional security measures and the progress that has been made by applying AI and ML to cybersecurity in pursuit of a greater solution to deep-rooted problems. Additionally, we will cover the risks and concerns of AI, how these concerns will be addressed, and look at an exciting new possibility for AI and cybersecurity paring in machine learning.

9.3 Cybersecurity in Its Current State

Cybersecurity is a very complex and important field. Businesses must be ensuring that they are looking carefully at cybersecurity issues to protect their enterprise. Cybersecurity is normally managed from a functional perspective. Cybersecurity needs to be analyzed by businesses because measuring coverage and utility is not enough anymore. It isn't about a looking at series of tasks. The enterprises must analyze the operational efficiency of actions, the resiliency of people and technology, maturity of practices, cost of ownership, gap analyses, and more. Although looking at cybersecurity like this is complex, there are methods to help access it.

Business owners should ask themselves whether their security program is operating efficiently to determine changes they may need to make. They should think about how effective, resilient, mature, and optimized their security system is. A solution should be cost-effective. Maturity and resiliency are related, and maturing capabilities are a path to building resiliency.

Enterprises should think about each aspect of their security system carefully and determine whether each aspect is up to standard. This mindset has many benefits, such as decreasing costs while increasing benefits, using resources optimally, making more informed decisions and advantage investments.

Some of the challenges that enterprises are facing are risk management, operational efficiency, skill sets and training, budgeting, negligence, and prioritization. By fixing these issues, enterprises can see a lot more success. Many businesses are not performing risk management correctly, so by looking at that aspect and learning about what can be done, companies can improve this aspect. Companies need to be aware of their due diligence to prevent problems from occurring. They also needed to analyze their systems to ensure efficiency. This starts by hiring people who are well trained to see these problems. This and prioritization can help with budgeting.

Cybersecurity does not have a defined identity that everyone in the field can agree on. Throughout my research into the topic, it became apparent that the term had many variables and often subjective definitions that struggled to truly capture all that this broadly complicated and diverse subject covers. It was only when I looked deeper that I found a satisfactory answer. Given to me by a research team that set out to do exactly what I was looking to close on, the definition they concluded was "Cybersecurity is the organization and collection of resources, processes, and structures used to protect cyberspace and cyberspace-enabled systems from occurrences that misalign de jure [rightful] from de facto [factual] property rights" [1]. Though no definition (not even this one) has the backing of all professionals, I will be referring to cybersecurity as defined in this way throughout this chapter.

All the parties involved in cyberspace make up a range of people from individuals, private organizations, and governments, all of which aim to protect their cyber assets, make attacks on others, or even both. These attacks find their source in financial, political, or militaristic purposes, and it is in everyone's interest that cybersecurity measures be fully developed [2].

The need for improving the cybersecurity environment cannot be understated, especially now when cyber threats find themselves as some of the highest risks on a global scale when it comes to the economy [3]. Cyberattacks may not seem that detrimental to a nation's economy. Still, when you look at how they affect entities such as banks, hospitals, power plants, and smart devices, suddenly the stakes are much higher and have the potential to run businesses for multibillions of dollars in losses.

As the cybersecurity community evolved, so did the mindset; it is no longer a question of "if we are attacked," but "when we are attacked," – then what? The realization has set in how likely it is for a system to be compromised if an attacker sincerely wants to and has the equipment to do so. In any massively connected chain of systems, the attack area is so vast and spread out that gaps and vulnerabilities are nearly impossible to defend. There is no such thing as forcing into a network since any intruders will get in by way of vulnerabilities in the system itself. An attacker merely needs to find one vulnerability to exploit one time to be successful, whereas defenders must always be on guard if they have any hope. Being prepared is preferable to being unrealistic and thinking cyberattacks are completely avoidable, because as soon as the inevitable is realized, focus and attention can be used more productively elsewhere [4]. Cybersecurity's main concern is to expose those vulnerabilities and threats so that countermeasures can be taken. This is a challenge that requires decisive, intelligible data for interpretation and effective use in detection and prevention.

Automation in data processing processes has been around for a while, largely because humans do not have the cognitive or sensory capacity to regulate enormous amounts of data. The human mind cannot logically manage these data streams in and out of their networks, produced by software and hardware alike that alerts system administration; automation is the only solution that makes sense. Most relatedly, we see this automation in anti-virus programs that utilize intrusion detection systems (IDS). These programs inspect networks looking for known signatures of malware that attach themselves to machines and cause harm. Over the years, we have seen this increase in malware protection and defending against worms, Trojans, spyware, adware, ransomware, rootkits, and other vulnerability-exploiting software. Despite its effectiveness in the past, the emergence of new polymorphic and metamorphic malware, which can change their appearance every second, can easily evade signature-based detection. It is impossible to recognize these threats if nothing is matching them to the detection database that will continuously be outdated.

Now that brings us to the discussion of possible solutions for such a troublesome problem in securing data. Anomaly-based capabilities are being developed now. The goal is to scan the network traffic for anything that seems out of the ordinary and then alert a human administrator for revision. But this, too, brings us right back to the original problem. Human cognitive and organizational abilities will falter at the increase in alerts and the certainty that something may be let through that is malicious. This only gets worse when you consider that this automation is subject to false negatives and false positives (good behavior labeled bad, and bad behavior labeled good), which only serves to exhaust more resources [5]. How do we go about correcting these problems? One very promising response is to layer in more automation to handle these big data sets using AI to process and interpret this suspicious behavior and feed it back into cybersecurity decision-making in short cycles. The history of AI has always been filled with distrust and fear of unpredictability. Still, now that technology development lays AI at our feet, dynamic learning and adaptability processes make it an ideal countermeasure scenario to equally be susceptible to dynamic threats.

9.4 AI as a Concept

Arguably the most famous words to ever be spoken regarding AI is that by Alan Turing, in his 1950 paper *Computing Machinery and Intelligence:* "We may hope that machines will eventually compete with men in all purely intellectual fields" [18, p. 460]. This foresight of the future surrounding the ability for machines to one day be so intelligent that they rival that of humans may very well be a reality with how advanced developments are coming along in the field.

AI is a software development focused on creating intelligent machines and projects. It has the motivation to imitate human intelligence and perform tasks like humans. In practice, this implies the ability of a machine or program to think and learn. Typically, the term *artificial intelligence* implies a machine or program that imitates human intelligence. Computer-aided intelligence has become an integral part of the technology business.

Artificial intelligence experienced three phases in the historical context of AI in the 1950s: the supposed period of neural systems from 1980 to 2000, the age of machine learning, and the current era of deep learning. A neural system is a type of programmed learning that consists of interconnected substances (such as neurons) that process the data as it responds to external information sources and transmits data between each element. The method requires several steps in the data to detect associations and to determine the meaning of indistinct data. Machine learning automates the development of exposure models and uses different strategies, – such as neural systems, research tasks, measurements, and materials science – to discover data that is excluded from the data without appearing clearly in the view or reaching what is to be achieved. Deeper learning is more like machine learning but in a deeper dimension. The goal of deep learning is the use of an algorithm to create a fearful system that can deal with the given problems. It is used in particular to solve problems in which arrangements with conventional techniques require exceptionally complex patterns. For example, through deep learning, speech, images, and content are distinguished or monitored [6].

An algorithm is a well-ordered strategy to tackle problems. It is constantly used for the management of data, estimates, and other numerical and PC activities. It is an accurate determination of the guidelines to perform a task or solve a problem. Algorithms can perform multiple tasks effectively and immediately if the right data is included in the framework [7].

The most important factors for increasing AI are currently the limit and the driving force, the data and the algorithms. The execution and the limit of the processors have been completely extended. Today, there is a huge amount of data about the atmosphere, online networks, and medicines, and machines can eventually use that data. In the meantime, data storage costs for executives and improved data storage have led to the faster investigation of a large amount of data.

As with cybersecurity, AI fails to hold a widely accepted definition and has been used in many ways since its coining at the Dartmouth meeting. The pioneers of this very concept attended: John McCarthy, Marvin Minsky, Allen Newell, and Herbert Simon. It was not only because these individuals attended the Dartmouth meeting but because they established three leading research centers, and their ideas largely shaped AI into what we know it today.

Newell and Simon said [8], "By 'general intelligent action' we wish to indicate the same scope of intelligence as we see in human action: that in any real situation behavior appropriate to the ends of the system and adaptive to the demands of the environment can occur, within some limits of speed and complexity" (p. 116). According to Minsky [9], intelligence usually means "the ability to solve hard problems" (p. 71). And McCarthy said [10], "It is the science and engineering of making intelligent machines, especially intelligent computer programs. It is related to the similar task of using computers to understand human intelligence, but AI does not have to confine itself to biologically observable methods" (p. 2).

Though these conceptions are revolutionary and essential to the early development of AI, they are admittedly vague and without much of a common foundation. The uniquely frustrating aspect of AI is that the scientific domain is notably disagreeable in targeting any objective, and many researchers subscribe to subdomains of AI than AI itself. Subdomains include knowledge representation, reasoning, planning, machine learning, vision, and robotics, for example.

Now to get to the point, what definition of AI do I most closely relate to and will be referencing throughout the rest of the chapter? That is a definition given by Dr. Pei Wang of Temple University: "Intelligence is the capacity of an information-processing system to adapt to its environment while operating with insufficient knowledge and resources" [11]. This definition, though simple, outlines exactly the important points in understanding AI and defines a clear objective as well, fulfilling a goal set out by my research into the broader purpose of AI usage in cybersecurity.

9.5 AI: A Solution for Cybersecurity

Recently AI has played a defining role in advances in many scientific and technological fields, including robotics, image recognition, natural language, expert systems, and much more. However, the advances that AI means for cybersecurity are probably the most profound. Cybersecurity augmented by AI will lead to increased awareness, and reactions in real-time, and overall make it a more effective system of protection for the data-connected world that so desperately needs it. It is known that even a small team of cybersecurity professionals can effectively protect networks supporting thousands of users. Just think of how this same expertise could be extended by AI. However, it takes a lot of time, money, and effort to generate and implement an algorithm used in cybersecurity. On top of that, training and employing specialists is not only expensive but very difficult. Especially in the coming years as the field will be stretched thin when you consider how fast the world of technology is evolved compared to how many people are joining the workforce, there is already a troubling shortage of talent, and it will only get worse if something is not done [12]. AI-based methods are expected to address these concerns.

The most relevant application of AI in cybersecurity is for IDS. This advancement is needed to offset the future deployment of offensive AI that will surely be seen very soon attacking organizations. As mentioned previously, anomaly-based detection strategies seem to be the most receptive and effective receivers of AI to bolster defenses. In addition to being well-equipped to combat low-level threats, leaving higher-level threats to dedicated human analysts, we are now looking at the ability of AI to mimic said analysts' techniques and then relay this data to learning algorithms that can replicate it and be used to shape AI cybersecurity entities [13]. Having AI monitor individuals within an organization will allow them to pick up on their cyber defense skills and glean from them the normality of the organization, meaning when anomalies come, they are that much more detectable [14]. AI also can monitor the mutation of accesses on a system, labeling it a potential threat. If it is malicious, validation will be reinforced and used to scan other similar actions and detect any malicious behaviors. That is the AI complex. It is subjected to normality within a system of a set of learning data. Over time, as the natural course of cyberspace takes place, it becomes more and more sensitive to the goings-on and thusly better prepared to detect threats. This, in its essence, is machine learning and holds a lot of promise for the future of cybersecurity efficiency.

We have talked about most in line with the supervised learning subtype of ML, which is much different from unsupervised learning. Suppose supervised learning involves parameters and data set by programmers before the learning begins. In that case, unsupervised learning involves algorithms purely doing all the categorization and labeling by themselves, subverting the biases of programmers. The success of these systems is very much dependent on large amounts of quality data. Traditionally, large quantities of data made it difficult to pinpoint relevant data and make connections. However, machine learning thrives off every bit of data, giving it much advantage. The more data AI must reference, the better off it can make connections and distinguish patterns. Machine learning is the most popular Ai-based solution for cybersecurity, especially when detecting network intrusions. But as the inevitability of more sophisticated and complex cyberattacks approaches, AI solutions need to be further explored to manifest their true potential. There is still much in the way of uncertainties and challenges that need to be addressed, and that is what will be tackled in the next section.

9.6 AI: New Challenges in Cybersecurity

Despite all the positive aspects of AI and ML, many challenges and concerns are involved with such a promising science. In conversations involving AI, the innate problems of uncertainty and transparency present questions of trust, safety, and control, and this is no different in the realm of cybersecurity. The truth is that all the advantages that AI and ML give to cybersecurity also present an equal and opposite advantage for those meaning to penetrate these systems for malicious purposes. This concern, along with the others highlighted, will be addressed here.

AI processes are by design intended to make decisions automatically without much human interaction, yet they are not entirely autonomous. Since they cannot completely replace human cognition, humans still need to keep a close eye on AI systems to make sure the job is being performed to expectation and no issues happen during this process. As previously states, humans control the application of learning algorithms necessary for machine learning and ensure its effectiveness.

Data privacy is also another major concern for AI. A big part of maintaining a secure cyber community is via information sharing. This will be done to form collective knowledge and experience databases to understand cyber threats better. Such sharing undoubtedly augments cybersecurity on a global scale, but this is a nightmare when considering data privacy. How can this data possibly stay secure, making so many leaps and bounds across countless data stores? [15] Especially challenging is when analyzing large quantities of data, the attention of private as well as public organizations will be concerned with how secure their personal data is being kept and may even refuse to share their information. What data is used, why it is used, and the conclusions AI makes are not always clear to translate, meaning that even when a threat is detected, it could remain opaque to analysts since the AI performs most of its operations without expressly giving a reason.

Something particularly concerning is that AI cannot discern morality in any aspect and, therefore, will make decisions indiscriminate of ethics. This means that AI might make decisions for humans that a human would never make. A big problem with an algorithm's fairness is the possible discrimination of certain groups of people because of the lack of data surrounding them used in the learning process or algorithm development [16]. Compound this with the uncertainty of AI decision-making, and things could turn out disastrous.

Overall, there is a lack of legality surrounding AI in cybersecurity. The prospect of humans losing control over their automation and the consequences being dire due to the unknown nature of AI is something existing laws may not necessarily be ready for. The fast-moving development of this science will disrupt a lot of things with positive and negative effects; one of the biggest actors that fall behind will be lawmakers. There will need thoughtful implementation of laws and regulations governing data privacy that needs to be updated regularly following developments. AI and ML systems will also face unique complications in the cybersecurity realm when it comes to cyberattacks and attackers adopting AI and ML techniques to improve their attacking efficiency. Many types of attacks affect ML performance [17]:

- *Poisoning:* This attack inserts hostile code into the training algorithm and data sets, causing availability and integrity problems with the machine.
- *Evasion:* Inserts many hostile codes into a system to reduce the overall security so that an attacker can evade detection more easily.
- *Impersonation:* This causes the existing machine learning system to use a different label than the impersonated code to misclassify it.
- *Inversion:* Basic information about a system is collected using its API (application programming interface) and the basic information is reversed to gain access to sensitive information.

New avenues of attack are not the only developments in the adversarial usage of AI and ML. Machine learning will be aptly used to train their means of identifying vulnerabilities in systems to gain access. Because AI algorithms are employing self-learning, they get smarter and smarter with each attempt, whether it's a success or a complete failure; just as cybersecurity will get continuously better, so too will the threats it defends against. Communities of hackers will most likely be sharing data and experience between their AI just the same way, and superior malware will come of it, certainly.

The challenges discussed in this chapter are the very basic and general issues that ML AI will be facing, and I will discuss now potential solutions. In the growing AI environment, it is paramount that lawmakers stay up to date on all the potentially troublesome problems so that many headaches can be avoided. Some companies will benefit most from having one AI algorithms handle their data but potentially having another AI or even an analyst check the processing of the first AI to ensure the system's integrity. This will, of course, require additional costs, but to avoid privacy and ethical hang-ups, this is one of the simplest fixes. It will not prevent all the challenging aspects of these problems, but it will provide much more security for an organization's systems. Another important note is that AI should not be used exclusively to deal with cyber threats; there still needs to be human analysts who are expertly trained to take over for the AI if something unexpected happens. Not only that but having someone knowledgeable to translate the operations of an AI system will be invaluable in reinforcing the entirety of any one organization's security. Being thoughtful about how your company will utilize AI will help plan expectations and make you better prepared to deal with issues.

9.7 Conclusion

I hope this chapter paved a path to comprehending artificial intelligence more extensively, notwithstanding other related issues identified with AI. It is not typical to trust that artificial intelligence has existed since the 1950s, in light of the way that we have been talking about it for the most part in the last 5–10 years. In the last decade, innovation has come a long way and is without a doubt one of the key clarifications for why man-made consciousness is such a convincing subject at this moment. At present, the advance has been quicker than at any other time, and manufactured brainpower was depicted in the nineteenth century as a dynamic improvement like power. A gander at the AI hypothesis unmistakably demonstrates that associations have just now started to comprehend the potential advantages of artificial intelligence. Toward the start of this decade, man-made reasoning undertakings were negligible, and this is astonishing. When we consider how cell phones has changed over the previous decade, we discover that enthusiastic innovation is not that astounding any longer. The greatest changes and hypotheses will arrive in a couple of years. It is anything but a surprising thing that China and the United States are driving the test, and I do not see that another country will in all likelihood move them a lot. Finland and other EU nations have communicated that they want to be among the primary countries in the AI race.

The impact of AI on the workforce is complex and still very much uncertain. A Deloitte study of automation and AI technologies in the United Kingdom found that 800,000 low-skilled jobs were eliminated, but 3.5 million new jobs were created, with an average of nearly $13,000 more per year than the ones that were lost. Therefore, there is, in fact, an important connection between automation and work. Automation technology can lead to a more intellectual and creative workforce for solving complex tasks which can improve our lives [18].

In both competitive sports personal game playing, players often push to be the best. Advances that can improve execution are great news for video gamers, athletes, and sports organizations. The reproduced knowledge brings every one of the inclinations of a player or relationship into the situation where it is utilized in the best way. The affiliations need better outcomes from the on-screen characters and a regularly expanding pay. As I would see, this recommendation indicates how the affiliations currently utilize man-made reasoning, in addition to how artificial intelligence can change the entire movement. For us, as observers, I discover it genuinely interesting how extraordinary the effect of AI will be on the transitory future, as the age of live video develops with new estimations.

Artificial intelligence is undoubtedly one of the most promising developments of the information age, and machine learning is the most promising development of AI. In this research chapter, we looked at the shortcomings and limitations of traditional security measures and the subsequent progress being made by AI and ML developments in the cybersecurity field. We also covered the risks and concerns that come with AI and how these concerns can and should be addressed.

Compared to traditional cybersecurity solutions, AI presents a more dynamic, adaptable, and efficient alternative. In the face of equally efficient cyberattacks, this is exactly what the world of technology and the world in general needs. Within the AI spectrum, the machine learning aspect is probably the most powerful, promising element that will push us that much higher into a new phase of cybersecurity.

Despite all the promises, AI is a global hazard for human existence by proxy because of the lack of AI's complete autonomy, data privacy problems, lack of legal framing, and the fact that machines do not abide by any moral code. With how fast things are moving in the world, the necessity to resolve these concerns is of the highest priority for those concerned with the integrity of the future. Nevertheless, until then, being responsible is the best course of action.

As the world grows and new technological advances emerge, you cannot forget the adversaries involved, and they, too, are taking notes and developing their means of attack. You bet AI and ML capabilities will be used against the law-abiding organizations, and the new challenges presented by these techniques will present new vulnerabilities to boot.

References

1 Craigen, D., Diakun-Thibault, N., and Purse, R. (2014). *Defining Cybersecurity*, 13–21. Technology Innovation Management Review (rep.).

2 Rehman, A. and Saba, T. (2014). Evaluation of artificial intelligent techniques to secure information in enterprises. *Artificial Intelligence Review* https://dl.acm.org/doi/abs/10.1007/s10462-012-9372-9.

3 Taddeo, M. and Bosco, F. (2019). *We Must Treat Cybersecurity as a Public Good. Here's Why.* World Economic Forum https://www.weforum.org/agenda/2019/08/we-must-treat-cybersecurity-like-public-good.

4 Libicki, M.C., Ablon, L., and Webb, T. (2015). *The Defender's Dilemma*. RAND Corporation https://www.rand.org/pubs/research_reports/RR1024.html.

5 Libicki, M.C. (2009). *Cyberdeterrence and Cyberwar*. RAND Corporation https://www.rand.org/pubs/monographs/MG877.html.

6 SAS (2018). Deep learning: What it is & why it matters. https://www.sas.com/en_us/insights/analytics/deep-learning.html#technical.

7 Rouse, M. (2021, October 7). Artificial intelligence. https://www.techopedia.com/definition/190/artificial-intelligence-ai.

8 Newell, A. and Simon, H. (1976). Symbols and physical symbol systems. In: *Computer Science as Empirical Inquiry Symbols and Search*. Essay. Associations for Computing Machinery, Inc https://iiif.library.cmu.edu/file/Newell_box00024_fld01660_doc0003/Newell_box00024_fld01660_doc0003.pdf.

9 Minksy, M. (1987). The Society of Mind. *Mind-Body East and West* 3 (1): 19–32.

10 McCarthy, J. (2007). *What Is Artificial Intelligence?* 2. Stanford, CA: Leland Stanford Junior University (tech.).

11 Wang, P. (1995). Non-Axiomatic Reasoning System: Exploring the Essence of Intelligence. PhD thesis. Bloomington: Indiana University.

12 Oltsik, J. (2020). *The Cybersecurity Skills Shortage Is Getting Worse*. CSO Online https://www.csoonline.com/article/3571734/the-cybersecurity-skills-shortage-is-getting-worse.html.

13 Bresniker, K., Gavrilovska, A., Holt, J. et al. (2019). Grand challenge: applying artificial intelligence and machine learning to cybersecurity. *Computer* 52 (12): 45–52. https://doi.org/10.1109/mc.2019.2942584.

14 Darktrace. (n.d.). *Autonomous Response*. https://www.darktrace.com/en/autonomous-response.

15 Buckner, B. (2019). *How Can We Trust Decisions Made by AI?* Leidos https://www.leidos.com/insights/how-can-we-trust-decisions-made-ai.

16 Middelstadt, B.D., Allo, P., Taddeo, M. et al. (2016). The ethics of algorithms: Mapping the debate. *Big Data & Society* 3 (2): https://doi.org/10.1177/2053951716679679.

17 Comiter, M. (2019). *Attacking Artificial Intelligence: AI's Security Vulnerability and What Policymakers Can Do About It*. Belfer Center for Science and International Affairs https://www.belfercenter.org/publication/AttackingAI.

18 Deloitte (2015). From brawn to brains: The impact of technology on jobs in the UK. https://www.sas.com/en_us/insights/analytics/deep-learning.html#technical.

10

Impact of Artificial Intelligence and Machine Learning on Cybersecurity

10.1 Introduction

Cybersecurity is the protection of systems, different software, and data shared over various hardware and equipment connected to an internet connection or are connected in a single network [1]. Cybersecurity ensures the protection of enterprises, organizations, businesses, employees working there, individuals or the public, and their confidential and credential information.

Cybersecurity protects unauthorized protection access to the valuable and confidential information stored on their computer systems, data centers, or databases. These devices are used personally, such as smartphones, tablets, and laptops. Not only this, but the routers, WLAN, and other such network devices to which all the identified systems such as computer systems, mobiles, laptops, data servers, or databases that are connected to are protected through cybersecurity [2].

With the increased adaptation of cybersecurity and beneficial outcomes, some threats emerged [3]. Over time, multiple risks are identified as they were carried from time to time and provided potential outcomes. This includes ransomware, which is malicious software explicitly designed to extract money from the target by blocking his/her access to the attacked system [4]. The target is forced to pay the amount of money required and then provide access to the information that is not guaranteed [5]. Malware is another major threat included in cybersecurity, which gains access in an unauthorized way to cause damage to the system [6].

Social engineering is associated with cybersecurity, a tactic used for accessing confidential information by revealing sensitive information about the system or computer [7]. For example, monetary payment can be solicited for accessing sensitive information to gain access, or an approach might include winning the trust of an employee or an individual, and using that person to gain access [8]. Besides this, phishing is one of the most used and threatening aspects of cybersecurity due to its ease of use and vulnerability in the system to human error. In phishing, fraudulent emails are sent to a network, and once the email is opened or the fraudulent link is clicked and opened, attackers are able to steal sensitive data [9].

The goal is thus to focus on enhancing and strengthening security. For this, different approaches and tactics are used to ensure the overall quality of cybersecurity. One of the focused aspects is machine learning (ML) and artificial intelligence (AI).

The term AI was coined in 1956 but was not considered important in ordinary aspects of life, such as organizational, business, and daily life aspects. AI has emerged as essential for the growth and success of organizations and countries to grow. It is because of the nature of AI, as it includes automation in different processes that might require a more significant number of human and higher human power to perform a task. However, AI has made its mark in various aspects and has contributed significantly to bringing automation to security aspects.

After its introduction in the first place, AI has kept evolving year by year and decade by decade. From the 1950s to the 1970s, AI contributed to neural networks as the focus was on the thinking machines [10]. From the 1980s to the 2010s, with the help of AI and its automation, ML was introduced and became popular. It is not so long that ML has been adopted in multiple aspects with AI utilization in a much more efficient way of producing deep learning (DL) [11].

Currently, the applications of AI are well implemented in sectors such as manufacturing robots, proactive management of healthcare, automated financial investment, natural language processing (NLP) tools, monitoring tools and software, social media monitoring, disease mapping, and virtual travel book agent conversational marketing bot [12]. These are the major examples of the evolvement of AI to this day, and they are the areas and aspects in which automation is significantly included.

From 5G to 6G: Technologies, Architecture, AI, and Security, First Edition. Abdulrahman Yarali.
© 2023 The Institute of Electrical and Electronics Engineers, Inc. Published 2023 by John Wiley & Sons, Inc.

Cybersecurity has focused on the mitigation of more significant threats that are arising due to advancements in technology. Therefore, bringing automation in cybersecurity and processes that are carried out in it included evolvement of AI. However, there are some advantages and disadvantages, potential risks and threats, and outcomes related to them. This chapter focuses on different aspects of artificial intelligence, such as the transformative of AI and the relationship of cybersecurity with AI. This discussion covers the outcomes due of the integration between AI and cybersecurity.

The risk of AI to cybersecurity is discussed. What impact has AI had in cybersecurity and related areas in which cybersecurity is implemented or used in the light of AI and bringing automation to its processes and activities? This chapter emphasizes threats and impacts on the national, local, and domestic levels of implementation. The reasons and requirement for prediction, prevention, and protection are included in our discussion. We will also discuss the role that predicting and prevention can play in providing protection from identified threats and risk factors.

The broad domain of AI security in the landscape of cybersecurity will also be discussed. This landscape includes the social, economic, political, and digital and physical security domains. Each of the areas will be focused on, providing significant information regarding security risks that are associated because of envisioning the beneficial future as a product of the relationship between AI and cybersecurity.

Technology continues to adapt as its potential abilities increase, including areas such as social media and the use of cybersecurity. This includes additional threats to global AI security. Privacy is at risk around the globe. Therefore, global aspects must be analyzed. This chapter discusses economic, political, physical, digital, ethics, data rights, and privacy aspects globally that are now affected or will be affected shortly.

10.2 What Is Artificial Intelligence (AI)?

Technology has been around for centuries, adopted differently for different purposes of life as new inventions expanded its reach. For example, typewriters, invented in 1714, significantly changed communication, affecting every aspect of life, from business use to government operations to sending letters to friends [13].

Technology has included significant variations as more enhanced, effective, and efficient technology has been produced based on the advancement in science and increased demands globally. Like the typewriter, breakthroughs such as smartphones and supercomputers, which perform tasks and operations at the highest possible speed, have revolutionized every aspect of life, including communications, education, engineering, and scientific operations [14].

AI is one of the most significant breakthroughs in this decade. Artificial intelligence is human intelligence that is simulated and processed by different computer systems and machines. It includes three major processes: learning, reasoning, and self-correction. The learning aspect deals with acquiring information and related rules for using the data for a specific purpose.

Reasoning relies on the rules that include reaching the most relevant conclusion; it is definite in terms of the required information or given instructions. Self-correction corrects any irrelevant information or any flaw in the process or information that is to be provided [15]. The technological advancements do not stop with artificial intelligence; they have made their way into development and improvements in four types of artificial intelligence.

10.2.1 Reactive Machines

Reactive machines are the most basic type of artificial intelligence. They cannot use past experiences and do not form memories for making decisions now or in the future [16]. An example is the chess-playing supercomputer of IBM [17].

10.2.2 Limited Memory

Self-driving cars are the biggest examples of the limited memory of artificial intelligence, including observation of the environment, speed of other vehicles, distances, and directions. This limitation in memory development prevents the vehicle from being able to store and search for memories from the past [18].

10.2.3 Theory of Mind

Another type is a more psychological aspect of AI machines, and it involves how people feel about the advancement of machines and their potential intrusion into things that have always been the realm of human beings. This includes detection of people, creatures, objectives, and expression of people. Therefore, it deals with the theory of mind that understands human social dynamics and normal norms in interactive, first person, and, second person and not like conventional scenario of the third person, and will not be achievable until sometime in the future [19].

10.2.4 Self-Awareness

Type III AIs, as discussed above, are included in this type. This type is a saner and more extensive form of type III AIs. It consists of a self-aware system, forming representations not only about the world but also about other entities to understand that each of us has thoughts, expectations, and feelings and that they have to cope and adjust their behavior accordingly [20].

10.3 The Transformative Power of AI

Considering artificial intelligence as an advanced technology that includes engagement based on cognitive abilities, machine learning, and automation, AI contains significant transformative power to shape the way people perform tasks and foster a future based on the interaction of artificial intelligence, automation, and cognitive engagement. Therefore, the transformative power of AI is determined by evaluating the current progress, changes, and enhancement it has provided in different aspects of life and majorly in businesses and organizational elements.

The cognitive automation of robotics has resulted in significant automation in manufacturing processes. This application is known as robotic process automation (RPA) for completing different tasks such as physical tasks and digital tasks [21]. Big Data analysis has been dealt with efficiently and effectively, made possible through cognitive insight [22]. For example, algorithms are used for Big Data analysis. This Big Data analytics has increased efficiency in detecting the significance of data and detecting the patterns in great volumes of data. Artificial intelligence will transform the way operations are performed by employees and make workplaces more and more automotive than they are right now.

Artificial intelligence will not wholly replace humans with the automation of different tasks because artificial intelligence can only bring automation to the processes or operation that is included or enlisted in a job rather than transforming a job entirely. Still, it requires human interactions. But this transformative power can still reduce the number of employees or time needed to perform a task significantly. It will enable employees to perform tasks more efficiently and provide more productive outcomes [23].

AI will transform the working environment and foster an environment in which better tools are provided for employees to make better decisions. Mundane tasks will be dealt with easily and will free the employees from such tasks. It will enhance the engagement with clients and customers. Major performance increases and will lead to performance gains in terms of having man and machine collaboration in organizational aspects [24].

The transformation of AI can be seen from the performances of several global examples. Grover Ocado is a British pioneer in the giant in e-commerce company known as Alibaba. He has implemented artificial technology in Alibaba's businesses and the processes involved. The results can be seen as Alibaba leads the way in transformative artificial intelligence [25].

Research projects have been initiated to identify and predict the results that AI implementation will have on business. Upon investigation, analysis, and evaluation, the reports and conducted research about the current progress of the companies and organizations globally have provided significant information regarding the transformative power of artificial in terms of the near future outcomes.

The PwC estimated the gross domestic product globally and identified in the report that by 2030, it will be 13.99% higher than it is right now. This growth is because of AI implementations in the relevant departments and the productivity gains from automated business processes. There will be a significant contribution to this transformation due to an increase in the demand, services, and products provided by technology enhancement [26].

10.4 Understanding the Relationship Between AI and Cybersecurity

There have been some major developments in cybersecurity. Techniques and tools are being redeveloped through AI and ML. This integration has paved and deepened the relationship between AI and cybersecurity. However, this relationship can be deeply understood by the roles that AI plays in cybersecurity and what outcomes it has provided or will offer soon.

The detection of threats is one of the significant concerns in cybersecurity and has emphasized it greatly. Officials, authorities, and organizations have always been keen on developing newer and more advanced ways to detect a potential threat or to be prepared for any advanced threat and attack that can take place. ML has played a vital role in this aspect of cybersecurity and strengthened the overall relationship as it has provided effective approaches and outcomes in terms of threat detection in cybersecurity [27].

It has been instrumental in detecting threats by analyzing data and identifying risks in the initial stages. AI has enabled the ways to detect the threat before it affects the system or produces potential vulnerability in the system [28]. Machine learning has made computers effective and enabled them to identify, use, and adapt to the received and/or perceived data, detect algorithms, learn from the observed data and its algorithms, and make changes if required. These changes include improvements. This enables the computer systems to predict the threats before they can appear or take place and observe the peculiarity that includes accuracy than humans [27].

Password protection and authentication have always been one of the major concerns in cybersecurity. AI has enhanced and fostered a much more secure environment in terms of password protection and authentication. In this aspect, biometric authentication has been implemented greatly. AI has been in use for the enhancement of biometric authentication. For example, the face recognition software of Apple's iPhone X devices includes AI-based biometric authentication. Infrared sensors and neural engines are used to detect the user's face [27]. The role of AI, in this case, is the creation of a user's face model that is much more sophisticated, which includes the patterns and correlations identification [29].

Besides this, artificial intelligence is helping cybersecurity in managing various phishing attacks as significant roles are being played in its detection and prevention. AI-ML detects a higher number of sources of active phishing and tracks them. It reacts to phishing sources in a much faster way than any human can. This results in the prevention and control of phishing attacks significantly. One of the significant aspects that AI-ML has unlocked and shaped a path toward it is identifying fake websites and other related sources, which plays a more substantial role in phishing attacks [27].

AI has enhanced the overall network security by contributing to its two major parts: the creation of security policy and identification of the topography of the organization. As time-consuming, AI has hit the right spot as both processes can now be expedited easily. This is done by observing and identifying network traffic patterns. With such identification and observation, effective and efficient security policies are suggested by AI. As a result, much of organizations' time, resources, and efforts are saved and can be invested in other innovations or technological developments [27].

10.5 The Promise and Challenges of AI for Cybersecurity

As technology reaches into different aspects of life, more concerns arise about how AI will enhance or intrude into personal and business life. Artificial intelligence has paved a path where technology is at the core of almost every organization – large and small business, nonprofits, governments, and the life of the individual in different aspects and forms.

Smartphones such as Apple's iPhones with artificial intelligence, to the extent possible, are included in their functions and specifications. But this is what artificial intelligence holds for people in the current era and generation.

Following the expectations of people and or requirements, many promises have also been made that AI will be secure. This is because of its adaptation to different aspects of cybersecurity and enhancing its operation. However, there are some areas and aspects of life where AI and cybersecurity are implemented and already benefit.

The current use of AI in different applications and software used in organizations and businesses has provided beneficial outcomes. These outcomes can be attributed to contributions made by AI-based applications and software such as Google. Google uses deep learning, which is considered the innovative area of AI. This includes image recognition and speech learning. It is widely used in different aspects and products to provide customer offerings and support for businesses in their performance [30].

AI and ML implementation have been growing dramatically in the last few years. Similarly, in the last few years of implementing machine learning and artificial intelligence in different areas of cybersecurity, there is much to expect. It is expected that the products that are based on machine learning will provide significant and productive outcomes due to their increased enhancement and efficiency.

The adaptation of machine learning and artificial intelligence will lead to more accurate, precise, faster, and effective, and overall results than human operators could accomplish. This should result in cost-saving, as will the reduction in employees in areas such as analysis.

Cybersecurity and having effective protective measures will be core concerns for almost everyone. AI will need to prove that it can hold its promise for delivering cybersecurity as it relentlessly deciphers the behavior of users and analyzes the used patterns and recorded information by AI and machine learning.

In case of any sort of irregularities in the pattern or behavior of users, such as suspicious activities and related factors, it makes it easier to detect threats. By analyzing and tracking such data and various models related to it, systems based on artificial intelligence detect any cyber threat and any vulnerability quickly, resulting in recording the whole pattern and behavior related to that attack or threat that occurred. As a result, a more sophisticated security system is formed considering the recorded information and performance. Furthermore, as a result of such functionalities, it makes cybersecurity more effective and efficient and gives faster and timely alerts in case of any suspicious activity or threat before it even proceeds with its plan. The promise of AI in cybersecurity includes even more effective, efficient, secure, and complex security and has fulfilled some promises [31].

The promise fulfilled by AI in cybersecurity is that AI can handle a large volume of data due to its increased efficiency in processing data using machine learning [32]. Machine learning has enabled learning behavior in AI as malicious attacks, and their behavior can be identified. However, with having some advantages and promises being fulfilled, there are some major challenges and threats of AI in cybersecurity. AI is being adopted by cybersecurity but is adopted by attackers [31].

Each year, cyberattacks are on the rise and are getting worse than their impacts in previous years. In this case, cyberattacks have been increasing, and more and more threats and vulnerabilities are occurring. The sharp increase of cyberattacks has resulted in more and more security threats and made it complex due to AI's evolution in cyberattacks or the use of AI by the attackers. As a result, more sophisticated cyberattacks that are more complex than ever are taking place, which is one of the greatest challenges in AI and its implementation and usage in cybersecurity despite all the promises it has fulfilled and is going to deliver [31].

In 2016, almost 36.9 million malware programs contributing to malware attacks were identified worldwide [31]. With the development of IoT and massive devices connected in such networks, the complexity, threats, and vulnerabilities have been increased. More and more hacks have occurred with too much extremity and complexity. The overall potential has been higher as they have become extremely mischievous that even intelligent security systems must be considered for dealing with such threats and preventing adverse outcomes [31].

Sandbox is being considered more and more for detecting and evading malware attacks. However, with the increased advancement and adaptation of AI and machine learning, just as security aspects are becoming sophisticated and advanced, so are cyberattacks due to an adaptation of the latest technology and new ways of carrying out attacks [33]. The rise of state-sponsored attacks is a serious challenge in which technology is part of almost every aspect of life. This issue not only puts individuals at threat but also the entire nation, which is the most concerning aspect of cybersecurity and related elements [33].

Besides the discussed aspects, there is a lack of skilled staff and equipment to implement more advanced cybersecurity devices. IT infrastructure that is spread thin geographically creates a significant challenge to adopt, implement, and make use of such advanced technology accordingly while keeping up with the demands, requirements, and system security. These are the major challenges that are being faced among the promising aspects of AI in cybersecurity [33].

10.5.1 Risks and Impacts of AI on Cybersecurity (Threats and Solutions)

There are also many aspects of cybersecurity that can be assessed with artificial intelligence. The escalation of artificial intelligence in the corporations' enterprises is causing the rise in the respective concerns as well.

Biometric access granted by providing eye prints or fingerprint login has leapt out of the pages of sci-fi novels into reality. It was impressive when users could touch a smartphone to open it; now they just have to look at it and the facial recognition software will bring up the main menu. These advances have enhanced security at multiple levels, previously compromised by hacking. They grant enhanced security to the organizations and cause the framework of the cyberworld to act more securely. These are the advantages of artificial intelligence that affect a more secure system framework for organizations. All these early advancements have granted secure horizons to the previous breathable security framework. These can gratefully be claimed as the proximities of cybersecurity that will be helpful for society and the corporate world.

With the advances in our community, there is an advancement in the world of cybersecurity too. Therefore it is highly essential to have access to the vulnerable spaces limited. Proximities are being worked on about the coverage of the aspects of viruses and malware. In this manner, it will be possible to escape the virus and malware intrusion before it even inflicts the system. These steps would ensure further security in the cyber world. In this manner, there will be enhanced proximities to induce better protection against ransomware and malware.

10.5.1.1 Domestic Risks

The domestic risks and impacts of cybersecurity that are undermined by artificial intelligence are the impacts of imploring AI to act as a watchdog, keeping civilians in a state of surveillance in the name of security. The people are experiencing a situation in which they feel in a state of inequity – they do not know what information governments or corporations are retaining about their personal lives. There must be a solution where enforcement of the law does not ignore people's privacy. There have also been privacy concerns about the issue associated with security cameras and consideration of their captured image as evidence, and more recently, with the massive DNA data voluntarily surrendered to companies that trace one's ancestry. Such concerns are going to grow as advancements in technology continue.

10.5.1.2 Local Risks

In the local debate, many people feel a sense of invasion of privacy. From smart appliances to phones to virtual assistants like Alexa and Siri, surveillance is being imposed on us in ways never experienced before. It is one of the most negative aspects of the advancements in artificial intelligence that are currently surfacing. The enhancement of the proximities of capabilities and development also gives us a whole spectrum of such concerns to deal with. With the advancement of artificial intelligence, there will be issues associated with such concerns.

Legal scholars are assessing the legal implications of some of these issues. Some cases have developed positively from this aspect, like those associated with security cameras and artificial intelligence, and others have spawned changes in the law to protect individual rights.

10.5.1.3 National Risks

AI is only as smart as the data that feeds it, a situation called *data diet vulnerability*. When it comes to national aspects, faulty data input can create cybersecurity risks. Lacking training factors or good data can result in falsehood and, therefore, some major attacks. In this case, the application of AI and cybersecurity in national security can open a new attacking vector based on the vulnerability of the data diet. The corruption resulting from Microsoft's AI Chatbot is the best demonstrator of such risk and its impact on the national aspect and security [34].

Based on the identified factors for national aspects and security, data used to train AI and AI-based systems must be viable enough. This can be done by testing data based on a proper schedule, testing, and simulations to identify any lacking factor to eliminate vulnerabilities and any underlying factor that is risky.

10.5.1.4 Why Prediction and Prevention

Prediction of possible threats via testing, simulating, training, and testing again must take place repeatedly to identify outcomes. This prediction is necessary because of the newly introduced era of automation, as it may hold potential negative and disastrous outcomes. With an emphasis on prediction, prevention of predicted issues, and identified underlying factors, risks, and impacts, prevention is necessary to ensure healthier and safe practices are carried out via AI-based systems. The fusion of AI and cybersecurity in a much more secure manner will exploit autonomous systems to their fullest. This should include the obligation to innovation as security and intelligence agencies (SIAs) are required to focus. It is necessary to predict and protect public expectations of privacy, or else it would include disastrous outcomes for the public and or nation [35].

Cyberattacks are taking place at a grander scale, and the prediction of various reports has already shown that. The attacks will occur in a much more outraged way if preventive measures and aspects are not in focus [36]. Therefore, it is necessary to focus on AI and ML in phishing detection and prevention and their control, use of AI and ML in the vulnerability management of businesses and organizations, and at a local scale, which is AI-based password protection and authentication [27].

10.6 Broad Domain of AI Security (Major Themes in the AI Security Landscape)

AI is interactive in different and major systems of today's world. This includes the economic system, political system, and social system. Based on this interaction, AI security can be defined as the resiliency and robustness of AI systems.

10.6.1 Digital/Physical

The physical domain of AI has expanded over the years and resulted in major advancements. This includes self-driving Uber cars and talking and walking robots. However, this digital and physical domain includes some potential threats as well. These threats are too critical and can result in the death of an individual. With such advancements, critical dangers are already being exposed.

In 2018, an Uber vehicle was testing its autonomous driving mode. Due to some malfunction in which Uber sensors did not detect the pedestrian, an Uber vehicle in autonomous mode hit and killed Elaine Herzberg, who was walking her bicycle across the street. Another case in which critical outcomes resulted was a robot that was designed at China Hi-Tech in Shenzhen for kids and related activities. The robot was named Xiao Pang.

The robot suddenly started to function automatically and rammed itself repeatedly in a display booth. The glass broke and went flying and injured a man. He was admitted to the hospital due to a severe injury. In 2017, Facebook revealed some information regarding their AI bots designed to communicate with people. Later, it was found that both of their AI bots had developed a complete language in which they started to communicate with each other. Both of the bots were turned off [37].

10.6.2 Protection from Malicious Use of AI and Automated Cyberattacks

As already discussed, with the advancements of AI, cybersecurity has advanced due to its adaptation, but cyberattacks and attackers have gained more significant advantages as well, which resulted in potential impacts and outcomes. AI and based tools and devices have provided ease to attackers as cracking complex security protocols, and the use of machine learning has enabled them to carry attacks in a more sophisticated way that is even more effective. This situation has been proved in the DARPA Cyber Grand Challenge as autonomous hacks were performed [38].

The AI-based defense system is much more advanced than ever, but cyberattacks are getting stronger and more complicated. Therefore, it is necessary to institute the right communication channels and processes to ensure the integrity and enhancement of AI-based security and defense systems.

10.6.3 Other Technologies with AI and Their Integration

With the development of different technologies and interactions of AI with those technologies, the convergences offer more significant opportunities and related threats, such as military and associated aspects. For example, the decision-making in the military has become more effective and efficient with potential and beneficial outcomes. However, the use of AI in decision-making will impact destabilized on the nuclear strategic balance as a result of a flaw in the AI system.

Other advancements of AI and their integration with different technologies include blockchain, robotics of various sizes, types, and nature, aerospace, and bioengineering. With the use of AI in such technological areas, overall outcomes are more integrated and beneficial. Still, they are hard to control and more dangerous due to the increasing involvement of technology [39].

10.6.4 Political

In the political domain, there are various areas in that AI is involved in one way or another.

10.6.5 Manipulation and Disinformation Protection

Social media networking and any other communication channel have included much misleading information or disinformation and manipulation. There can be greater support by AI technologies and systems based on AI for the campaigns of disinformation. They can be enabled more and can become more productive, centric to objective, narrowing down required outcomes in a much more potent way, and overall scale can be increased. Undermining of democratic aspects can result from such campaigns, which can also result in targeting the overall population of minorities in a country by including a more effective and efficient process. This process can result in many harmful outcomes due to the spread of disinformation and include greater loss of or widespread media communication trust [40].

There are already some examples of this, such as the presidential election of 2016. It was discovered using social media bots generated a higher number of posts, and the percentage was surprisingly high. In this case, nearly a third of pro-Trump tweets were identified to be the fake generation, while a fifth of pro-Clinton tweets were the result of fake production, *Atlantic* reported on 1 November 2016 [41].

Based on the discussed democratic and political aspects, the overall relationship of people or nations can be disrupted by such interference of AI technologies. However, AI technology has contributed significantly to detecting and providing the right, valid, and required information that chases it to its core. Therefore, it is necessary to focus on using AI to enhance engagement and promote the positive and trustful relationship of citizens with their governments as AI systems are helping in managing different constraints. Constraints such as resources and administrative aspects encourage more robust behaviors and gain attention in a positive manner and aspects.

10.6.6 Infrastructure Based on AI and Digital Expertise of Government

There is increased attention toward AI technologies and systems adaptation in governmental aspects by governments worldwide. It has been proved helpful as management aspects of governments and their operations, required resources and removal of constraints, administrative burdens that are easily managed, and other such aspects are enhanced and improved. Optimization of planning and scheduling are the major outcomes that support different governmental organizations and result in better and more effective customer services with increased, faster and better support. An example of such enhancement and improvement is the automation in data related to governments' operations and customers [42].

The infrastructure is based on digital and AI technology greatly, and governments are focused even more on balancing the goals and benefiting from the advancements of technology; however, due to such nature of infrastructure, which is more digitalized and is based on AI. Threats related to AI and related technologies also wander around it. Governments are also considering standards and policies to secure the environment.

10.6.6.1 Economic
The economic domain of AI security includes some benefits as well as some full threat outcomes.

10.6.6.2 Labor Displacement and Its Mitigation
AI and machine learning usage enhance security in the work environment. Unfortunately, AI-based technology and systems have been affecting one of the major security aspects, which is job security, and is affecting the economy differently, in this case. With the adaptation of AI technology in different organizations and companies, many complex tasks that once required a greater labor force are now carried out more easily and with fewer people. It includes automation in complex processes as well.

The introduction of automation in complex business processes and operations has endangered the labor force and increased job loss. The Organization for Economic Cooperation and Development (OECD) has estimated that automation through technological advancements and AI-based robotics will result in a 13.99% job loss in advanced economies [43].

A study that Oxford carried out has identified that automation will take over greatly in the next few decades, such that almost 46.99% of jobs will be at risk in the United States. A report by McKinsey, which emphasized economic aspects in upcoming years, has identified that almost 799 million jobs will be affected by automation by the end of 2030 [44]. The World Economic Forum estimated a loss of 85 million jobs worldwide by 2025 – but at the same time, 97 million new jobs would be created [45]. AI will create millions of jobs in robotics, Sean Chou, former CEO of AI startup Catalytic, says that as more and more things become automated, "you move from worrying about the impact of high technology to actually helping to create the technology. When you look at AI, there's this nonstop need for training, for data, for maintenance, for taking care of all the exceptions that are happening" [46]. It is important to monitor and train AI, to make sure it is not "running amok." These could all be sources for new jobs [46].

10.6.6.3 Promotion of AI R&D
AI technology and systems based on AI and their advancements are considered the promise and fuel of substantial economic growth. However, a significant amount of investments must be included to gain such advantageous outcomes. In this case, research and development are considered the foundation upon which such economic growth can occur. China has increased its investment in research and development for AI significantly. The overall increase in investment in AI is

200%. This increase and growth in investment occurred from the year 2000 to the year 2015 and was expected to surpass the investments made by the United States [47].

10.6.6.4 Education and Training That Is Updated

For propelling AI and related technology development, countries became education and training centric to have more effective development and implementation of AI and related technologies. Therefore, educational aspects included more and more focus on providing opportunities related to AI aspects and developed a deep connection to different areas of education: science, technology, engineering, and math, also known as STEM.

AI education and learning have not been limited to just STEM but have also focused on humanities and social sciences and are more transparent and apparent, such that it is now being called the "soft skills." There is some limitation in conducting AI research and including it in education which is the lack of AI researchers and experts of AI. Therefore, the government focused on and increased the number of people trained over the years in AI and are experts in AI and related areas for helping in educational courses and programs.

As discussed, the elimination of jobs will take place on a much greater scale in upcoming years due to automation, so there is a stronger need for retaining and adjusting accordingly to meet the requirement of a job linked with AI and areas. Therefore, education and training in these growing fields will be essential [48].

10.7 Transparency of Artificial Intelligence and Accountability Societal Aspects

Just like other domains, AI has been proved beneficial and threatening in societal aspects as well.

With the increased adaptation of AI systems, integration has resulted in various areas of society. In this case, decision-making and integration are one case. Capacities of decision-making have been increased in different areas like legal scrutiny, individual screening for police, medical diagnosis, different loan applications, and their reviewing. The accountability of AI in society has been improved as transparency in decision-making has been improved in the identified areas.

There are various reasons for discussing accountability improvement as an AI system. Their programming has made it easier to describe the whole decision-making process to humans in a much more quick and understandable way. Therefore, the overall accountability of AI and AI-based systems has been improved. Not only this, but the responsibility is also yet in consideration to be enhanced even more. However, there are some complications and challenges as well in the social domain.

One of the major challenges that are included in the social domain is the legal aspects, as questions are raised regarding AI accountability in terms of having legal standards, policies, and regulations to make it understandable by the AI system. How can the alignment of objectives of policy and legal purposes take place? These are significant areas of concern and are considered vital for the near future of AI security in the social domain.

Besides the discussed aspects of the social domain, AI has made decision-making free of the biases of humans. Every decision that the AI takes to be the right one and use is not biased based on sex, religion, color, norms, and cultural differences. However, the doubts regarding AI systems affect the overall decision-making and its outcomes significantly. The dominant mechanism of AI diminishes the administrative burden as well. Such utility of AI-based systems can only be useful and productive if society starts to trust the decision-making of AI and its results.

10.7.1 Rights of Privacy and Data

When it comes to data flow throughout the organizations and different online platforms such as Facebook, the privacy of data and rights to it has been a significant concern. AI security has interfered with the rights of privacy and data much. With the adaptation of AI-based systems in such platforms and organizations, new laws and regulations have been passed in recent years for ensuring that rights are protected, citizens are saved from data theft and misuse.

A European privacy law, The General Data Protection Regulation (GDPR), has enabled the right to access the information from the companies and organizations and how data is used, as well as how and for what logical reason the automation or AI has been used in different processes.

Privacy acts are considered to deal with the questions that have been resulted due to advancements in AI-based automation systems and increased concerns about consumers' privacy. California Consumer Privacy Act of 2018, implemented on January 1, 2020, provides four rights to citizens regarding their personal information. This act will grant rights and permission to customers to inquire how their data is being held, used, and via what platform.

The rights and permissions include requesting companies to delete personal information or opt out of data being sold to third parties. In any case, repulsion of these rights or discrimination for using and exercising these rights must not occur. AI has resulted in more sophisticated attacks and potential threats; however, it has also been used to increase social protection by securing their rights, resulting in regulations, laws, and policies and increasing overall security.

10.8 Global AI Security Priorities

An analysis performed by McKinsey & Company, a US state-run consulting firm, of more than 400 use cases across 19 industries and 9 business functions highlights the broad use and significant economic potential of advanced AI techniques to create $3.5 trillion to $5.8 trillion in value annually [49]. An additional incentive for advertising and business is expected to be $2.6 trillion and $2 trillion in-store network, stores, and meetings. Artificial intelligence will have an incidence of 11.6% in the income of the motor industry and up to 10.2% in fluctuations in innovative income. In retail, AI will increase the value of valuation and development and customer management by $100 billion. McKinsey reports that the AI's assessment is not based on the models themselves, but on the ability of organizations to use them. Despite the fact that the use of AI strategies has monetary potential, concerns such as data security, protection, and potential tilting issues need to be addressed [49].

10.8.1 Global Economy

AI and its security aspects have been identified as a game-changer in different global aspects. One of the major ones in all global aspects is the global economy. PwC analyzed the role of AI security and its priority in the global economy. The reason that AI is considered the game-changer in a global economy is because of its value potential. AI contributes greatly to the global economy as the outcomes can be almost $15.8 trillion by 2030 throughout different aspects of the economy.

The study showed that almost $9 trillion would be resulted from the side effects of consumption, while $6.7 trillion will be due to the enhancement and increase in productivity. However, the impact of AI on the economy globally will be different in different countries and regions based on the potential to gain more than others. In this case, North America will have 14.5% of its GDP, which is estimated to be $3.7 trillion, while North Europe will have an impact, which will increase by 10% of GDP. This estimation is almost $1.8 trillion.

China will have almost $7 trillion due to the impact, which will be almost 26% of its GDP. Asian, Oceania, and African markets will impact almost 5.5% and will result in $1.2 trillion. The study identifies that by 2030, all of the region's global economy will benefit from the adaptation of AI and increased security. In this case, China and North America will have greater economic gains through AI adaptation, enhancement, and improvements.

The global impact of AI on our economy would perhaps be more than the industrial revolution. Robots and inexpensive advanced technologies create a challenge to human labor value. AI will improve the economy by speeding workflow, reducing costs, increasing accuracy and precision, and enhancing the efficiency and performance of businesses. Figure 10.1 [50] and Figure 10.2 [51] show the potential of AI in general and particularly in healthcare (hospitals, healthcare payers, providers, and pharmaceutical) where North America is the advanced region and key contributor of AI to healthcare.

10.8.2 Global Privacy and Data Rights

Privacy concerns for the consumers and users of such advanced technology have made it a more debating aspect globally. Therefore, efforts are being made worldwide to provide increased security to consumers and protect their privacy. For this, the GDPR has been spotlighted and went into effect in 2018 [52]. This act focused on consumer privacy and legalized the protection greatly. Similarly, the California Consumer Privacy Act implemented in 2020 will allow people to manage personal information used by organizations in their automated processes and specify for what purposes the data is used [53].

The standards that are focused globally to ensure the protection of data and privacy of people worldwide, including related aspects. This includes that AI systems must be transparent in performing operations. There must be rights for the information collected by the AI. More clearly, AI systems must hold rights to the information that is to be collected. On the other hand, consumers will also be provided the right to opt out of any information from the system they would not like to

AI CANDOUBLE GROWTHRATES

A comparison of baseline annual gross value added growth (%) in 2035 to a scenario where AI has been absorbed.

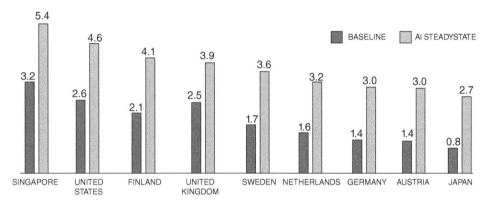

Figure 10.1 Economic growth rate across various countries.

ARTIFICIAL INTELLIGENCE IN HEALTHCARE MARKET, BY REGION
(USD BILLION)

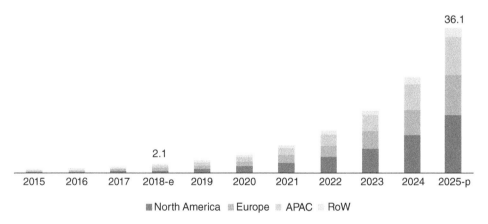

Figure 10.2 AI and healthcare across various regions.

share or to be collected. The aim and purpose of AI will be made limited in terms of data collection and its use to be more specific [54].

Citizens can easily make requests, as they will be granted the right to delete some data they would not like to include. A study conducted in 2018 measured the cost of data breaches globally, which is almost $3.64 million due to negligence in the security aspect, which costs the privacy of millions and millions and even greater cost as identified globally. Therefore, the emphasis has been laid on privacy aspects worldwide, emphasizing making policies, regulations, and laws [54].

10.8.2.1 AI and Ethics

Ethical concerns have been raised according to the adaptation of AI and globally. With such greater adjustment and bringing automation in different aspects of life, the ethical aspects of each element cannot be simply ignored just for successive outcomes as it may impact the societies significantly. In this case, questions have been raised about which ethical aspects are in focus by AI systems and what ethical aspects are ignored. There have been programs and initiatives introduced worldwide to ensure that ethical standards and regulations comply with the use of AI and its purpose correctly.

Different countries are participating in assuring ethical standards and regulations to be followed properly and effectively. There are already different governance, policy, and existing laws that will be applied to AI and its implementations. These features include fundamental rights, administrative and industrial rights, privacy laws, intellectual property, and data protection. Europe and China are leading in such initiatives, programs, standards, policies, and regulations as compared to other countries that are focused on identifying ethical aspects or addressing any relevant ethical issues linked with AI systems, technologies, and their implementations [55].

10.8.3 Automation of Cyberattacks or Social Engineering Attacks

With the involvement of AI and machine learning, cyberattacks have become more advanced. There is a rise in such attacks around the globe. The Internet Security Threat Report of 2018 of Symantec showed that phishing or spear-phishing attacks have increased over the years. It was identified as the most used attack as it was held responsible for three-quarters of the attack and making it the most successful and used attack [56].

Automation in cyberattacks was identified as a successful vector for attacks. It was identified that automation would increase the vulnerability of digital security and increase the problem in a greater manner. This issue can be seen worldwide as digital security is under more threat than ever before [56]. Social engineering attacks are more automated with the development of AI, as it has created Chatbots. Such bots elicit the trust of humans as they engage with people for longer chats and can result in obtaining personal, confidential, and credential information.

The automation resulted in the creation and generation of automated attacks such as malicious links, emails, and websites are generated that mimic the official and professional look, and language and engagement are made from addresses that look like real contacts. As a result, automation reduces any error in such attacks, and social engineering takes place in a much more proper, effective, and potential way.

10.8.4 Target Prioritizing with Machine Learning

When it comes to using the internet, online behavior was used to identify the interest of an individual to offer related products, ads, and price ranges. This practice was used to increase customer satisfaction, improve their experience, and increase sales for organizations and businesses. However, this online behavior is used as a digital threat as it determines and prioritizes targets with such machine learning.

Machine learning is used to track the online behavior of people and determine them as a target. This identification of online behavior helps prioritize the target for attack based on his/her online behavior of purchasing products, financial activities related to transactions made, and products purchased. In other words, the identification of wealth takes place and identifies the target to be prioritized accordingly. As a result, demands are made based on their online behavior and condition of wealth.

10.9 Automation of Services in Cybercriminal Offense

AI technology and its techniques are significantly used to automate the processes and services included in cybercriminal activities. Cybercriminals and their implementation of AI techniques are used to automate the tasks that make up their attack pipeline. This includes processing payment-generation dialogs with the victims of ransomware attacks. With the experience of more human-like interaction throughout the attack in a much more precise manner, digital security has been on edge globally as such attacks are carried out more successfully.

10.9.1 Increased Scale of Attacks

With the increase in machine learning, human interaction with AI has been enhanced more and over the period. Autonomous systems have increased the amount of damage that even small groups of attackers can do. For example, the usage of drones has grown around the globe and is considered a breakthrough in technology.

There is an exponential growth in unmanned aerial vehicles (UAVs) and drones greatly around the globe. The studies show that the predicted that the use of drones worldwide would increase and will worth more than $124.99 billion [57].

With the extensive usage of drones equipped with AI systems worldwide, these unmanned vehicles have turned into killer robots.

With the development of autonomous vehicles, some positive aspects were spotlighted and identified as beneficial shortly. But, with such greater good, there are some disadvantages and drawbacks perhaps more than the benefits. In this case, the availability of AI systems can be exploited by attackers in different ways. A report in February 2022 predicted that terrorists would use autonomous vehicles for the deliverance of explosives that will be used for attacks. Terrorists that have low skills on their own will have access to high-end technology and products based on it, and the attacks carried out by such people will result in more physical damage around the globe [58].

However, the use of AI in different security scenarios has proved very beneficial for enhancing overall aspects of military, recreational, humanitarian, and commercial applications worldwide. Across different countries, the application of robotics and enhancement resulted in supply chains, production, and distribution worldwide in a more secure and precise manner compared to past years [59].

The use of drones also makes it easier to identify, track, and capture criminals and stop terrorism. However, not all drones and systems are AI-based. To some extent, human interaction is required. A greater autonomous era is on the horizon, though, including greater application to increase security and different other applications [59].

10.10   The Future of AI in Cybersecurity

Cyberattacks are one of the major threats to businesses, government, and institutions. Data breaches have become common in today's world, and most of the misused exposures have occurred because of overreliance on firewalls as a defense. Because the firewalls cannot seize the determined hackers, artificial intelligence can only be worthy of defending against the hackers. Cyberattacks can simply be recorded in the system to learn and check the patterns and recognize their eccentricity [60].

It is very difficult to stay safe online, which is why artificial intelligence will replace the cybersecurity site in the upcoming years. This artificial intelligence will eliminate the human element in gatekeeping, and it will be the technology that will control cybersecurity in the networks. It simply seems to be a hide-and-seek game among the criminals who are the reason for online threats and the people who provide security to the networking sites and web pages. However, artificial intelligence can provide better identification of cyberattacks and cyberthreats. Every time people connect their devices to the internet, the risk of cyberattacks increases. Data breaches are no more difficult for the attackers because if the threat is aimed at a single workplace, the whole organization can be affected [61].

The major part of legal and cybersecurity tools need human interaction or configuration to some extent. Still, due to increased cyber threats, companies will be compelled to rely on smart tools and IoT devices. This is the event monitoring and incident responses of this bulk are not possible without technology. In the few upcoming years, firewalls will possess machines learning technical skills that will allow the software to identify the recorded arrangements of codes in the web request and automatically block if some kind of threat or deviation is identified [62].

AI will play an important role in the future. Many researchers have presented their theories that for the scanning of enormous data across the internet, only artificial intelligence can know about the method of generating cyberattacks. Theorists believe that only AI can provide a solution to the decision-makers of the organizations through the frameworks, and the security products of artificial intelligence are not accessible and easily available [63]. This is an issue for small or moderate business organizations because it is difficult for them to hire and provide support to ML experts in building custom cybersecurity solutions. It will take a huge investment in purchasing the hybrid tools that are embedded by artificial technology.

Loopholes still exist in today's world as hackers can easily find the methods to beat the machines by sneaking into the system and cracking the passwords. Most people do not know about a data breach because it requires a few months for an organization to recognize the data breaches in their company [64].

10.11   Conclusion

Several aspects are covered in this chapter about AI and cybersecurity and their impact on each other. This includes social, domestic, local, economic, and political aspects globally. The fusion of AI and cybersecurity has contributed greatly to increasing and enhancing productivity and efficiency. It has been proved beneficial for organizations as the operations they perform or are carried out have become more time-efficient, cost-effective, and productive. For example, automation in the processing and packing of products.

There are already some promises that AI has met about cybersecurity and have a significant impact on economic, social, local, and political aspects worldwide. The signs it has fulfilled already are the automation in vehicles, processes, and activities performed in daily life, including smartphones with enhanced and increased security due to AI-based face detection, authentication, and biometric password protection. Many software platforms with AI are used in businesses, organizations, and enterprises that have been around for an extended period. For example, Google, as it includes deep learning, and identifies online behavior, enhances overall experience according to the interests of individuals, and other related processes are included.

The use of deep learning by Google is the cutting edge of AI in such an area, such as image recognition and speech learning. The relationship between AI and cybersecurity has promised more productive and effective outcomes due to machine learning. Organizations and enterprises will have more cost savings with increased efficiency, accuracy, and precision and will have more effective protective measures. AI has promised greater enhancement in cybersecurity as it will include automation in its processes and activities. As a result, threat detection will take place at a very early stage, and attacks can be prevented before they even begin. This is made possible by bringing automation to cybersecurity processes and providing more complex security compared to current security protocols, measures, and frameworks, making cyberattacks ineffective and unsuccessful.

There are some risks and impacts of AI on cybersecurity that are identified. These impacts are discussed in on social, domestic, local, and national aspects. Privacy concerns are identified at the individual level or in communities that are local or domestic. National security may be at stake due to the increased advancement of AI and cybersecurity despite advancements in different fields. In conclusion, with more significant improvements and profitable outcomes to this date, AI and cybersecurity hold some threatening aspects and wander as a cloud of fear in each area discussed.

The relationship between AI and cybersecurity has great promise, as national, local, domestic, and social security is considered and significantly focused. The vital role of prediction will enable the prevention of threats and risks before they can occur or overall impacts can be mitigated. The future of AI and cybersecurity discussed briefly here identifies that significant beneficial advancements are still to come; as the partnership improves productivity and increases global revenues by billions and trillions over the next decade, some of these revenues can be invested in a more secure and sound infrastructure.

References

1 Rouse, M. (2019, July). Cybersecurity. https://searchsecurity.techtarget.com/definition/cybersecurity#targetText=Cybersecurity%20is%20the%20protection%20of,centers%20and%20other%20computerized%20systems.

2 Cisco (2019). What is cybersecurity? https://www.cisco.com/c/en/us/products/security/what-is-cybersecurity.html#targetText=Three%20main%20entities%20must%20be,software%2C%20and%20email%20security%20solutions.

3 Abomhara, M. (2015). Cybersecurity and the internet of things: vulnerabilities, threats, intruders, and attacks. *Journal of Cyber Security and Mobility* 4 (1): 65–88.

4 Kharraz, A., Robertson, W., Balzarotti, D. et al. (2015). Cutting the gordian knot: A look under the hood of ransomware attacks. *The Conference on Detection of Intrusions and Malware, and Vulnerability Assessment* (July 9): 3–24.

5 Everett, C. (2016). Ransomware: to pay or not to pay? *Computer Fraud & Security* 4: 8–12.

6 Porter, K. (2019). Malware. https://us.norton.com/internetsecurity-malware-malware-101-how-do-i-get-malware-complex-attacks.html#targetText=Written%20by%20Kim%20Porter%20for,usually%20without%20the%20victim's%20knowledge.

7 Kromholz, K., Hobel, H., Huber, M., and Weippl, E. (2015). Advanced social engineering attacks. *Journal of Information Security and Applications* 22 (C): 113–122.

8 Atkins, B. and Huamg, W. (2013). A study of social engineering in online frauds. *Open Journal of Social Sciences* 1 (3): 23–32.

9 Palmer, D. (2019, 23rd July). These are the most common types of phishing emails reaching your inbox. https://www.zdnet.com/article/these-are-the-most-common-types-of-phishing-emails-reaching-your-inbox.

10 SAS (2019). Neural networks. https://www.sas.com/en_us/insights/analytics/neural-networks.html.

11 Marr, B. (2018). What is deep learning AI? A simple guide with 8 practical examples. *Forbes* (October 1): https://www.forbes.com/sites/bernardmarr/2018/10/01/what-is-deep-learning-ai-a-simple-guide-with-8-practical-examples/#44e050d58d4b.

12 Daley, S. (2019). 19 examples of artificial intelligence shaking up business as usual. *BuiltIn* (September 24): https://builtin.com/artificial-intelligence/examples-ai-in-industry.

13 Xavier (2019). A brief history of typewriters. https://site.xavier.edu/polt/typewriters/tw-history.html.

14 Rouse, M. (2019). Supercomputer. https://whatis.techtarget.com/definition/supercomputer.

15 Burns, E., Laskowski, N., and Rose, M. (2019). AI (artificial intelligence). https://searchenterpriseai.techtarget.com/definition/AI-Artificial-Intelligence.

16 Granda, J.M., Donina, L., Dragone, V. et al. (2018). Controlling an organic synthesis robot with machine learning to search for new reactivity. *Nature* 559 (7714): 377–381.

17 Sharples, J. (2017). *A Cultural History of Chess-Players Minds, Machines, and Monsters.* Manchester University Press.

18 Joshi, N. (2019). 7 types of artificial intelligence. *Forbes* (June 19): https://www.forbes.com/sites/cognitiveworld/2019/06/19/7-types-of-artificial-intelligence/#2263cc8233ee.

19 Hintze, A. (2016). Understanding the four types of artificial intelligence. https://www.govtech.com/computing/Understanding-the-Four-Types-of-Artificial-Intelligence.html.

20 School, S. (2018). Machines demonstrate self-awareness. https://becominghuman.ai/machines-demonstrate-self-awareness-8bd08ceb1694.

21 Aalst, W.M., Bichler, M., and Heinzl, A. (2018). Robotic process automation. *Bus Inf Sys Eng* 60 (4): 269–272.

22 Lohr, S. (2012). The age of big data. New York Times.

23 Franklin, T. (2018). f5 reasons AI won't replace humans. . . it will make us superhuman. https://hackernoon.com/5-reasons-ai-wont-replace-humans-it-will-make-us-superhuman-413c499e1e68.

24 Moreno, A. and Redondo, T. (2016). Text analytics: the convergence of big data and artificial intelligence. *IJIMAI* 3, Special Issue on Big Data and AI (6): 57–64.

25 Lancefield, D. (2019). The transformative power of artificial intelligence. https://www.strategyand.pwc.com/uk/en/transformative-power-ai/transformative-power-artificial-intelligence.html.

26 PWC (2019). The transformative power of artificial intelligence. https://www.strategyand.pwc.com/uk/en/transformative-power-ai/transformative-power-artificial-intelligence.html.

27 Ramachandran, R. (2019). How artificial intelligence is changing cyber security landscape and preventing cyber attacks. https://www.entrepreneur.com/article/339509.

28 Ahmad, N. (2019). Artificial intelligence: A tool or a threat to cybersecurity? https://readwrite.com/2019/08/21/artificial-intelligence-a-tool-or-a-threat-to-cybersecurity.

29 Madhavan, R. (2019). AI in biometrics and security – current business applications. https://emerj.com/ai-sector-overviews/ai-in-biometrics-current-business-applications.

30 Marr, B. (2019). *Artificial Intelligence in Practice: How 50 Successful Companies Used AI and Machine Learning to Solve Problems.* United Kingdom: Wiley.

31 Ghanchi, J. (2019). The possibilities of AI and machine learning for cybersecurity. https://thenewstack.io/the-possibilities-of-ai-and-machine-learning-for-cybersecurity.

32 West, D.M. and Allen, J.R. (2018). How artificial intelligence is transforming the world. https://www.brookings.edu/research/how-artificial-intelligence-is-transforming-the-world.

33 Banafa, A. (2018). Challenges facing using AI in cybersecurity. https://www.bbvaopenmind.com/en/technology/artificial-intelligence/challenges-facing-using-ai-in-cybersecurity.

34 Osoba, O.A. and Welser, W. (2017). *The Risks of Artificial Intelligence to Security and the Future of Work.* RAND.

35 Babuta, A. (2019). A new generation of intelligence: National security and surveillance in the age of AI. https://rusi.org/commentary/new-generation-intelligence-national-security-and-surveillance-age-ai.

36 Dixon, W. and Eagan, N. (2019). Three ways AI will change the nature of cyberattacks. https://www.weforum.org/agenda/2019/06/ai-is-powering-a-new-generation-of-cyberattack-its-also-our-best-defence.

37 Hebbar, P. (2017). Balls have zero to me: What happened when Facebook's AI chatbots Bob & Alice created their own language. https://www.analyticsindiamag.com/facebook-ai-chatbots-created-their-own-language.

38 Solon, O. (2016). These engineers are developing artificially intelligent hackers. https://www.theguardian.com/technology/2016/mar/03/artificial-intelligence-hackers-security-autonomous-learning.

39 Sharikov, P. (2019). Artificial intelligence, cyber attacks, and nuclear weapons: A dangerous combination. https://www.eastwest.ngo/idea/artificial-intelligence-cyber-attacks-and-nuclear-weapons-dangerous-combination.

40 Feldstein, S. (2019). How artificial intelligence systems could threaten democracy. https://www.govtech.com/products/How-Artificial-Intelligence-Systems-Could-Threaten-Democracy.html.

41 Guilbeault, D. and Woolley, S. (2016). How Twitter bots are shaping the election. *The Atlantic* (November 1): https://www.theatlantic.com/technology/archive/2016/11/election-bots/506072.

42 Eggers, W.D., Schatsky, D., and Viechnicki, D.P. (2017). AI-augmented government. https://www2.deloitte.com/us/en/insights/focus/cognitive-technologies/artificial-intelligence-government.html.

43 OECD (2019). *OECD Employment Outlook 2019: The Future of Work.* Paris: OECD Publishing.

44 Meyer, D. (2017). Robots may steal as many as 800 million jobs in the next 13 years. https://fortune.com/2017/11/29/robots-automation-replace-jobs-mckinsey-report-800-million.

45 World Economic Forum (2020). *The Fugure of Jobs Report 2020.* World Economic Forum https://www3.weforum.org/docs/WEF_Future_of_Jobs_2020.pdf.

46 Thomas, M. and Powers, J. (2023). Robots and AI taking over jobs: What to know about the future of jobs. https://builtin.com/artificial-intelligence/ai-replacing-jobs-creating-jobs.

47 Vincent, J. (2019). China is about to overtake America in AI research. https://www.theverge.com/2019/3/14/18265230/china-is-about-to-overtake-america-in-ai-research.

48 European Council on Foreign Relations. (2019). Harnessing artificial intelligence. https://www.ecfr.eu/publications/summary/harnessing_artificial_intelligence.

49 Chui, M., Chung, R., Henka, N. et al. (2018). Notes from the AI frontier: Applications and value of deep learning. https://www.mckinsey.com/featured-insights/artificial-intelligence/notes-from-the-ai-frontier-applications-and-value-of-deep-learning.

50 Williams, A. (2017). Singapore's economic growth rate could nearly double by 2035 with artificial intelligence: Accenture. https://www.straitstimes.com/business/economy/singapores-economic-growth-rate-could-nearly-double-by-2035-with-artificial

51 Baiju, N. (2018). Artificial intelligence in healthcare worth $36.1 billion by 2025 – Report. https://markets.businessinsider.com/news/stocks/artificial-intelligence-in-healthcare-market-worth-36-1-billion-by-2025-exclusive-report-by-marketsandmarkets-1027814602.

52 Satariano, A. (2018). What the GDPR, Europe's tough new data law, means for you. *The New York Times* (May 6). https://www.nytimes.com/2018/05/06/technology/gdpr-european-privacy-law.html.

53 Morgan Lewis (2019). California's consumer privacy act and other state privacy & security laws. https://www.morganlewis.com/topics/ccpa-and-state-privacy-security-laws#targetText=California%20has%20passed%20a%20comprehensive,Jerry%20Brown%20the%20same%20day.

54 Forbes (2019). Rethinking privacy for the AI era. https://www.forbes.com/sites/insights-intelai/2019/03/27/rethinking-privacy-for-the-ai-era/#229dcce37f0a.

55 Daly, A., Hagendorff, T., Li, H. et al. (2019). *Artificial Intelligence, Governance, and Ethics: Global Perspectives.* The Chinese University of Hong Kong Faculty of Law.

56 World Economic Forum (2019). AI raises the risk of cyberattack – and the best defense is more AI. https://www.weforum.org/agenda/2019/04/how-ai-raises-the-threat-of-cyberattack-and-why-the-best-defence-is-more-ai-5eb78ba081.

57 Choudhary, M. (2019). What are the popular uses of drones? https://www.geospatialworld.net/article/what-are-popular-uses-of-drones/#targetText=Thanks%20to%20their%20ability%20to,and%20firefighting%2C%20has%20only%20multiplied.

58 Ware, J. (2019). Terrorist groups, artificial intelligence, and killer drones. https://warontherocks.com/2019/09/terrorist-groups-artificial-intelligence-and-killer-drones.

59 Brundage, M., Avin, S., Clark, J. et al. (2018). The malicious use of artificial intelligence: Forecasting, prevention, and mitigation. https://doi.org/10.48550/arXiv.1802.07228.

60 Dilek, S., Çakır, H., and Aydın, M. (2015). Applications of artificial intelligence techniques to combating cybercrimes: A review. 15–19. https://arxiv.org/abs/1502.03552.

61 Link, J., Waedt, K., Zid, I.B., and Lou, X. (2018). Current challenges of the joint consideration of functional safety & cyber security, their interoperability and impact on organizations: how to manage RAMS+ S (reliability availability maintainability safety+ security). In: *12th. International Conference on Reliability, Maintainability, and Safety (ICRMS)*, 185–199. Shanghai, China: IEEE https://ieeexplore.ieee.org/abstract/document/8718942.

62 Demertzis, K. and Iliadis, L. (2015). A bio-inspired hybrid artificial intelligence framework for cybersecurity. In: *Computation, Cryptography, and Network Security*, 161–193. Cham: Springer https://link.springer.com/chapter/10.1007/978-3-319-18275-9_7.

63 Kott, A., Wang, C., and Erbacher, R.F. (2015). *Cyber Defense and Situational Awareness*, vol. 62. Springer https://link.springer.com/book/10.1007%2F978-3-319-11391-3.

64 Rehman, A. and Saba, T. (2014). Evaluation of artificial intelligence techniques to secure information in enterprises. *Artificial Intelligence Review* 42 (4): 1029–1044. https://link.springer.com/article/10.1007/s10462-012-9372-9.

11

AI and Cybersecurity: Paving the Way for the Future

11.1 Introduction

In the twenty-first century, cybersecurity and artificial intelligence (AI) have presented numerous opportunities for the world while facing challenges behind the scenes. There is a battle to counter cyberthreats, turning to AI and machine learning (ML). The US government employs methods to combat a new type of terrorism in a completely different environment. While it may seem that there are many challenges cybersecurity and AI present, there are also many opportunities for advancements in all types of technology.

As technology grows and becomes more connected to the physical world, the need for cybersecurity grows exponentially. Everyone from the average person to governments to criminals needs to take aware of cybersecurity. Cybersecurity is within a gray zone of conflict between war and peace in the global world. AI and ML are being used to help this.

Many people are attempting to find gain, whether that be economic or political, through the internet. This makes cybersecurity in a gray-zone conflict below the armed conflict threshold. Governments are using the internet to gather information from other countries. AI and ML are planned to be the essential solution to handle cybersecurity threats. AI can identify cybersecurity threats quickly using algorithms.

The key aspect of cybersecurity is monitoring the access to and protection of data, especially as more devices such as the internet of things are becoming more common. The attack surface is being increased dramatically. If one very popular device is created with cybersecurity issues, then it can be deviating for all people involved. Therefore, it is essential to ensure that a device is put out with cybersecurity being considered forefront. Although vulnerabilities can be fixed with patches, some people do not update their devices. This is where forced, or automatic updates are important to implement. This is especially important for governments because if their information is leaked, it could be very dangerous and even potentially start a war.

A key part of cybersecurity is making vulnerabilities and threats visible so they can be countered. AI will be very important to cybersecurity as humans do not have the cognitive capability to handle huge amounts of data to identify problems. The shift from signature-based detection to anomaly-based detection is an important part to keep up with new vulnerabilities. Although AI will be a good thing for combating threats, it can also be used offensively. This is a huge concern for governments. This means that AI is in a battle between people using it for bad and good.

AI will allow for overcoming human constraints while increasing the speed of decision-making. It will also provide better protection to organizations and end users. It prioritizes operational efficiency. ML and AI have the potential to detect and stop attacks with no human intervention. This allows humans to focus on more important things that machines cannot deliver, like writing code or performing physical activities. AI also has the benefit that it can catch things that humans may overlook. AI systems are able to learn about the threat environment within and outside of the organization it is monitoring. An important thing is that as AI continues to be used, it will become smarter. It will start to learn patterns. AI will combine the capabilities and decision-processing abilities of a human brain.

Many countries now consider AI to be essential for their militaries. If a country isn't using AI, it is behind the curve. UK is developing human–machine teaming to help promote AI, robotics, and humans into warfighting systems. The US is using centaur warfighters. These allow for automation without losing human intelligence. AI has potential advantages and disadvantages in warfare. AI has the possibility to eliminate human warfare as machines could fight battles. It may be that we see a race for who can create the best AI machines to fight in wars. This has the potential to be lifesaving, as people

From 5G to 6G: Technologies, Architecture, AI, and Security, First Edition. Abdulrahman Yarali.
© 2023 The Institute of Electrical and Electronics Engineers, Inc. Published 2023 by John Wiley & Sons, Inc.

would not have to physically fight in war anymore. But not sending soldiers into the field does not mean that AI warfare could not cause devastation. Bombs dropped from a drone are no less deadly. It will be incumbent on politicians and generals to recognize this, and not escalate disputes just because they have the technology.

There is one thing certain about cybersecurity; it cannot proceed without data. The access and protection of data are core practices for cybersecurity professionals. With the world of IoT growing, more devices are being connected every day. This creates an easy entry point for hackers to gain access to critical data. As Libicki notes [1], there is no forced entry in cyberspace; whoever gets in it enters through pathways of the network itself. There are vulnerabilities in every system, and no shortage of hackers looking for ways to exploit them. Cybersecurity is spotlighting vulnerabilities and threats to counter them.

Since the early years, AI has been around cybersecurity, but it wasn't brought to light until recently. AI deployments are now part of the emerging arms race between world governments. They have also been created to detect vulnerabilities in target systems and assist in their exploitation and subversion. Cybersecurity professionals are worried that some militaries and governments could use an offensive AI for their purposes.

AI plays a huge role in the landscape of cybersecurity. So many pathways to different technologies and algorithms are being created to revolutionize how data is collected. This will come with my challenges, such as militaries and other agencies wanting to use offensive AI. The world will be very different in the next few decades, and the role of AI and cybersecurity will be involved in many daily aspects.

Through recent advances in artificial intelligence, we have seen what it is truly capable of. Image recognition can analyze an image with a precision that would take hours for a human to do. New AI applications will bring challenges to the community that they have not previously faced. There are various security concerns with these applications because they are still vulnerable. Security will be the most important factor in the future for artificial intelligence systems and applications.

There are many complexities involved with these systems, and they need to operate in all environments. There are four components in integrated AI systems: perception, learning, decisions, and actions. Each of these components must react independently from its counterparts, and each has unique vulnerabilities.

Since these AI systems can have high risks, trustworthiness must be ensured. There are a few areas that need to be addressed when it comes to making this assurance. Areas that need to be addressed for trustworthy decision-making include defining performance metrics, developing techniques, making AI systems explainable and accountable, improving domain-specific training and reasoning, and managing training data.

Even though AI does well on many tasks, vulnerabilities are still produced from corrupt inputs that produce inaccurate responses. These are current challenges facing the community that need to be solved. Modern AI systems are vulnerable to surveillance where adversaries query the systems and learn the internal decision logic, knowledge bases, or the training data. This is often a precursor to an attack that attempts to extract security-relevant training data and sources or to acquire the intellectual property embedded in the AI.

With these reasons, you can see how cybersecurity can be affected by AI. Cybersecurity can use AI to increase its awareness, react in real time, and improve its effectiveness. This means that it uses self-adaptation and adjustment to face ongoing attacks and the ability to alter its course of action. It can help highlight weaknesses in an adversary's strategies by analyzing and reacting to them in real-time.

11.2 IoT Security and the Role of AI

With the prospect of global digital connectivity disturbing our entire understanding of what traditional war and peace are, we are entering a sort of gray zone in which cybersecurity is a persistent conflict. Within the "cyberwar" environment, there is much to be gained in the form of political, strategic, and economic value, and its growing effect on the world should not be ignored. Just think about how omnipresent the internet is in daily living; electronics literally share and shape our lives. To deny the power of cybercrime would be a massive blunder. To combat this threat, artificial intelligence and machine learning are aptly suggested as solutions to all things cybersecurity. This chapter addresses the identification of cyberthreats through AI and the evolution thereof, the implications of new human-non-human subjective, and even goes into the military and intelligence aspect that questions the stake of cybersecurity automation.

The evolution of cybersecurity from signatures to anomalies is hinged on the fact that access to and protection of data are paramount. As the IoT becomes more and more real, the foundation on which can be attacked and exploited in terms of global information is expanding continuously. As such, it is a near-impossible assignment to prevent exploitation of all facets of this massive system composed of numerous vulnerabilities casually and professionally overseen [2]. Thus, as stated in the introduction, a key facet of cybersecurity is about making those vulnerabilities and threats visible so they can be countered.

This brings us to the idea of automation, which has always been a part of the cyber world; data has always been automated by machines because of human cognitive and sensory limitations. Considering how notoriously bad humans have been in the field of cybersecurity, the ability to automate this process seems a reasonable fix. The systems that are currently in place that detect and alert you of malware are employed based on signatures of known malicious entities and as such, cannot defend against these rapidly mutating threats, which have no signature recognition. In direct response, anomaly-based capabilities that continuously scan network traffic for normal behavior are being brought into consideration. However, as the number of alerts increases, which require human attention, we're right back to human biases – not to mention the creation of false negatives and false positives that the anomaly system presents. Now, though, the possibility that AI can learn and adapt in dynamic ways makes machine-learning AI an attractive counter-measure to adapting to threats.

AI is a cognitive technology, and the growth of mathematical models on a large collection of data with different bases, values, and purposes allow greater advances in applications such as health, economic, political, and social fields. AI has different bases, values, and purposes, depending on the field to which they apply.

AI is data-driven and not a model-driven tool, and it is not a solution for every problem; it is an approach for highly scalable neural networks. There are no "good" or "evil" mathematical functions. How this technology is used depends on governments, international agreements, and companies who create profit from it. When AIs are able to discover vulnerabilities, they will favor hackers on less-complicated but still-effective ways to exploit loopholes and vulnerabilities in security systems by decomposing software. Even now, that software has become more adept and secure at defending against malware; the threat has shifted from this to phishing via multiple paths such as web pages, mobile devices, and broad diverse apps.

AI is also distributed, and the design and implementation of it are complex, hidden, and proprietary. It is too easily accessible to any company or organization; hence, it will be almost impossible to govern without serious, enforceable international agreements on the appropriate use and principles for AI. AI prioritizes business needs over end users, and the owner of AI builds it to maximize profits over people. Ethical behaviors can be hard to define and even more difficult to build into AI systems for the ever-evolving universe and global cross-cultural.

AI will likely be developed and used with ethical intent, but as a simple fact, American, European, and Chinese governments, and Silicon Valley companies have different ideas about what is ethical. How AI is used will depend on the hierarchy of values for economic development, international competitiveness, and the social impacts of governments. There are two motivations for AI functionality, namely to collect more and more personal data and to employ Big Data for total control of the citizenry. The main drivers of AI owners and developers are geopolitical and economic competition with little or no incentives addressing ethical behaviors or concerns where we have seen cases of abuse of AI for privacy violations and personal gains.

Although AI and its subtechnology of ML and deep learning (DL) are still in their infancy, they have been used by corporations to evaluate tasks in terms of revenue generation, which can lead to better productivity, creation of economic value, and efficiency. But these tools could get out of control as they become autonomous, working against society. That could lead to segmenting, segregation, discrimination, profiling, and inequity. The impact of AI and robotics on human evolution has not been very significant, but over the course of decades or more, that will change.

The standardization of AI ethics has been a major concern. An ethical AI must place humans at the center with global collaboration, common standards, and transparency. AI is promising in many ways and even more so when we realize that the adversaries have the same idea, initiating a sort of arms race between cyber attackers and defends. Back to the anomaly idea, if you combine this with machine learning, it becomes a sort of immune system that grows and evolves based on the norms of what a network should be. We can even go in manually and adjust things or speed up the process for a better overall system. An interesting aspect is AI learning from monitoring cybersecurity professionals and emulating how they secure a network.

The search for anomalies shifts the idea of cybersecurity from preventing known threats to predicting and preventing yet unknown threats. This brings up an important question: If AI is learning from humans, are we not just going to fall into the same traps and biases that we've already been fighting? This is where the idea of unsupervised learning comes in, which is the idea that if we let the algorithms do the labeling and categorizations themselves, they may become unbound by the bias of specific localities. But at the same time, you cannot rationally have machines determine proper cybersecurity, for at least three reasons:

1) Anomaly-based analytics are geared to actionable information rather than what is good or truthful information, showing that uncertainty is embedded within algorithmic reasoning.
2) Anomalies are divergences from normality rather than set-in-stone rules.
3) Categories would make for comparing things based on similarity, putting a lot of things into the realm of invisibility.

With all these questionable qualifications of AI, the only solution is to either double down on AI or figure out how humans fit into AI decision-making. In the hybrid system, human and nonhuman experts contribute to expert knowledge, which will be distributed rather than individual, and at that point, are we a part of this process if choices are based on paths of action presented by machines? On the other side, leaving machines to learn unsupervised means humans have no say whatsoever. Either way, it is just a circle of the agency.

It is true that AI has a role in automating the trivial parts of life and provides a safe environment for everyone; however, when you think of it on a political and economic scale, you get something of devastating mass and impact. Think of how these systems constitute new avenues of political influence and control, for example, surveillance capitalism. Another avenue of concern is how integrated the future of the military is with ML and the prospect of combining people and machine capabilities into warfighting systems that seek to outperform each other. The role of AI, concerned with the conflict of offense-defense, as with anything, is blurred at the lines. When we consider AI in cybersecurity, maybe we should think twice before starting something we do not have the power to stop once it gets going. What agency will be accountable for making cybersecurity a tool for good?

In a society where high-speed data transfer is not just a luxury but also a way of life, we have come to realize that the need and demand for our wireless infrastructure have come under immense pressure to expand. I will cover several characteristics and trends in the mobile industry that will show that the need to move from a 5G system to a 6G system is on the horizon [3]. This will be a very condensed summary due to the large amount of information involved. I will briefly cover the prospects, applications, specifications, requirements, and technologies that will enable growth, industry standardization, and future challenges. All the topics listed can be vastly expanded on and can occupy pages that cover changes in the network systems.

We will start with the global traffic volume of mobile devices in 2010 showed that 7.5 EB/month of data was being consumed. This was not the beginning but gives a quick reference to the large jump that will be made in the next decade. The predicted traffic in 2030 was estimated at a whopping 672% increase, making data traffic consistent at 5016 EB/month with expectations to increase even greater in the future. This will not only put a strain on the 5G network that is not fully functional in all areas of North America but will cripple it if the network structure does not change quickly. The 5G network is rapid including new techniques of managing data, including but not limited to new frequency bands. These changes are being implemented at a rapid pace, but not fast enough to keep up with the trends.

We are already reaching the limits of 5G – it is expected to max out in 2030, only a few years from now. However, the new 6G network will include all the bells and whistles of the 5G network, such as network densification, higher throughput rate, increased reliability, lowered energy consumption, and massive connectivity. This will pave the way for new services and devices to automate much of our lives. The most important aspect of the 6G network is its capability to handle massive amounts of data with very high data rate connectivity per device.

With the increased data rates and communication speeds, the 6G systems will foster a new revolution in the digital age of industrial manufacturing. Some of the key prospects and applications, such as the super-smart society, will accelerate the quality of life for all humanity with smart AI-based communication. The ushering in of extended reality (XR) includes augmented reality (AR), mixed reality (MR), and virtual reality (VR), which all use 3D objects within the real-life environment. This will also lead to advancements in robotics and autonomous systems, wireless brain-computer interaction, haptic communication, smart healthcare, automation in the manufacturing industry, full-sense information transfer, and the ever-growing IoT that has been placed under an umbrella term already going by the name IoE (Internet of Everything), which will interconnect the world.

Some of the specifications and requirements that will be changing on the new horizon are the service requirements of the 6G network. This will include enhanced mobile broadband (eMBB), ultra-reliable low latency communications (URLLC), massive machine-type communication (mMTC), AI communication, higher throughput, increased network capacity, higher energy efficiency, low backhaul congestion, and a hot topic the last few years – increased data security. This change will also include newer integrated networks with connected intelligence, seamless integration of wireless information, and energy transfer of super 3D connectivity. This will require fewer general requirements in the network characteristics, including small-cell networks, ultra-dense heterogeneous networks, high-capacity backhaul, and mobile technologies integrated with radar technology and software with visualization.

Some of the key enabling technologies that will help push this high-speed age are AI and its ability to learn without the need for human interaction, terahertz communication that is needed for large data transfers, free space optical (FSO) backhaul network for constant communication with devices and sensors, MIMO technology to increase data connection while reducing latency, the ever-growing blockchain ledger system with its decentralized data system, speedy retrieval speeds, and quantum communication. Big Data analytics has to be added to this list in order to market the expanding technologies so that the companies can process the growing data retrieval.

There will have to be standardization in protocol and research activities. This has begun with Samsung Electronics leading the way, as well as Finnish 6G Flagship programs that will help maintain consistency in the rapidly changing environment. The International Telecommunication Union has also reared its head while in the discussions of the new 6G wireless network at the summit by the same name.

Some of the future challenges and the headed directions include but are not limited to areas in high propagation and the atmospheric absorption of THz to show the loss as frequency travels through free space. The added complexity in the resource management required for the 3D environments while minimizing hardware constraints. The required need to model sub-mmWave (THz) frequencies to be managed with higher wavelength capabilities. This will include needed spectrum management to prevent overlap or interference as well as beam management in the THz communication range.

What this all boils down to is that there is a rapidly changing technological environment coming very soon. It leaves the consumers oblivious to the amount of change needed behind the scenes to bring this to fruition. Looking on the bright side, this will make our lives easier and even better in the future. Most of all, it will bring increased opportunities in this industry for students taking this class as I write this report.

In the world of IoT, people are wondering whether they should adapt to the trend of connected products because of security concerns. These devices are more vulnerable to security threats because they are connected over a network with many open paths that can be attacked. Practices are in play involving strong authentication and tracking IoT devices. But artificial intelligence (AI) and 5G may address some of the concerns people are having about IoT devices [4].

One positive effect that 5G is having on the security of IoT devices is the low latency and faster speeds it offers. There is one thing worrying developers about 5G, though: the protocol flaws discovered. Hackers have demonstrated that they can find certain connected devices in a certain area on 5G networks. One can see that 5G has many advantages but also has been shown to have its faults. The future for 5G and IoT devices is still largely unknown, but engineers expect it will continue to pave the way for innovative capabilities.

One thing about AI and IoT that has amazed everyone is how advanced it has become in the last decade. Thirty years ago, people could not imagine the advancements we can make today. However, one thing is halting advancement in a certain AI field that has yet to be solved, and it is in the autonomous car industry. There is too much data that needs to be processed and stored for these cars to run. If we want to fill the roads with autonomous cars, implementations need to tackle these problems.

The fusion of IoT, 5G, and AI has a long way to go, but developments are continuing to release that will help pave its future in technology. We will soon be living in a world run by connected devices, and AI will pave this road [5].

11.3 Cybercrime and Cybersecurity

There are potentially many problems in the future if there are hidden flaws in the network. When a business hires an ethical hacker, often called a *penetration tester,* it can uncover hidden vulnerabilities in its network [6]. They have a unique toolset that lets them completely test and observe the network.

The first thing a penetration tester will do is survey the network. There are two types of survey, passive and active. When you are passive, you are not creating a footprint or alerting the network to be an intrusion. This includes sniffing and information gathering. The other type, active, might alert the host if certain detection systems are enabled.

Active includes probing the network for information such as open ports, details about the operating system, accessible hosts, and the locations of routers. Reconnaissance allows the tester to gain information on the system that they can use later to find exploits for vulnerabilities in the host's systems. With reconnaissance, the tester uncovers that information, so it is a key part of the process.

The next step in the process is scanning. This is where the tester scans the network with the specific information that they gathered in reconnaissance. This is where port scanners come into action; when the open ports are scanned, you can gather information. Once you see which ports are vulnerable to attacks, you can then plan how you will exploit the vulnerabilities [6].

After the tester completes the scanning step, the object is ready to access. Data can be exploited over the internet, LAN, locally, and even offline. The penetration tester will uncover any security problems and try to gain the highest access possible. This way, it exposes how a real hacker might attempt to break into their network.

The tester has many different attacks that they can use on the network – buffer overflows, distributed denial of service (DDoS), session hijacking, password filtering, and even social engineering. Their arsenal is limitless, which gives them a good chance of finding a vulnerability in your system and then being able to exploit it [6].

The next thing the tester must do is maintain access while they are still in the system. This is where the tester will try to create a backdoor or install a rootkit to stay logged on to the system and access the network until they're discovered, which may never happen on some networks. While access is maintained, files and documents can be sent to the owned system. After this comes the most important part, covering your tracks. Good hackers will always cover their tracks, and a tester needs to know what to look for.

This will hide any information on the system that you may have left connected to a device. Hackers will often edit the log to delete their tracks to prevent them from pursuing a legal matter. The tester will test this to see what hackers would use, and the tester will be able to circumvent that action from happening.

Anyone can see that it is vital for a business or organization to hire a consultant to test their network. This has created an explosion of ethical hackers trying to become certified to get hired by an agency that can legally hack systems [7]. A new field has arisen solely based on network security, which keeps getting more advanced as time progresses.

Millions of dollars each year are saved by firms patching security flaws that could have been exploited. In the past, companies like Sony, Target, and multiple crypto exchanges have been targeted and lost millions because hackers found vulnerabilities in their systems – no telling how hard it was for the hackers to gain access. It could've been something simple or a complex attack. This is why it's important to test every aspect of your security and hire professionals to do it.

Another huge thing that companies and organizations need to pay attention to is social engineering. This is one of the easiest tools a hacker has in their arsenal because they can access your system with little or no work. People have been doing it for centuries, and other people are always the easiest targets because they're not as hard to crack as computers; people make mistakes. This can be accomplished with defined skillset such as shoulder surfing, identity theft, and gaining unauthorized access to closed-off building areas such as the server room.

The hacker acts like an employee for shoulder surfing and hovers over workers to see what password they enter. You would be surprised how many people openly type their password or leave it somewhere on their desk, such as a note. People do not realize there is always a chance that someone can be looking for anyone's password to gain access to critical information [8].

The next way is by using someone else's identity. They act like an employee within the organization you work for, and they will send you an attachment or a link in an email that has a virus bound to it or they take you to a phishing site. This is where many people make mistakes because they do not pay attention to who the sender is or whether they are on an encrypted site.

This is another common attack to gain employee credentials because some people are always easy targets. The hacker can gain access to the system through social engineering by gaining access to an unauthorized room such as a server room. They can gain root access if they have access to the server room, which essentially means game over. Many companies do not cover all aspects of social engineering, which leaves them very vulnerable.

Thanks to ethical hackers, they can find and point out these vulnerabilities, often circumventing such actions from even taking place. Ethical hackers are starting to make a name for themselves and are being snatched up by great companies [6]. With 5G, AI, and IoT products being launched daily, there is a need now more than ever for security experts and certified ethical hackers.

With the pace of technology never slowing down, the industry needs to keep up with the latest information on systems. Every time a system is patched or something new is added, it potentially opens doors to new vulnerabilities. That is why it is important to stay up to date with the security industry to learn about new vulnerabilities and exploits so they can prevent intrusions into their network. With AI, automated systems can detect most of these problems before they even happen.

11.4 How Can AI Help Solve These Problems?

With the introduction of AI in cybersecurity, there are various ways to tackle the problems mentioned above. The world is changing, and so is how devices are connecting. IoT is revolutionizing the world of connected devices. One problem that has arisen from connected devices is the security behind the technology. Many companies are seeing increased threats because more devices must be managed. This is not hurting adoption, but it is certainly a concern within the community.

One fix has been suggested, and that involves the 5G network and AI. One example is an AI company that focuses on user authentication. This provides an extra layer of security that is harder to crack than the application or device's traditional security. The 5G network brings faster speeds and low latency, which provides better management for devices that go a step further than user authentication. Both implementations address the concerns people have about IoT-connected devices and will help adoption in the future.

However, as always with benefits, there will also be cons. The 5G network has protocol flaws that could cause security flaws. Security experts have agreed that the 5G network has better security than the previous versions but also has some of the same flaws ported over to the new network. 5G is also vulnerable to new attacks that we have not been concerned about with the previous versions. With the increase of data, attackers can go unnoticed attacking the network because there is more traffic.

There is a huge factor at play with all these implementations growing with each other; they are gaining traction together. 5G, IoT, and AI are all helping each other out by implementing their features under one roof. AI used to be something straight out of a comic book, and now we see it used more every day. IoT is making this possible by allowing data to be shared through devices. AI also enables data to be processed with much greater efficiency than ever before and will only grow as we progress. This is creating smart devices that are essentially learning.

The innovations in these technologies are creating new platforms for both individuals and organizations. It brings new functions in transport, entertainment, medical, and public services. From a technological standpoint, the applications are limitless, and so are the number of devices that can communicate with each other.

The only problem I see with this is where exactly the data will be stored and how much we will be able to store. So far, this has not been much of a problem, but it will need to be addressed in the future. However, the benefits outweigh the cons and will have a lasting positive impact on people and the world. We already see this effect in the world of virtual reality. We can fly drones and view the world at our leisure and watch live concerts from the comfort of our homes. This technology is creating a great future for everyone and will only get better.

IoT and its implementation in AI will continue to expand how we use technology and how we are connected to the world. There are some security concerns, but the technology itself is helping collect data on how we can fix these problems. The applications are limitless with what IoT, 5G, and AI have in store, and we will see life-changing technology appear faster than ever before.

11.5 The Realm of Cyberspace

With cybersecurity being at the forefront of the technological revolution, there are many challenges they must overcome. More satellites are being deployed, and they also face cybersecurity threats. These threats are relatively new to the cybersecurity landscape because more people are figuring out satellite vulnerabilities. Threats include compromised launch systems, communications, telemetry, tracking and command, and mission completion. Therefore, security is a huge aspect of satellite communications and their networks [9].

There are three major types of satellites: scientific satellites, application satellites, and communication satellites. As their name implies, scientific satellites, also referred to as space science satellites, carry instruments to study an object or phenomena. Application satellites survey the Earth's resources, whether on the ground, beneath the ground, or inside water bodies. Communication satellites relay telephone calls and television signals down to Earth and relay voice communication between the astronauts orbiting on Earth's space shuttle and mission controllers.

Satellites provide so many benefits that the day-to-day person probably never even thinks about. They provide information on Earth's resources, provide communications, medicinal purposes, remote work, and so much more. Research is currently being done to address concerns regarding not only security but also space debris.

Security is discussed at all phases of building the satellites so they can maintain their integrity throughout completion. Recently, the Turla hacking group attempted hacks on satellite communications by exploiting satellite-based internet links. This has raised questions about the international laws behind these attacks. These attacks have caused physical damage to ground stations and satellites by jamming signals and disrupting computer guidance systems. Several new states that have launched satellites, such as North Korea, have also originated some attacks.

Cyber-enabled destruction of satellites and their signals can cause major disruption. These satellite communications control critical infrastructures such as military systems, banking, and financial systems, air traffic control, electricity grids, traffic and transport systems, and early-warning weather systems. It has been said that the space environment is becoming congested – the more satellites in space, the more threats they face.

We will continue to see attacks carried out on satellites during the next few decades as technology continues to advance and more satellites are launched into orbit. A new approach is needed to protect our cyberspace, and the UN is currently trying to combat this. International law is being considered to map out how the laws regarding space are governed in the future. Since these threats are new, the countermeasures must also be new. Sanctions are one approach for countries or states that violate these laws or guidelines.

The range of hostile disruptions can vary from kinetic, virtual, to hybrid. Even at the beginning of being built, satellites are vulnerable to certain threats. Hostile disruptions can occur at any phase, from the prelaunching testing phases, through launch into orbit, during the satellite's active lifespan, and through its deactivation. The attacks can be distinguished between being hostile or through just error and negligence.

One hostile intention can be to cause a direct impact with a satellite to disable the former. Another intention is to block or jam signals to disrupt the satellite's communication. They can also be attacked by electromagnetic pulses (EMPs), which will disable everything electronic [9]. A proposed topology lists responses available to affected satellites. There are legal ramifications if the attacker is discovered and evidence backs up the find.

Companies must stay up to date with current vulnerabilities and protect their satellites at the planning and the entire production stage. Deployment of AI to combat these threats will be necessary to maintain a stable environment. Data management will need to be maintained by AI programs to keep the systems in shape. Everything in the cybersecurity realm in some way revolves around the use and deployment of AI.

11.6 Connected Devices and Cybersecurity

IoT is transforming the world as we know it. As technology advances, so do the limitless applications that are built on top of it. Data is being collected from thousands of sensors that help improve products' quality, management, and security. IoT brings innovation to the health care industry, cities and communities, energy and construction industries, and the insurance industry. These developments come solely from IoT, and anyone can see how much of an impact it has made [10].

IoT has been great for businesses, but its practices are being deployed slower than expected due to issues of security, interoperability, and costs. Many IoT products have not been pushed to production and are still in the concept phase. It is also hard for them to hire experienced staff that can work on these complicated projects. For now, companies are relying on simple IoT projects because they generate value quickly, thus helping the company retain a consumer base.

Even though IoT is slowly growing, great innovations have been made to cost, security, and technology. There have been a lot of great products developed through IoT, one being by the pharmaceutical industry. IoT solutions are able to track the recordkeeping of medicines from production to the patient. Introducing these new products offers competitive rates to the consumers. This will essentially develop into more use cases.

These are just a few examples of limitless applications and products from IoT. I believe the future is through IoT and the optimization it brings to the table with artificial intelligence. We will continue to see this market grow, expand to other industries, and create more competitive pricing models [10].

IoT-connected products are being implemented in homes, wearables, the industrial internet, connected cities, and cars. One product that is playing a huge role in its expansion is the use of sensors. They can be used for security, optimization, and autonomous technologies. Data from sensors is even being used to enhance the safety of first responders. Anyone can see how limitless the applications can be and how important they will be for the future.

The use of sensors is creating the next generation of public safety services. Sensors can track data in real-time and send alerts to responders. A high-level architecture is used, allowing various sensors to be interconnected, allowing more than one application is the same architecture. This is huge for optimization in any industry because it creates less hands-on work, and the data is even being processed and stored more efficiently.

The current state of technology is readily developed for a future state. It will be connected over broadband internet, have multimedia/supplemental data, and provide situational awareness in the future state. There are many considerations for this future technology, such as an open ecosystem for applications and developers, data security and privacy, and a national movement toward open government data.

This technology provides way more benefits than cons. At first, developing some of these products could be pricey, but as technology advances, the pricing will get more competitive, reducing the cost. IoT-connected devices are here to stay, and we will see many applications, giving us a better quality of life [10].

In the world of IoT and connected devices, the challenge of security is ever so apparent. Devices are connected to almost everything nowadays. They are usually mobile devices and PCs, but now we have automated cars and radio frequency identification (RFID). This presents many challenges in the security field. With the emergence of cloud-based operations, there is an increase in challenges, making analyzing the data a lot harder.

Cloud computing makes it much harder for forensics professionals to analyze what is on the hard drive because it is sometimes potentially unrecoverable [11]. More complex procedures for investigating must be used because it is much harder to analyze this type of device data. Sometimes it is not necessarily harder to come across the data, but it becomes cumbersome. Often there is just so much data that needs to be analyzed, and it takes a lot of extra work.

The devices that are connected come with weak security because these devices are new to the market. It is hard to test for all vulnerabilities if the item hasn't been made public to be tested in real time. The attack surface for hackers is becoming a lot wider with all the new products coming to market. Sensors, automated systems, and infrastructures can be easily targeted because of their new, unknown vulnerabilities. Attention must be paid to all of these new devices because of the high workloads they are creating and the data they're processing.

All of this can be combated by properly seizing data, storing data, extracting, and analyzing it. As of now, there isn't a procedure that covers the world of IoT. This must change as we continue to push out connected devices and build the world of IoT.

11.7 Solutions for Data Management in Cybersecurity

For IoT to go well, IoT devices need a solution for data management. Currently, the data generated by the devices is hard to process, and data speed may be slow. 5G is having problems with network latency and bandwidth usage. It must also deal with data security and keeping track of data lineage.

Cloudera Dataflow has come up with a solution to these problems. It tackles major problems that most IoT devices are experiencing. Its management, processing, and analysis all help make this possible. It is also open-source, which is a game-changer for products in the IoT field [12].

With all the devices being connected, something must be done with their data. This is where data management comes into play, which has a key focus on automation. By using data management, businesses will become more productive, accelerate IoT initiatives, and help deliver a comprehensive view of IoT-related data.

Many vendors are tapping into this new market and already offer data-management solutions. Google, Amazon, and Microsoft all offer data-management solutions using a vast computing and storage infrastructure. Data-management vendors are constantly releasing open source projects so other companies can help build on the technology. Several hardware vendors have designed systems used for IoT.

11.8 Conclusion

We have discussed the role of artificial intelligence in cybersecurity and its impact on our everyday lives. From security to smart devices, there are multiple uses for AI. Most of the problems in cybersecurity can be addressed by the use of AI. We have discussed how intrusion detection systems can provide more security for your network and how AI can prevent and deter hackers. Artificial intelligence is even used in cyberspace when dealing with satellites and other systems such as GPS. With all of the pros discussed, we have also discussed some cons. Data management is a problem that will continue to be discussed until we find a permanent solution to the problem. Another problem it faces is constant connectivity to smart devices and sensors. With everything being connected to the internet now, it is more vital than ever to have artificial intelligence deployed in more places than one.

References

1 Libicki, M. (2009). *Cybersecurity and Cyberwar*. RAND Corporation https://www.rand.org/content/dam/rand/pubs/monographs/2009/RAND_MG877.pdf.

2 Gupta, S. (2019). Ethical hacking – learning the basic. Video. https://link.springer.com/video/10.1007/978-1-4842-4348-0.

3 Ramachandran, V. (2021). 6G network – a detailed guide to sixth generation network technology. https://expersight.com/6g-network-detailed-guide-to-sixth-generation-network-technology.

4 Porter, M.E. and Heppelmann, J.E. (2014, November). How smart, connected products are transforming competition. Harvard Business Review. https://hbr.org/2014/11/how-smart-connected-products-are-transforming-competition.

5 Yarali, A. (2021). *Intelligent Connectivity: AI, IoT and 5G*. Wiley-IEEE Press. ISBN: 978-1-119-68521-0 https://www.wiley.com/en-us/Intelligent+Connectivity:+AI,+IoT,+and+5G-p-9781119685180.

6 Ashvin, G. (2016). *Ethical Hacking*. TutorialsPoint https://www.academia.edu/32432762/Ethical_hacking_tutorial.

7 EC-Council. (n.d.). Ethical hacking. https://repo.zenk-security.com/Magazine%20E-book/EN-Ethical%20Hacking.pdf.

8 Deepika, S. and Pandiaraja, P. (2013). Ensuring CIA triad for user data using collaborative filtering mechanism. In: *2013 International Conference on Information Communication and Embedded Systems*. ICICES https://doi.org/10.1109/icices.2013.6508262.

9 Housen-Couriel, D. (2016). Cybersecurity threats to satellite communications: towards a typology of state actor responses. *Acta Astronautica* 128: 409–415. https://doi.org/10.1016/j.actaastro.2016.07.041.

10 Yarali, A., Ramage, M.L., May, N., and Srinath, M. (2019). Uncovering the true potentials of the internet of things (IoT). 2019 Wireless Telecommunications Symposium (WTS), New York, IEEE, pp. 1–6, doi: https://doi.org/10.1109/WTS.2019.8715545.

11 MacDermott, A., Baker, T., and Shi, Q. (2018). IoT forensics: Challenges for the IoA era. In: *2018 9th IFIP International Conference on New Technologies, Mobility and Security (NTMS)*, 1–5. Paris, France: IEEE https://doi.org/10.1109/ntms.2018.8328748.

12 Cloudera (2019). Cloudera dataFlow: IoT data management from edge to cloud. https://www.cloudera.com/content/dam/www/marketing/resources/analyst-reports/forrester-now-tech-iot-data-management.pdf?daqp=true.

12

Future 6G Networks

12.1   Introduction

As is glaringly obvious, as each new generation of technologies comes forward, the needs to be met get bigger and bigger. For example, what we are seeing with 6G is the prospect of high-fidelity holograms and immersive reality, tactile/haptic-based communications, and the support of critical applications for connecting all things, as we know it. However, to meet these goals even larger bandwidth and capabilities are needed to support such advancements. To fully understand what systems will be capable of in the future, it is very important to analyze the requirements of daily life, which naturally direct the requirements of 6G.

Backward compatibility is essential for 6G technologies because devices will be multimode and multiband. Therefore, in certain circumstances, there may arise a moment when a natively 6G device must fall back on 5G and possibly even 4G, depending on the coverage conditions. Undoubtedly, there will be design and other tradeoffs needed to achieve this, but ultimately the integrity of these devices must be ensured.

As we look further into this topic, we have established in this book that according to ITU-T, the three most important characteristics spearheading the future of lifestyle and societal change that directly affect 6G are:

- *High-fidelity holographic society:* If you have spent any meaningful amount of time on the internet, you would begin to notice just how powerful and central video is to today's communication. Just look at the methods by which we have learned to deal with the onset of COVID-19; video has allowed people, businesses, governments, medical professionals, and their patients to remain in contact and establish a new normal that allows functions to keep on going despite the risks of in-person communication. As has been previously mentioned, holograms are a present possibility for society to look forward to. In 2017, Stephen Hawking was able to give a presentation via hologram. Admittedly, it was not fully realized yet, but with 6G, the potential for a high-enough transmission rate to make this possible is within the realm of possibility.
- *Connectivity for all things:* With interconnectivity we have already seen with 5G, 6G will need to bring it on a much bigger scale to keep up. Infrastructure that is necessary to society, like water supplies, agriculture, power plants, transport, etc., will need to operate multiple network types and go well beyond the networks of today. Furthermore, considering how essential these operations are, security will need to level up alongside.
- *Time sensitive/time engineered applications:* The importance of timeliness cannot be overstated in the onset of a world interconnected. The new technology that relies on information delivered will demand timeliness. Within a network of massively connected sensors that endpoint communication, late arrival of information could spell catastrophe given the right infrastructure.

The above changes will ultimately provide the need for a 6G network in the first place, and as such, they are extremely important.

The critical and principled approaches to follow when constructing a 6G network design are:

- *Super convergence:* To foster an ecosystem that will allow for dramatic changes, an easier and more scalable convergence between various technology families is going to be paramount. In addition, non-3GPP-native wired and radio systems is the backbone that allows for this.

From 5G to 6G: Technologies, Architecture, AI, and Security, First Edition. Abdulrahman Yarali.
© 2023 The Institute of Electrical and Electronics Engineers, Inc. Published 2023 by John Wiley & Sons, Inc.

- *Non-IP-based networking protocols:* Decade-old IPv6 has long been researched for an upgrade, but the onset of a 6G connected society will make it that much more justified that more than 50% of networking traffic does not originate or become terminated at the wireless edge.
- *Information-centric networks (ICNs) and intent-based networks (IBNs):* Like the above NGP (next-generation protocols), ICNs are actively in research. They are a step toward separating content from the location by which it originates. To make the connection between networking design and operational management, intent-based networking and service design have popped their head up. It is a lifecycle management spearhead for networking infrastructure, constituting continual network monitoring and real-time optimization processes able to adapt to any change in network/service state.
- *360-cybersecurity and privacy-by-engineering design:* 6G requires that end-to-end solutions be secure as well as top-down. Privacy-by-engineering will ensure that mechanisms are natively built into the protocols and architecture, thereby preventing the forwarding of packets uncertified.
- *Future-proofing emerging technologies:* As technologies continue to improve day to day, it will be necessary that 6G adhere to technologies that will be invented in the future so that they may be embedded quicker and more efficiently into the architecture. Such technologies include but are not limited to artificial intelligence (AI), distributed ledger (DLT), and quantum computing.

12.2 Vision, Challenges, and Key Features for Future 6G Networks

We can subdivide the telecommunication industry like this:

- Wireless
- Communication equipment
- Processing systems and products
- Long-distance carriers
- Domestic telecom services
- Foreign telecom services
- Diversified communication service

The amazing changes this industry has brought and continues to bring have changed the lives of the global population, establishing new problems and benefits. As one of the largest and fastest-growing industries, advancements will be coming in great volume and quality in the coming years. It continues to evolve with each new generation toting flexibility and reliability that consumers desperately look for in their communication.

No one is refuting how prevalent in society wireless communications have been. Look at radios and televisions; they provide people with entertainment, local news, global news, and more. With the advent of smartphones, this connectivity has become cheaper in the last years and can now be taken with you outside of the confines of a stationary radio or TV. The implication of advancement is beyond spectacular and can be brought to developing countries to promote economic advancement.

It's crazy to think that wireless communication in its entirety is only a little over 50 years old. And throughout that time, wireless networks have undergone many specific developments: wireless personal area networks (WPAN), local area networks (WLAN), ad hoc networks (Adhoc), or mesh networks, metropolitan area networks (WMAN), wireless wide area network (WWAN), cellular networks that span from first generation (1G) to 5G, and space networks. The author dives deeper into the specifics of each wireless system, starting with 1G and ending with 5G.

1G was introduced in 1970 and only offered the capability to call and stayed in use until 2G was released in the early 1990s. 2G was extremely advanced compared to 1G, primarily flexing digitized signals. The efficiency of 2G allowed for it to utilize spectrum and penetrations offering data service. Not only that, but short message systems (SMS) were enabled, allowing people to communicate through text.

3G marked the biggest communication milestone by being better in almost every possible way: wider data bandwidth, faster internet connectivity, and faster data transfer. It enabled the wireless telephone, internet access, wireless internet service, and mobile streaming. However, it did not deliver on its promises and ultimately failed.

3.5G was released to correct some of the initial problems with 3G by performing more efficiently and offering faster uplink/download packet access.

4G came about in 2010 via the International Telecommunications Union-Radio communications (ITU-R) through extensive research and testing. It was released mainly for mobile cellular network connection with the intention of lasting through many more generations to come. There are many reasons why it is so accessible, the first being how wide a range of network frequencies it has access to, from MHz to GHz. Second, a wide bandwidth is 1 Mbps to 2 Gbps. A very interesting feature of 4G is that its network bandwidths are assigned to different technologies. For example, 100 Mbps or higher includes cars and trains, while 1 Gbps transmission is for low mobility connections like pedestrians and stationary devices. The overall goals set out for 4G are:

- To be a completely IP-based integrated system
- To provide communication everywhere with speeds of 100 Mbps to 1 Gbps
- To combine Wi-Fi and WiMAX (Worldwide Interoperable Microwave Access) technologies utilization

12.2.1 Fourth Generation Long-Term Evolution (4G-LTE)

4G LTE is meant for mobile devices and consists of many advancements and features compared to 3G and earlier LTE, giving smartphones the speed and power they need to function properly. However, it only has that great integration in metropolitan areas. To overcome this, an ingenious idea of combining 3G and 4G LTE was engineered to meet these requirements:

- Highly advanced bandwidth and data transfer speeds of up to 1 Gbps
- Faster response times (shorter round-trip delay)
- Enhanced quality of service control mechanisms
- Lower mobile implantation cost due to simplified and simple architecture
- Orthogonal frequency division multiplexing (OFDM) for downlink
- Single carrier frequency division multiple access (SC-FDMA) for uplink
- Multi-input, multi-output (MIMO) for enhanced throughput
- Reduced power consumption
- Higher radio frequency (RF) power amplifier efficiency (less battery power used by handsets)
- Increased ARPU (annual revenue per unit)

5G is nothing like what we have seen so far. It sets out to revolutionize the telecommunication industry, and not only that, it will be built for other systems that need higher speeds, unlimited to the broadband world. Its primary goal is to trailblaze for networks and internet connections. It has features such as:

- Simultaneous seamless connections for great cloud deployments
- Improved signal efficiency
- Significantly lower latency compared to 4G
- Enhanced spectral reliability and efficiency
- Improvements to browse, download, and upload data files from any place to anywhere
- Reduced network energy usage, which will, in turn, increase the battery life of the device
- Increase in the users' density over the unit area many times, which will help users use high bandwidth for a longer period
- Improved internet of things (IoT), machine-to-machine communication, and device-to-device communication
- Increased object-oriented works and data management

Through the deployment of 5G technology, users will develop frameworks to utilize a machine-to-machine communication system. The onset of this architecture will increase and improve business and consumer demands. It will always establish these characteristics:

- *Improved reliability and security:* With the onset of things like cars and planes connected via 5G, the need for reliability and security is obvious.
- *Traffic prioritization:* Within dense populations, actions like calling for an emergency will take priority over networks.
- *Internet of things*: A major point is automation, and seamless communication amongst various devices is a big part of 5G.
- *Radio interoperability performance:* An enhancement over the existing 4G radio will provide capabilities that manage various single physical cases.

It is blatant that 5G brings so much to the table but, at the same time, will bring its own set of new challenges to overcome. However, that's exactly what it means to advance and succeed.

Sixth-generation (6G) wireless replaces 5G cellular technology. 6G networks look to use increased frequencies compared to 5G networks. 6G networks will also offer comparatively increased capacity and reduced latency [1]. One of the objectives of the 6G network is to back microsecond-latency communication. This kind of communication is almost 1000 times faster compared to one millisecond.

In more ways than one, technology specialists maintain that 6G intends to be an addition to what 5G already offers. 6G will offer additional devices, higher rates in data, immersive extended reality, increased use of wireless cognition, and e-health [1]. In other ways, the same specialists look to see a more central move from human-centered communication toward machine-centered communication. There are advances in the use of technology exemplified through artificial intelligence and machine learning (ML). These advances and integration of sensing communication will ensure that 6G networks can arrive at additional decisions minus the need to involve human thinking [1]. There will be optimized results founded on various applications in the long run.

Several companies of global capacity invested in realizing the 6G network dream. An example of such a company is National Instruments. The company is actively dedicated to working in several areas linked to 6G research [2]. The company has research involvement in over-the-air (OTA) measurements, wider bandwidth test answers, ML-trained systems testing, and increased frequency. The research by the company also points out that the spectrum needed to deploy the coming 6G networks involves sub-6 GHz frequencies as the mainstay, particularly with the mid-band frequencies employed [2]. At the same time, the rollout and adoption of mmWave will keep up. Even though sub-THz bands have gained traction, it is still not yet clear for which cases the bands are commercially and technically feasible.

12.3 Rationale for 6G Networks with Prevailing and Future Success of 5G

The 6G technology market will likely enable significant enhancements in imaging, location awareness, and presence technology in the future. 6G network will work closely with AI. As such, the computational infrastructure of the 6G network will independently decide the optimal computing location. Such decisions will include data processing, storage, and sharing [1]. Aside from that, it is also highly likely that 6G wireless sensing abilities will use various frequencies to measure the absorption of frequencies and make necessary adjustments to the same. This is attainable because molecules and atoms give off and take in electromagnetic radiation at particular frequencies, and the release and absorption frequencies are identical for whichever substance.

6G promises considerable effects for several industries and government techniques to critical asset protection and public safety. These implications include health monitoring, the measurement of air quality, detection of any threats, facial and feature recognition, sensing toxic substances, and making crucial decisions in social credit systems and law enforcement [2]. Improving aspects of the mentioned sectors using a 6G network means a plus for mobile technology. In addition to that, autonomous vehicles, smart cities, and augmented and virtual reality representing emerging technology will also benefit.

6G networks aim to bring together a collection of past incongruent technologies like Big Data analytics and deep learning. With the introduction of 5G, the convergence between such technologies is more attainable. The necessity to implement edge computing to enable general throughput and reduced latency for low-latency, ultra-reliable communication is a crucial motivator for the development of 6G [2]. Aside from that, another motivator of 6G network development is the necessity to back machine-to-machine communication in the Internet of Things. A formidable relationship is evident between high-performance computing and 6G networks. Edge computing technologies aim to take care of some of the Internet of Things and data from mobile devices even though most of the same will need centralized high-performance computing resources to go through processing.

With 6G, we have the expectation of fully global connectivity, where the physical object is mapped with high detail in the digital domain for analyses and acting upon. The network would establish the links among the domains by sensors and massive devices embedded everywhere, as well as provide the platform and the cognitively of the digital domain. Humans' position with connected bodies and intelligence in this cyber-physical continuum would be placed in the middle, with our bodies as well as our intelligence connected. As shown in Figure 12.1, three classes of interactions – human, physical, and digital – will converge with the twinning of systems between domains:

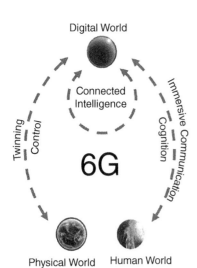

Figure 12.1 Convergence of personal domains, digital, and the physical world [3].

- *Control:* Sensors and actuators can achieve digital twins of cities, factories, and even our bodies. This will enable rich data mining and highly efficient control, but data integrity and security must be ensured.

- *Connected intelligence:* This offers high capacity, low end-to-end latency, and secure compute functionality. The trusted AI functions can operate in and on the network. Information in the digital domain is exchanged between virtual representations of persons and physical devices. Immersive communication is where people extend their senses through the digital domain.
- *Cognition:* Humans represented as physical objects in the digital domain, but it is also important to be aware of their intentions, desires, and mood [3].

12.4 Missing Units from LTE and 5G That 6G Will Integrate

6G aims to offer support to data rates of one terabyte for every second. There will be access points to enable service to more than one client simultaneously using orthogonal frequency-division multiple access. This degree of capacity and latency aims to lengthen the performance of 5G applications and improve upon the capability scope to offer to back to innovative applications in imaging, cognition, sensing, and wireless connections [1].

6G will offer increased frequencies to enable faster sampling rates and offer improved throughput and higher data rates. 6G networks will integrate sub-mm waves and frequency selectivity to enable relative electromagnetic absorption rates [1]. There ought to be potential advances in wireless sensing technology in the long run. 6G networks will incorporate mobile edge computing already existing in 5G networks. Core and edge computing will be seamlessly combined in a computation and communications framework structure by the time 6G networks are implemented [1]. There will be several possible advantages that 6G will have with these techniques the moment it becomes operational. One of the advantages is improved access to what artificial intelligence offers.

12.5 Features of 6G Networks

12.5.1 Large Bandwidth

6G mobile devices will have the advantage of using highly directive antennas through millimeter-wave and terahertz frequency bands and massive bandwidth. As such, applications will benefit from seamless coverage [4]. The Federal Communications Commission (FCC) recently commercialized frequency bands. Ultra-high-precise positioning grows to be accessible with 6G due to high-end imaging and direction-seeking sensors, compared to human ears and eyes. It is possible to equip mobile phones with capable, intelligent algorithms and robots [5].

Obtaining a secure transmission infrastructure with an efficient range and without power hunger is a viable solution. The accessibility of wide bandwidths can change the importance of spectrally enhanced solutions to upgraded coverage solutions [6]. The tradeoff from power efficiency, spectrum performance, and coverage in new frequency spectrums ought to play a determining role in advancing devices. The result will be a design of contemporary air interface with additional regard paid to single-carrier systems. 6G communication networks should cover a large bandwidth from 3 to 60 GHz [6]. For communication to occur at whichever frequency in this bandwidth, the comprehensiveness of the hardware circuitry must increase.

12.5.2 Artificial Intelligence

With the rapid progress of smart terminals and infrastructure – coupled with diverse applications like remote surgery, augmented and virtual reality, and holographic projections – current networks are finding it hard to keep up with sufficient image resolution. It is hard for 4G and 5G to match the increasing traffic demands [7]. For that reason, scholars and members of the telecommunication industry have delved into research concerning 6G networks. In recent times, AI has been used as a novel paradigm for configuring and optimizing 6G networks with increased intelligence. Consequently, there are propositions on AI architecture for 6G networks to attain knowledge finding, automatic network adjustment, smart resource management, and intelligent service provisioning [7]. The architecture is distributed into four layers: data mining and evaluation, smart application layer, intelligent sensing layer, and intelligent control layer.

12.5.3 Operational Intelligence

Operational intelligence is real-time intelligence that, in most instances, is obtained from technical approaches and transmitted to ground troops dealing with opposing forces. Operational intelligence is instant with a short lifetime. The real-time nature of operational intelligence needs that evaluation has immediate access to collection systems and can bring together financial intelligence in a setting with increasing pressure [8]. Nonetheless, there is a need to update operational intelligence, and 6G networks promise to do so. Modifications in operational intelligence can significantly affect resource allocation in organizations.

12.6 Wireless Networks

Smartphones and IoT polarization contribute to the progress in generating mobile data through wireless networks. The most recent research indicates that 54% of generated traffic from Wi-Fi and cellular connectivity devices got offloaded using Wi-Fi back in 2017 [9]. In 2022, the projections were that the figure would go up to 59%. With wireless mobile networks, it is possible to include space–time data on consumers and network conditions to offer the system end-to-end intelligence and discernibility. There is an improved understanding of long-term network dynamics in the long run.

Wireless networks allow geolocation and acquiring of user positioning data [10]. These two functions enable identifying user patterns and recognizing any irregularities. In addition, evaluating data provided by the wireless networks enables self-coordination of network functions and units. As such, it is possible to develop additional effective and proactive networks.

The use of wireless networks is made possible through machine learning approaches. Using machine learning techniques gives rise to a whole research perspective termed Wi-Fi analytics [11]. Wi-Fi analytics in big mobile data permit wireless networks' characterization through identification from gathered data obtained from environmental factors and actors. This characterization comes in handy when proposing solutions for positioning devices in outdoor and indoor contexts and enhancing spectrum management [11]. In the same light, there exist many techniques like the characterization of the network topology through keeping track of access points and the characterization of the spectrum.

In most instances, it is important to employ particular tools to gather data. In wireless networks, stream processing is the most optimal model for Big Data processing because of its dynamic features and the unlimited volume of data [11]. In such a scenario, new prospects come up for research and industry to fuel intelligence and knowledge. Nonetheless, Big Data management comes with more than one computational hurdle. The processing capacity of one machine cannot match up to the growth in the volume of data. It is important to have a distributed processing system [12]. In such a case, countless computational cluster programming frameworks emerge, intending to treat more than one workload [13].

12.6.1 Beyond 5G and Toward 6G

The effort to realize the 6G network dream warrants attention from several industries. As noted before, there are many companies devoted to realizing this dream. Another one of these companies is Keysight Technologies [1]. The race to 6G may be easier than in other countries that do not have the same devotion.

There are significant projects in the pipeline regarding the development of 6G networks. For instance, the University of Oulu, located in Finland, recently started a 6G research project to progress a 6G vision with a 2030 timeline [1]. In addition to that, the university has signed a partnership with Beyond 5G Promotion Consortium from Japan to synchronize the tasks in the Finnish 6G Flagship research on 6G technologies.

The efforts to attain the 6G network dream can also be witnessed in other countries. The Ministry of Industry and Information Technology in China invests and keeps track of 6G research and development. Ericsson from Sweden and Nokia from Finland are communication companies spearheading a consortium, Hexa-X, located in Europe [1]. The consortium comprises industry and academic leaders who come together to progress 6G standards from a research point of view. The Electronics and Telecommunications Research Institute in South Korea is currently studying the terahertz frequency band for 6G networks [1]. The research also pictures data speeds of 100 times more than 4G LTE networks and five times faster than 5G networks. In 2020, the US FCC started a 6G frequency to test the spectrum for frequencies over 95 GHz to 3 THz [1]. There are vendor commitments to 6G, including significant infrastructure corporations like Nokia, Huawei, and Samsung, with evident commitments to the course.

12.6.2 Visible-Light Communications

There is a fast upcoming trend of setting up the IoT in smart cities, homes, and next-generation wireless frameworks. It is expected that there will be a connection of the plurality of devices like sensors, cameras, machines, and tablets, among many others [14]. When this happens, the radio frequency spectrum currently in use proves inefficient to satisfy the needs of huge data-heavy wireless applications. Operating at the radio frequency spectrum would fail from data jamming and interloping. To eliminate the challenge of spectrum insufficiency and fulfill high data-rate requirements beyond 5G frameworks, visible light communication (VLC) for Internet of Things applications in indoor settings through effectively routing data in VLC nodes [14] will be a method to consider.

Operating at the unrestricted optical band in the visible light area uses simple light-emitting-diodes (LEDs). LEDs are a health-friendly communication technique [14]. VLC comes with the upper hand by offering ultra-high bandwidth, intrinsic physical security, and robustness to electromagnetic interloping. On the downside, VLC has the shortcoming of suffering from severe short communication ranges and restrictions regarding line-of-sight [14].

6G networks promise to offer significantly increased capacity and fulfill the needs of emerging applications. However, the existing frequency bands might prove inadequate [15]. What is more, 6G will offer improved coverage through the integration of underwater, space, and air networks with terrestrial networks keeping in mind that traditional wireless communications cannot offer high-speed communication data rates for networks existing outside the terrestrial space. VLC is a high-speed communication approach. VLC has an unrestricted frequency range from 400 to 800 THz [15]. VLC can be taken up as a substitute technique for dealing with the mentioned challenges. There are prospects that VLC will grow to be an essential part of 6G considering its high-speed-transmission upper hand [15]. VLC will also work with other communication techniques to improve daily living.

Even though VLC is not close to radio in terms of maturity, it is prospectively better in most cases, including underwater communication, V2X communications, and communications with a need for high security and low interloping [16]. VLC is an increasingly appealing communication technique, especially for short-range exchange. This quality of VLC acts as a complement to what radio communication already offers. VLC uses reconfigurable optical-radio networks to generate increased performance and highly flexible communication frameworks that can function in the harshest contexts of 6G [17].

Aside from providing a huge and unrestricted bandwidth to counter the packed radio spectrum, the VLC technology also has many additional advantages, like being easily available with no electromagnetic interloping and radiation [18]. In addition to that, the free-space optical (FSO) communication system falls under line-of-sight technology. FSO attracts considerable attention for being a high bandwidth last-mile transmission approach. FSO also functions as a viable substitute for optical fibers because it needs less primary deployment [18]. Additionally, FSO can be installed in places with a challenge with deploying a wired connection.

12.6.3 E-MBB Plus

6G promises the dawn of e-MBB-Plus, which will come in as a replacement for e-MBB already existing in 5G. e-MBB will offer a high-quality experience in data standards and utilization [19]. Outstandingly, additional integral parts of the wireless communication of optimizing networks, interference, and handover ought to use the ideas of Big Data to enable such operations. Making available other add-ons like precise indoor location and a well-matched worldwide connection in different mobile operating networks comes without question [19]. However, there ought to be a strategy meant for e-MBB-Plus communication minus compromising on network subscribers' confidentiality, security, and privacy.

12.6.4 Big Communications

5G communication networks were primarily deployed in 2019. Research in communication science now pays attention to 6G communications [20]. Nonetheless, with the fast progression of 5G data and communications technology, it is challenging to accept that close to 3.7 billion people worldwide count as underconnected or not connected at all [21]. As a result of this fact, ICT research agrees that the benefits of the industry need to share all over the world as opposed to serving developed and urban regions alone. In the same light, while 5G aims to fulfill the application cases in densely populated urban areas with the compact distribution of telecommunication infrastructure, 6G instead intends to offer its services to distant regions which are less populous, with restricted financial resources, and less physical infrastructure [22]. In this way, 6G networks will contribute to closing the digital divide by the 2030s.

Big Communications looks to provide a framework for sharing the benefits of the ICT industry and achieving global connectivity with 6G. Come the 2030s, the ICT industry is hopeful that Big Communications will play a significant role in attaining several goals [21]. One of the goals is to share the advantages of ICT throughout the world and work on reducing the digital divide. Another objective is linking the unconnected part of the population worldwide while paying extra attention to underrepresented factions with demographic, economic, and geometric drawbacks. The third goal is to enhance worldwide and universal connectivity as part of the sustainable development goals (SDGs) of the United Nations [21].

Big Communication is not the sole solution intended for resolving technical challenges. Aside from that, it also offers a comprehensive framework for evaluating the service demands, technological enablers, socioeconomic feasibility, and long-term applications for particular areas. From value engineering evaluation, political-economic, technological, environmental, and legal factors analysis, and additional business models, Big Communication can function as a bottom-up structure [21]. This framework will help link people living in regions not fully developed with restricted financial resources and infrastructure.

Living up to the suggested Big Communication structure demands that the local consumers go through analysis for better comprehension [21]. After that, the holdups averting residents from benefiting from telecommunications will be pointed out. The same can be attained through value engineering evaluation and create the basis of offering nontechnological and technological solutions. With this basic information considered, a comprehensive political, economic, social, technological, environmental, and legal (PESTEL) analysis can be done to fish out the foundation of societal challenges, economy, technology, policy, legality, and environment. The severe and all-inclusive results will aid in enhancing proposals, generating possible solutions, and arriving at important decisions at a high hierarchy. Before this happens, a suitable business framework can be chosen to define the value chain better, uphold profitability, and pinpoint matching stakeholders. With the groundwork laid, engineers and researchers can delve into progressing and putting into practice new communication technologies to attain global and universal connectivity [21].

For connectivity to be attained, it is also important to have appropriate promotion techniques and marketing strategies. Commercial institutions can take up pertinent content distributed concerning enabled connectivity [21]. Coupled with the comprehensive structures and inclusive considerations, nontechnological and technological factors are regarded as critical. It is easier to realize global and universal connectivity to close the digital gap in the long run. Putting Big Communications into practice is not easy. There are principal challenges in content and marketing and service provisions. These are intricate and founded on profits [21]. Nonetheless, these challenges can be solved because content providers, governments, and mobile network operators keep on realizing how important access to the internet is.

12.6.5 Secure Ultra-Reliable Low-Latency Communications

In 5G and beyond systems, ultra-reliable and low-latency communications are the principal promoters of diverse mission-critical services like inaccessible health care, industrial automation, and intelligent transportation [23]. Nonetheless, the two strict requirements of ultra-reliable and low-latency communications present considerable hurdles in designing systems.

In the prospective 6G networks, ultrareliable and low-latency communications will set the pace for upcoming mission-centered applications with strict end-to-end reliability and delay requirements. Prevalent progress in ultra-reliable and low-latency communications comes from theoretical frameworks and assumptions [24]. The solutions founded on frameworks offer crucial insight. However, it is impossible to implement the same direction. To take care of such challenges, it is important to develop a multi-level architecture that permits device intelligence, cloud intelligence, and edge intelligence for ultrareliable and low-latency communications. The foundational concept is to bring together theoretical frameworks and actual data to evaluate the reliability and latency and train deep neural networks [24]. Deep transfer learning is taken up in the architecture to advance the pretrained deep neural networks in mobile networks. With additional regard for the computing capacity of every user and every mobile edge computing server operating with limitations, federated learning is applied to enhance learning efficiency [24].

12.6.6 Three-Dimensional Integrated Communications

The implementation of 5G networks has progressed in many major cities worldwide, and this progression is owed to the ability of the network to sustain prevailing industrial use. Nonetheless, the prospects of 6G indicate the emergence of technological evolution and societal requirements that need researchers to explore past 5G. Currently, terrestrial communication is part of the principal needs in 5G networks [25]. Nonetheless, terrestrial communication is set to progress in 6G networks. This progress will move from terrestrial communication to aerial and underwater communication [25].

6G promises the desired level of freedom in comprehensive connectivity that will enable the emergence of various innovations like ultra-real human-computer relationships, flying cars, underwater recreations, and holographic telepresence. However, these mentioned innovations come with significant challenges. However, there is a way to address these challenges by making communication networks additionally intelligent. Intelligent communications mean improved chances of attaining multidimensional videos, improved bandwidth, 3D connectivity, and high data rates [26]. The prevailing 5G network cannot sustain such needs. With 6G, there will be new KPIs that will come with integrated performance requirements. Real-time remote robotic surgery needs concurrent ultra-low latency, security, and massive data rates to attain efficiency [27]. Figure 12.2 shows key performance indicators (KPIs) for different technologies.

It is highly expected that 6G networks will come with various heterogeneous network constituents that connect through many numerologies to sustain verticals, process Big Data, and need various levels of QoS [28]. There is a need for optimization, analysis, and decision-making capabilities to attain the same. At each communication level in the system, intelligence needs to be extensively embedded and integrated to attain end-to-end services.

One of the most important needs of 6G networks, which sets them apart from past generation networks, is to attain comprehensive worldwide connectivity in terms of terrestrial space, in high altitudes, and underwater [29]. With such a vision, 6G networks intend to attain prolonged and incessant communication between smart devices and people. In a decade or so, communication between ships, terrestrial planes, satellites, and aircraft will become evident [26]. 6G will prospectively affect network densification by increasing the number of user terminals. As such, consumers will network with other terminals on the same level, like in the terrestrial space of through different levels from terrestrial to underwater or from aerial to terrestrial using multiple radio access networks. Figure 12.3 depicts a 3D communication scenario in 6G [28].

The 3D perspective of integrated communication can be understood by explaining underwater communication, space communication, and unmanned aerial vehicle (UAV)-based communication representing terrestrial communication.

12.6.7 Underwater Communication

This part of integrated communication can be explained through the existence of deep-sea exploration. Exploration on such a scale requires video content and data transmission to a station on the ground. In some instances, communication during critical situations like rescues and warfare fails in its efficiency; however, unpredictability and challenging communication because of the distance from the coast can be eased through the use of VLC, RF, and acoustic signals. The coming of 6G networks promises a chance to integrate the mentioned solutions into terrestrial communication applications [30].

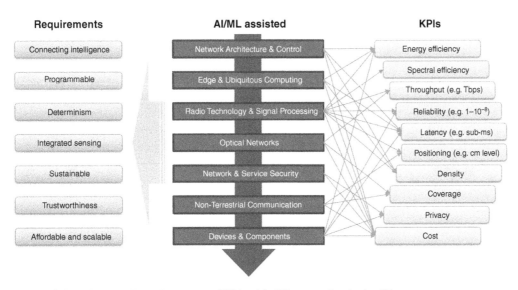

Figure 12.2 Different 6G requirements and KPIs with different technologies [3].

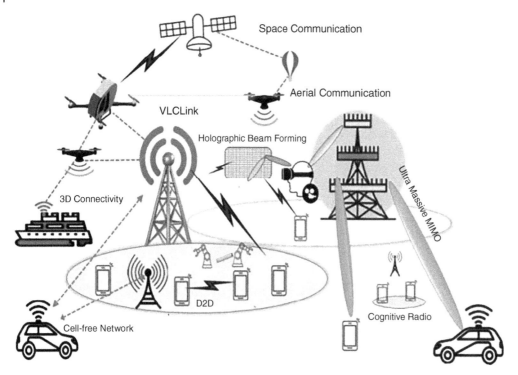

Figure 12.3 3D communication scenario in 6G.

12.6.8 Space Communication

Three principal patterns characterize space communication:

- UAVs, 6G base stations, and airplanes
- Correspondence between low Earth orbit (LEO) and Earth satellite transporters
- 6G cellular base station, UAVs, and cellular consumers

Prevailing communication satellites mostly belong to geostationary orbits, resulting in a large backlog for terrestrial mobile correspondence. As a result, theoretical research shows mobile broadband integration with LEO satellites [31]. With the coming of 6G, several countries have already made plans for satellite expeditions. The United States of America, China, and the United Kingdom are some countries making such plans.

12.6.9 UAV-Based Communication

UAVs have been comprehensively employed as relay nodes to progress cellular coverage in recent times. UAVs function at lower altitudes than satellites as mobile base stations stay afloat in space. This ability enables them to cover an increased footprint of terrestrial mobile clients and act as hotspots to progress coverage of base stations, especially in places with challenging mobile infrastructure [27]. With 6G networks, UAV communication will offer rural connectivity at a lower cost.

12.6.10 Unconventional Data Communications

Unconventional data communications are present in various technological advancements. An example is the use of smart sensors. Smart sensors are primarily used to track and regulate networks in different application domains. The principal feature of these types of devices is that they can send data using a wireless communication channel [32]. There are several standards and technologies progressed to enable the transmission of data, and these developments are featured in different protocols, frequency bands, and modulations. The selection of communication technology relies on network requirements regarding latency, bandwidth, cost of deploying infrastructure, and range of coverage. There is an option to employ a different technique considering the applications where smart devices like sensors are located near steel pipeline networks [32]. These exceptional data communications are exemplified through various scenarios like tactical communications, human-bond communications, and holographic communications.

12.6.11 Tactical Communications

Tactical communications mostly come in handy in the defense ministry. With 6G networks promising to improve connectivity and strengthen the quality of communication, it is definite that tactical communication will also be improved. Understanding how 6G networks contribute to enhancing tactical communications is exemplified through a prototype. The main aim of this prototype is to come up with a material solution to bring together network capability meant to pool modern telecommunications and internetworking ideas and 6G networks with military and tactical radio wireless communications technologies like wideband high-frequency (WBHF) networks, NATO narrowband waveform (NBWF), and barrage relay networks [33].

The projected implementation intends to bring together military-particular wireless communications technology and contemporary supporting network technology. Software-defined perimeter ought to be particularly explored to enable prevailing network configurations for military multilevel security services, black core networking services, and black/red cryptographic separation [33]. Contemporary military wireless communications offer progress goals, and their convergence with additional improved wireless communications solutions is essential, particularly with the coming of 6G networks.

People in war need to get ahold of precise data in real-time and at the right place. At the moment, some networks are radio aware coming up to address this concern. A new internet engineering task force, RFC 8175, describes a dynamic link exchange protocol technique for bringing together mobile radios and IP routers [33]. This integration enables faster convergence, improved performance for delay-conscious traffic, radio aware routing, and efficient route selection. Nonetheless, data networking services like WBHF, NATO NBWF, and Single Channel Ground Airborne Radio System (SINCGARS) were developed way before the mentioned innovations came into the picture [33]. With such a projection reaching successful completion, the commander will be able to integrate these data networking services into converged IP-based data networks, thereby leveraging contemporary convergence techniques.

The tactical communications project can be delivered in different phases. In phase one, the objective is to attain a comprehensive document describing how to achieve a converged SDx-founded network leveraging prevailing military tactical radios and 6G cellular technology [33]. The description will also evaluate the techniques and prospects for an SDx-founded tactical network. Phase two will develop and demonstrate a prototype solution of an SDx-founded network leveraging prevailing and upcoming PM Tactical Radio platforms and prevailing accessible technology [33]. The objectives of phase two are to offer a prototype solution pertinent for backing an operation, demonstrate the prototype with Army tactical radio systems, document products with highlights of operations and functions of the prototype through regular progress reports, and offer test reports with solution performance. The reports might highlight technical hurdles, risks, and developments [33].

In phase three, there will be demonstration and development of solutions that offer feasible technology insertion to move to an SDx-founded configuration. The objectives of phase three include a demonstration of the prototype with radio systems that are emerging or already owned by the army, product readiness reports comprising of implementation and design challenges, regular progress reports, prototype solutions pertinent for backing an army operation, product documentation with highlights of functions and operations, and test reports with the performance of the solutions [33].

12.6.12 Holographic Communications

Holography is an approach supporting the EM sector, generally, as the eventuality of a signal source spread out from objects, to be documented founded on the interference principles of the EM wave. The documented EM sector can be used to reconstruct the primary field founded on the code of diffraction. It should be clear that wireless communications through a continuous aperture come from optical holography. The holographic communication process contains a training phase where the generated training signals come from an RF origin. The training signals split the beam-splitter into two waves, the reference and object waves. The object wave goes to the object, and part of the reflected wave mixes with the reference wave beam where the latter does not impose on the object. The object wave is then fed to the Holographic MIMO Surface (HMIMOS). Aside from the training phase, there is a communication phase where the transmitted signal transforms into the needed beam to the target consumer over the spatially constant aperture of the HMIMOS.

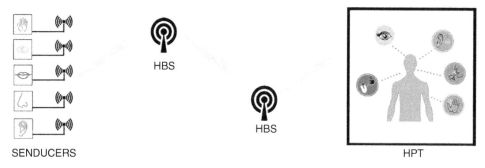

Figure 12.4 HBC system architecture [37].

There have been developments in the making of programmable metamaterials. As such, reconfigurable intelligent surfaces display the prospects to attain the almost unattainable vision for 6G networks and realize seamless connections and intelligent software-founded regulation of the environment in wireless communication structures when set up on surfaces of different objects [34]. By leveraging this progress, HMIMOS have the objective to strive past massive MIMO founded on weight, reduced cost, size, and reduced power consumption hardware architectures that offer a transformative solution for transforming the wireless setting into a programmable smart sector [35].

Constant prototyping advancements are increasingly needed to assert that HMIMOS idea with the novel holographic beamforming technologies and to realize possible matters that straightaway need research [36]. The instance of incessant HMIMOS, intelligent holographic beamforming is a technique to smartly target and monitor lone and small clusters of devices and offer high-fidelity beams and smart radio communication. Nonetheless, self-enhancing holographic beamforming technologies that rely on intricate aperture synthesis and lower-level modulation have yet to make it to the scene [36].

12.6.13 Human-Bond Communications

In coming times, the ability of people to communicate sensations will improve by incorporating the five sensory characteristics into the messages and permitting additionally expressive and wholesome sensory data exchange through communication methods [37]. Human-bond communication (HBC) is just a vision, but scientists are hopeful that it will be attainable. HBC is a newly described idea that brings together gustatory, olfactory, and tactile sensations to achieve an additional human emotion-centered communication for coming networks.

For HBC, there is a proposed baseline architecture considered. HBC comprises seducers (sense transducers) responsible for the sensory transduction of stimuli to electrical indicators [37]. The second component of HBC is human perceivable transposer, responsible for transposing data. The third component is human-bond sensorium, which gathers, processes, and transmits the data [37]. Figure 12.4 explains the proposed HBC architecture.

12.7 Challenges for 6G Networks

12.7.1 Potential Health Issues

With research work focusing on developing and launching the use of 6G networks, there are debates concerning how 6G technology will affect human life. Other implications bring negative consequences aside from the many advantages of realizing the 6G network dream. Some of these negative implications fall on health concerns. Exposure to high-frequency radiation connects to specific medical consequences like attention deficit hyperactive disorder, autism, nausea, blurred vision, dizziness, obsessive-compulsive disorder, and post-traumatic stress disorder [38]. Additionally, exposure to RF through smart devices like mobile phones connects to the rising incidence of cancer. However, there are studies to refute this claim. Many specialists and scientists raise concerns that using cell phones over an extended period poses an impact on specific sections of human cells [38].

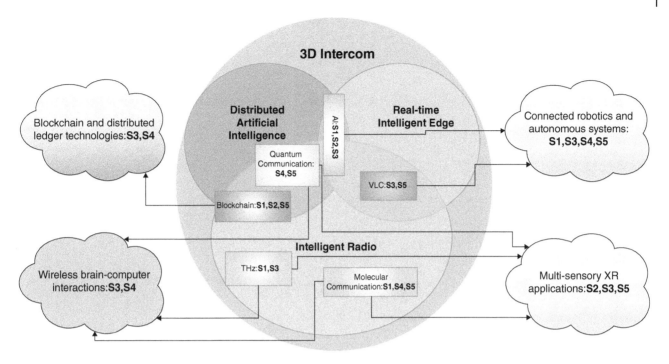

Figure 12.5 Security and privacy issues in the 6G network.

12.7.2 Security and Privacy Concerns

Figure 12.5 shows areas affected by the five principal security and privacy challenges: authentication (S1), access control (S2), malicious behavior (S3), encryption (S4), and data transmission (S5). Most parts of the diagram display susceptibility to malicious behavior, authentication, and access control [39]. Nonetheless, some technologies display specific sensitivity to some of the mentioned challenges. Case in point, VLC with real-time intelligent edge and intelligent radio display weaknesses in the face of malicious activities and data transmission procedures [39]. Molecular communication and THz technology both go back to intelligent radio. The molecular communication technology links to privacy and security issues regarding encryption, authentication, and communication. On the other hand, THz technology is particularly vulnerable to malicious acts and authentication security challenges [39]. The quantum communication and blockchain technology overlay with distributed AI and intelligent radio. The principal privacy and security considerations link access control, data encryption, authentication, and data transmission.

6G applications might display particular susceptibilities. Robotics and autonomous systems like self-driving cars from Tesla rely on AI and VLC technology where data encryption, data transmission, and malicious acts can cause challenges. The multisensory XR applications employ molecular communication technology, THz technology, and quantum communication technology, which is vulnerable to malicious acts, control attacks, and data transmission exposure [39]. Wireless brain-computer relations use the same approaches as the multi-sensory XR application. However, they possess specific security and privacy challenges.

12.7.3 Research Activities and Trends

5G network has already been commercialized. Stakeholders in the initial stages of 6G research need to function in partnerships with the mutual goal of generating momentum in research activities to attain next-generation technology [40]. The government, academics, and industries need to come together at this initial stage. 6G is yet to establish its standards and products. Instead, 6G needs a long runway with adequate room for maturity, discussion, and investigation [40]. Going in for the long haul, it is important to pool efforts at the initial stage.

In other instances, it is normal to see stakeholders compete against each other with no thought of sharing unless it is industry standards. However, with 6G, if stakeholders fail to engage in mutual discussions and efforts, with no public

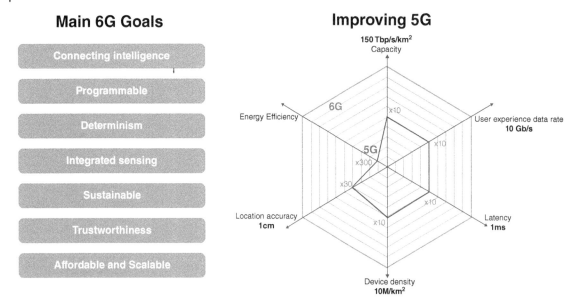

Figure 12.6 The main 6G goals, plus the KPIs where 6G will be improving 5G [3].

publications, other stakeholders will spend resources, making the same attempts meaningless [40]. Corporations ought to generate momentum and energy to back 6G research at such an initial point. At a later stage, different companies can then select specific technologies and products for additional development.

Several countries have already made significant steps toward 6G research. An example is Europe. In December 2021, Europe witnessed a newly founded Joint Undertaking on Smart Networks and Services with 6G [41]. This new venture took up its initial work program from 2021 to 2022 with public funding of approximately €240 million [41]. The Work Program, 2021–2022, intends to financially support activities from evolving 5G and inclusive of large-scale pilot ventures with vertical sectors to leading research in 6G systems.

The European collaboration on Smart Networks and Services is pioneered by the Commission, the industry, and close participation from the member states. This joint undertaking lays the R&I course for Europe [41]. This venture capitalizes on the EU's contribution of €900 million through the coming seven years, set to be matched by the private industry in an equal figure. The objective is to make it possible for European stakeholders to strengthen their R&I ability for 6G systems and grow lead markets for 5G infrastructure as groundwork for the green and digital revolution.

Major factors affecting research paths toward 6G networks and services include the future sensitive applications and usages, dynamic changes with time and space distribution, and deep technologies.

Security in 6G systems must explore many new phases and strategies such as virtualization (from basic functions to end-to-end virtual perimeters using slices), softwarization (making smart usage of flexibility and programmability of orchestrated security closely integrated with orchestrated systems and services), concepts such as deception or moving target defense, cloudification (delivery of security as a service, define/operate service attributes for SNS and integrated into the service-based overall paradigm) and holistic approaches encompassing the whole life cycle, from original component/code developments to introduction of technologies (e.g. AI Quantum). Security quantification in a 6G context should deal with complexity and fragmentation, and mandates the exploration of mechanisms to identify, evaluate, certify, and monitor the level of security. Based on such quantification, trust can be given to providers and services, enabling the growth of the sector by answering the diversity of demand from the mass market to specific B2B verticals [3]. Figure 12.6 shows the main 6G goals.

12.8 Conclusion

The sixth generation of wireless systems will support many features such as massive machine-type communications (MTC), ultra-reliable low-latency communications, high-fidelity holograms, immersive reality, and tactile/haptic-based communications. The 6G of systems will require new physical layer techniques and higher-layer capabilities.

According to the ITU-T, the next decade will see three disruptive changes in the wireless network design. Video communication is now the mode of choice for communications, and as a result, it requires a data rate of 15.4 Mbps per UE per user. In addition, it has an increase in viewing time, making it the norm for end users to watch complete television programs, live sports events, or on-demand streaming. The next frontier for a virtual mode of communication is holograms and multisensory communications. In 2017, Stephen Hawking gave a lecture via a hologram to an audience in Hong Kong, showcasing the growth potential of this technology. Communication involving all five senses, including smells and tastes, is also being considered for the future of technology. In addition, an order-of-magnitude or even higher number of planned interconnectivity will become a defining characteristic of future society. Humans and machines both need to react in real-time to the arrival of information in the future. Timeliness of information delivery will be critical for the vastly interconnected society of the future. Since the need for 6G is driving the research on 6G systems, we review the progress in the literature on 6G systems. Several papers provide applications and solutions for 6G systems. Some of these papers discuss possible use cases and enabling technologies for 6G research and development. Many studies have been published on 6G, including those that propose new waveforms for 90–200 GHz [31], that propose an internet of BioNanothings [38], that propose efficient wireless powering of IoT networks [30], and that discuss the role of collaborative AI for network management [8]. To realize the development of 6G systems, technologies that are not yet available need to be developed. These technologies will need to be combined with improvements in the network architecture, and the current transport network will need to be redesigned to support 6G services.

This chapter presented a detailed breakdown of the strengths and weaknesses of 6G technologies. In this chapter, we discussed the vision for 6G, seven use cases to be supported by 6G, and their technical requirements. It also contains a comparison with 4G and 5G systems and a discussion of the new frequency bands and deployment scenarios. 6G use cases require extremely high-speed wireless connectivity and other requirements from the ITU-T. Holographic displays are the next evolutions in multimedia experience, providing an immersive 3-D experience for the end user. Interactive holographic capability requires a combination of very high data rates and ultralow latency. The holograms' data rates vary from tens of Mbps to 4.3 Tbps, depending on the hologram display and the number of synchronized images. There are several challenges in realizing holograms and multisensory communications, especially their widespread adoption. We note that several standards are needed to supply data to a holographic display, and specialized optical setups are required for digital recording holograms. Data rates depend on the application, latency depends on the application, and high-definition 1080p video requires 1–5 Mbps. Other applications require sub-ms latencies and are used for autonomous driving. Due to the fast reaction times of the human mind and the fast reaction times of machines, the inputs need to be strictly synchronized, and the system must be able to prioritize streams based on their criticality. Mobile edge will be deployed as part of 5G networks, yet this architecture will continue toward 6G networks. An assortment of edge computing sites may be involved for computation-intensive applications, and the computing resources must be utilized in a coordinated manner. Information shower kiosks in metro stations and shopping malls may provide fiber-like speeds with a data rate up to 1 Tbps. Information shower kiosks in metro stations and shopping malls may provide fiber-like speeds with a data rate up to 1 Tbps. The network could support the Internet of Bio Things in real-time monitoring of buildings, cities, environment, cars and transportation, roads, critical infrastructure, water, power, and so on. On-chip, inter-chip, and inter-board communications use wired connections, which become bottlenecks when data rates exceed 100–1000 Gbps. Nano networks represent a promising area for 6G. This chapter presents an integrated system based on the seamless integration of terrestrial and space networks. This system provides internet access from space via a global constellation of LEO satellites.

The chapter also indicates that AI does indeed have the potential to be a key enabler for the next generation of wireless networking, 6G, by providing intelligent and robust security solutions. Given the expansive future that is 6G capability to connect in so many massive ways, bringing on more demanding and powerful technology, so too will the security that protects it all. AI will provide the key to autonomously identify and respond to potential threats based on network inconsistencies rather than the outdated cryptographic methods of the past. Given the deep reinforcement capabilities of AI to be proactive and immediate in its detection through machine learning, cybersecurity measures will be able to effectively defend against the main players in system destruction: overloading attacks, DDoS attacks, control plane saturation attacks, and host location hijacking attack. This is very significant and will hold a definite precedent for the impending advancements that need to, and will, come, even after this.

However, as with most things, when it comes to the internet and security, it is a double-edged sword. Wherein there is much to gain for the side of securing networks, there are many issues presented in the way of security, privacy, ethics, and its use as an attack. All of which were appropriately addressed in this book, but it remains to be seen how much more can be done and should be done. Thus, the author calls for more research regarding this extremely exciting and most certainly world-changing topic.

References

1 Kranz, G. and Christensen, G. (2021). What is 6G? Overview of 6G networks & technology. https://www.techtarget.com/searchnetworking/definition/6G

2 Tomas, J.P. (2021). How softwarization will be key to future 6G networks: National Instruments. *RCR Wireless News* (December 20). https://www.rcrwireless.com/20211220/5g/softwarization-play-key-role-future-6g-networks-national-instruments.

3 European Vision for the 6G Network Ecosystem (2021). White paper-6G-Europe- in new6g-noton desltop. *The 5G Infrastructure Association*, Version 1, https://doi.org/10.5281/zenodo.5007671.

4 Raja, A., Jamshed, M., Pervaiz, H., and Hassan, S. (2020). Performance analysis of UAV-assisted backhaul solutions in THz enabled hybrid heterogeneous network. In: *IEEE INFOCOM 2020 –IEEE Conference on Computer Communications Workshops*, 628–633. IEEE.

5 Viswanathan, H. and Mogensen, P. (2020). Communications in the 6G era. *IEEE Access* 8 (1): 57063–57074.

6 Akhtar, M.W., Hassan, S.A., Ghaffar, R. et al. (2020). The shift to 6G communications: vision and requirements. *Human-Centric Computing and Information Sciences* 10 (53): https://doi.org/10.1186/s13673-020-00258-2.

7 Yang, H., Alphones, A., Xiong, Z. et al. (2020). Artificial-intelligence-enabled intelligent 6G networks. *IEEE Network* 34 (6): 272–280. https://doi.org/10.1109/MNET.011.2000195.

8 Liska, A. (2015). What is intelligence? In: *Building an Intelligence-Led Security Program* (ed. A. Liska). New York: Elsevier.

9 Cisco, V. (2019). Global mobile data traffic forecast update 2017–2022. *White Paper* 1–33. https://www.cisco.com/c/en/us/solutions/collateral/executive-perspectives/annual-internet-report/white-paper-c11-741490.html.

10 Alessandrini, A., Gioia, C., Sermi, F. et al. (2017). Wi-Fi positioning and Big Data to monitor the flow of people on a wide scale. In: *European Navigation Conference (ENC)*, 322–328. Lausanne: IEEE.

11 Medeiros, D.S., Neto, H.N., Lopez, M.A. et al. (2020). A survey on data analysis on large-scale wireless networks: online stream processing, trends, and challenges. *Journal of Internet Services and Applications* 11 (6): 1–48. https://doi.org/10.1186/s13174-020-00127-2.

12 Zaharia, M., Das, T., Li, H. et al. (2012). Discretized streams: an efficient and fault-tolerant model for stream processing on large clusters. In: *Proceedings of the 4th USENIX conference on Hot Topics in Cloud Computing*, 1–10. Boston: USENIX Association.

13 Dean, J. and Ghemawat, S. (2008). MapReduce: Simplified data processing on large clusters. *Communications of the ACM* 8 (1): 107–113.

14 Faisala, A., Alghamdi, R., Dahrouj, H. et al. (2021). Diversity schemes in multi-hop visible light communications for 6G networks. *Procedia Computer Science* 182 (1): 140–149. https://doi.org/10.1016/j.procs.2021.02.019.

15 Chi, N., Zhou, Y., Wei, Y., and Hu, F. (2020). Visible light communication in 6G: advances, challenges, and prospects. *IEEE Vehicular Technology Magazine* 15 (4): 93–102. https://doi.org/10.1109/MVT.2020.3017153.

16 Božanić, M. and Sinha, S. (2021). Visible light communications for 6G. In: *Mobile Communication Networks: 5G and a Vision of 6G* (ed. M. Božanić and S. Sinha), 155–188. Springer https://doi.org/10.1007/978-3-030-69273-5_5.

17 Marcos, K. and Iqrar, A. (2020). Opportunities and challenges for visible light communications in 6G. In: *2020 2nd 6G Wireless Summit (6G SUMMIT)*, 1–5. Levi https://doi.org/10.1109/6GSUMMIT49458.2020.9083805.

18 Gupta, A., Sharma, N., Garg, P., and Alouini, M.-S. (2017). Cascaded FSO-VLC communication system. *IEEE Wireless Communications Letters* 1–4. https://doi.org/10.1109/lwc.2017.2745561.

19 Gui, G., Liu, M., Kato, N. et al. (2020). 6G: opening new horizons for integration of comfort, security, and intelligence. *IEEE Wireless Communications* 27: 1–7.

20 Dang, S., Amin, O., Shihada, B., and Alouini, M. (2020). What should 6G be? *Nature Electronics* 3 (1): 20–29. https://doi.org/10.1038/s41928-019-0355-6.

21 Zhang, C., Dang, S., Alouini, M.-S., and Shihada, B. (2022). Big communications: connect the unconnected. *Frontiers in Communications and Network* https://doi.org/10.3389/frcmn.2022.785933.

22 Yaacoub, E. and Alouini, M. (2020). A key 6G challenge and opportunity-connecting the base of the pyramid: a survey on rural connectivity. *Proceeding of the IEEE* 108 (1): 533–582. https://doi.org/10.1109/jproc.2020.2976703.

23 Feng, D., Lai, L., Luo, J. et al. (2021). Ultra-reliable and low-latency communications: applications, opportunities, and challenges. *SCIENCE CHINA Information Sciences* 64: 120301. https://doi.org/10.1007/s11432-020-2852-1.

24 She, C., Dong, R., Gu, Z. et al. (2020). Deep learning for ultra-reliable and low-latency communications in 6G networks. *Electrical Engineering and Systems Science* 1–8. https://doi.org/10.48550/arXiv.2002.11045.

25 Loven, L., Partala, J., and Pirttikangas, S. (2020). Privacy-aware blockchain innovation for 6G: Challenges and opportunities. In: *Proceedings of 6G Wireless Summit*, 1–5. Levi.

26 Zaho, Y., Yu, G., and Xu, H. (2019). *6G Mobile Communication Network: Vision, Challenges and Key Technologies*. Scientia Sinica Information.

27 Dang, S., Amin, O., Shihada, B., and Alouini, M. (2020). From a human centric perspective: what might 6G be? *Nature Electronics* 3 (1): 20–29.

28 Bhat, J.R. and AlQahtani, S.A. (2020). 6G ecosystem: current status and future perspective. *IEEE Access* 20 (1): 1–9. https:// www.researchgate.net/publication/348799624_6G_Ecosystem_Current_Status_and_Future_Perspective/ link/6010d86545851517ef19e8ff/download.

29 Chen, S., Liang, Y., Sun, S. et al. (2020). Vision, requirements, and technology trend of 6G: how to tackle the challenges of system coverage, capacity, user data-rate, and movement speed. *IEEE Wireless Communications* 27 (2): 218–228.

30 Khan, L., Yaqoob, I., Imran, M. et al. (2020). 6G wireless systems: a vision, architectural elements, and future directions. *IEEE Access* 8 (1): 147029–147044.

31 Huang, X., Zhang, J., Liu, R. et al. (2019). Airplane-aided integrated networking for 6G wireless: will it work? *IEEE Vehicular Technology Magazine* 14: 1–9.

32 Caro, D.D., Leo, G.D., Pietrosanto, A., and Paciello, V. (2020). Unconventional communication channels for smart sensors networking. In: *2020 IEEE International Instrumentation and Measurement Technology Conference*, 1–6. IEEE https://doi. org/10.1109/I2MTC43012.2020.9128581.

33 SBIR.STTR (2021, February 10). Software-defined everything (SDx) and 5G/6G cellular design prototype for tactical radios. https://www.sbir.gov/node/1695963.

34 Huang, C., Zappone, A., Alexandropoulos, G. et al. (2019). Reconfigurable intelligent surfaces for energy efficiency in wireless communication. *IEEE Transactions on Wireless Communications* 18 (8): 4157–4170.

35 Wu, Q. and Zhang, R. (2019). Intelligent reflecting surface-enhanced wireless network via joint active and passive beamforming. *IEEE Transactions on Wireless Communications* 18 (11): 5394–5409.

36 Huang, C., Hu, S., Alexandropoulos, G.C. et al. (2020). Holographic MIMO surfaces for 6G wireless networks: opportunities, challenges, and trends. *IEEE Wireless Communications* 99 (1): 1–9. https://doi.org/10.1109/MWC.001.1900534.

37 Re, E.D., Morosi, S., Mucchi, L. et al. (2016). Future wireless systems for human bond communications. *Wireless Personal Communication* 1–14. https://doi.org/10.1007/s11277-016-3240-x.

38 IP Look (2021, October 12). Top 10 Challenges for 6G. https://www.iplook.com/info/top-10-challenges-for-6g-i00096i1.html

39 Wanga, M., Zhua, T., Zhanga, T. et al. (2020). Security and privacy in 6G networks: new areas and new challenges. *Digital Communications and Networks* 6 (3): 281–291. https://doi.org/10.1016/j.dcan.2020.07.003.

40 The Guardian Nigeria (2022, January 12). 6G research needs collaborative efforts in the initial phase. https://guardian.ng/ technology/6g-research-needs-collaborative-efforts-in-the-initial-phase.

41 Directorate-General for Communications Networks, Content and Technology (2021, December). Europe launches first large-scale 6G Research and Innovation Programme. *Shaping Europe's Digital Future* 17: https://digital-strategy.ec.europa. eu/en/news/europe-launches-first-large-scale-6g-research-and-innovation-programme.

Index

From 5G to 6G: Technologies, Architecture, AI, and Security, First Edition. Abdulrahman Yarali.
© 2023 The Institute of Electrical and Electronics Engineers, Inc. Published 2023 by John Wiley & Sons, Inc.